GANGES WATER MACHINE

Designing New India's Ancient River

ANTHONY ACCIAVATTI

I0805467

APPLIED
RESEARCH
+DESIGN
PUBLISHING

Published by Applied Research and Design, San Francisco

Publisher: Gordon Goff

www.appliedresearchanddesign.com
info@appliedresearchanddesign.com

Ganges Water Machine: Designing New India's Ancient River / Anthony Acciavatti

Author: Anthony Acciavatti
Design: Station
Design Director: Nicholas Rock
Designer: Julie Ostrow
Editing by: Aleksandr Bierig and Duncan Corrigall
Production: Usana Shadday
Production Assistance: Meghan Wright-Martin
Proofread by: Kristin Lacey and Ryan Buresh
Typeset in Bryant Compressed and Elena
by Process Type Foundry

Portions of this work were funded by a J. William Fulbright Fellowship as well as grants and fellowships from the Ford Foundation and Harvard University.

10 9 8 7 6 5 4 3 2 1 First Edition

Library of Congress data available upon request.
World Rights: Available

ISBN: 978-0-982622-61-2

Color Separations and Printing: ORO Group Ltd.
Printed in Hong Kong.

APPLIED RESEARCH +DESIGN PUBLISHING

International Distribution:
www.appliedresearchanddesign.com/distribution

for Anne Donahey,
Patricia Sandage, and
Margaret Ellen Acciavatti

TABLE OF CONTENTS

TABLE OF CONTENTS

FOREWORD

The Ganges is fundamental to any imagination of India. The Ganges River Basin has defined the culture of India and nourished an entire cultural landscape with settlements along the river's edge. Cities in this river basin, which sustained most of north India, were historically located on one bank of the river to accommodate its volatile seasonal movements. The Ganges—fed by the melting snows of the Himalayas in the summer and the intensity of the southwest monsoon—doubles its size as it sweeps across the northern plains of India, often changing its course as it floods huge tracts of land each year, leaving behind a rich alluvial soil. It is perhaps one of the most fertile river basins in the world and, throughout much of its history, settlements were judiciously built respecting the volatility of the sacred river. But this phenomenon is hard to capture (let alone represent) in ways that might be useful for study or that could inform patterns of occupation and existence on the ground. And it is due to this lack of clear representation that the unpredictability of the river has often had terrible impacts on the lives of the people that live along it and are dependent on this massive resource.

It is precisely to unearth, capture, and represent the fluid and dynamic condition of this mighty river, which *Ganges Water Machine* has set as its central ambition. This is an admirable project that started in the early 2000s as a "Dynamic Atlas" and has now, as a result of material that surfaced in the research, morphed into a hybrid between an atlas and an almanac. An atlas by nature captures a moment. It is represented literally as a collection of maps – a static moment in the life of regions and terrains. In contrast, an almanac is a collection of calendars and astrological positions: a representation of rhythms, of implied movement, of trying to predict (based on past data) projected behaviors of nature. The almanac is somewhat speculative—it essentially aims to predict the potential effect of nature's rhythm on people who live and depend on its cycles.

The combination of atlas and almanac is, for me, the genius of the book—the static and the dynamic, the given and the ever-changing are all collapsed simultaneously into the research, study, and representation. This is a book that recognizes the dynamic nature of weather and juxtaposes this with the rhythms of people who occupy and construct the terrain of the Ganges River Basin. The book maps the cyclical changes produced by the monsoon and agriculture against the historic changes made by people at different scales of organization to the river—from the construction of the Ganges Canal in the nineteenth century to the proliferation of uncountable tube wells in the twentieth and twenty-first centuries. It is only through the book's hybrid format that one is able to simultaneously understand the nature of the river itself and its ever-changing and reconfiguring relationship with the people that share space with it, depend on it, and worship its presence.

In the past two decades, many attempts and projects have been imagined to "clean the Ganga." These imaginations have often been conjured by politicians wanting to leverage the symbolism inherent in purifying a sacred river, but have lacked substance in terms of understanding the true ecological and geographical dimensions of the problems—of dealing with a territory that extends from the Himalayas to the delta at the Indian Ocean at the eastern edge of India and Bangladesh. At other times, environmentalists have raised the boogey of pollution and impurity to push for the closing down of industries that pour effluent into this ecologically sensitive landscape. And while the activists' premises are well-grounded, they too do not have the mechanisms to embrace this problem holistically. To imagine and articulate a narrative that includes all the dimensions of this ecologically and culturally sensitive landscape is a real challenge, a challenge to capture and represent the very cross-section of India through its different geographies, geologies, hydrologies and the hundreds of culturally distinct societies and their settlements that colonize the river's edge.

To construct this nuanced and inclusive reading of the Ganges, Anthony Acciavatti immersed himself in travel. He traversed and experienced as much of this landscape from Gaumukh (Uttarakhand) to Patna (Bihar), as he could. He travelled on foot, by car, in boats (and even slept in them sometimes), visited sarais, towns, villages, historic sites, and agricultural landscapes. He spoke with and interviewed villagers, city dwellers, farmers, industrialists, academics, activists, government officials, and even the minster of environment and forests in the central government of India. In sum, he engaged the range of constituencies that are intrinsic to the formation, the occupation, and the future of this landscape. It is this deep excavation that informed his insightful understanding of the territory of the Ganges.

Limited by the fact that there has not been a systematic mapping of this terrain since the 1950's, Acciavatti developed tactics to capture the fluidity of the river he experienced: he created maps that worked on a temporal scale, allowing information to intersect itself and be visually layered. He worked through methods of mapping that showed correlations among otherwise disparate pieces of information. These techniques of superimposition and mapping based on time rhythms allowed Acciavatti to truly create a hybrid between almanac and atlas. In short, a unique instrument to capture this very particular condition, producing visualizations and representations that will prove useful for all those concerned with the future of this landscape.

More specifically, this book, as is revealed in its title, is not nostalgic. It is about the future of the Ganges and interested in the productive dimension of the river. The research is focused on infrastructure and the transformation of the sacred river and its basin into a large machine—with canals, dams, and means of irrigation. Since the era of the British East India Company, irrigation has been seen as a modernizing phenomenon. This infrastructure has formed the backbone of development in the Ganges basin. And over time, this obsession to harness the water as a resource has transformed the river into a giant water machine! Thus by carefully mapping and registering the entire range of instruments, from canals to tube wells, Acciavatti also documents the cultural and spatial consequences of such monumental (as well as minute) changes to the landscape. This is not only a useful inventory, but also makes us aware of the components that alter and continuously transform the landscape of the Ganges River Basin and their possible future implications.

The message of Ganges Water Machine resonates beyond the river and makes the reader aware of the interconnectedness of the emergent urban landscape along the river basin and in its hinterland. The understanding of this ecology of urban centers that grow, merge, and depend on each other is an important lesson as we imagine new futures for urbanism in South Asia. The fact that the book gives us samples of many such urban centers will hopefully fuel a broader debate on urbanization in the region and, most critically, situate natural landscapes as part of that discussion. For too long, planners, designers, and policy makers have treated the natural and manmade as separate domains. This approach has created a condition in which human occupation is understood as in conflict with nature. In this book, Anthony Acciavatti makes vivid for us the implications of constructing in and around a natural landscape. But, more crucially, he draws and represents this landscape in order that we not only understand its relationships more clearly, but are also sensitized to its complexity and feel compelled to engage with its future.

RAHUL MEHROTRA,

Professor and Chair of the Department of Urban Planning and Design, Harvard University, Graduate School of Design

PRINCIPAL, RMA ARCHITECTS, MUMBAI AND BOSTON

PREFACE

I made a mistake.

I sat patiently, while my passport and visa were looked at with suspicion. "What are you, C.I.A?" asked the director of the geography department at a well-known university in Varanasi. All I asked was to see the department's collection of maps. I assured him I was a J. William Fulbright Fellow from the U.S.—with no ties to acronym-based espionage organizations of any kind. A copy of my proposal, along with my letter of affiliation at Allahabad University laid before him on his desk, outlining my research interests in the development of cities and infrastructure along the Ganges River. Fairly straightforward. Or so I thought. Rather than answer whether or not he had such maps, he told me that maps of this territory were in fact sensitive information and, furthermore, sharing them with a foreigner was out of the question. When I naively asked why, he said it posed a security risk. Baffled by the chair's response, I took a sip of tea to think of what to say next. (No doubt maps may contain sensitive and detailed information; however, I was only hoping to see when the last time the Ganges was mapped.) Explaining how I planned to make my own set of maps, I showed him a low-resolution satellite photo of Allahabad—the best that I could find in the early days of Google Earth. Pointing at the small-printout, I noted the profile of the Ganges and the edges of the urban form in relationship to it, but the patchwork of beige, green, and grey made it hard to tell what was a building and what was a field. "This is what I plan to map, and how it changes over time," I said. He raised his eyebrows and said little more. Puzzled at my inability to see the maps I requested, I went on my way, sure that the official's anxieties over what seemed a quilt of farms were overblown.

Three weeks later, while traveling along National Highway 7, some 12km southwest of Varanasi, near the village of Narainpur, I came across an unusually wide canal passing beneath the roadway. I asked the driver to stop so that I might get out and photograph it. By now, having traveled along Grand Trunk Road (NH2) from Varanasi to Allahabad and now making a full loop on the opposite side of the Ganges, he was used to my insistence of stopping at each and every hydrological feature. But this one, with its unequalled volume of soupy, brown water, demanded greater attention. We traveled to its source: a large pumping station known as Narainpur Pump Canal system (said to be the largest in all of Asia). After watching hundreds of liters of Ganges water gushing from this great structure, all I could think to do was see where it flowed. After much protest, the driver agreed to follow the canal. Unaided by any map, we set off along it, passing through hamlets and fields. Without warning, he pulled over and said he would go no further because it was too dangerous. He said we were entering an area full of *dacoits*. Dismissing his warning, I got out and walked along the canal for some time, keeping an eye out for snakes (a longstanding phobia of mine). Focused on the ground and the line of the canal, I paid little attention to anything else. As I was getting ready to make a panorama, I noticed three women not too far away. Dressed in *salwar kameezs* and *dupatas* (common attire of many women in South Asia), they began walking in my general direction. I thought little of it and kept moving around to find the best spot to photograph from. Once they were within four meters of me, they each whipped out light machine guns.

They pointed their guns at me, and I immediately put my hands up, letting my camera swing about my neck. Unsure what to say or do—I had only been studying Hindi for just one month—I found myself saying the only thing that came to mind: "Aapka desh bahut sunder hain (Your country is very beautiful)." Fingers on the trigger, they looked at me and then at one another, then back at me, and then one another. They burst out laughing. They thought I was hysterical. I thought I was a goner. Two lowered their guns and looked in all directions, while the third asked, "Kya tum sarkar se ho (Are you from the government)?" Bumbling in my broken Hindi, I said no and that I was photographing the Ganges River. They motioned with their guns to the direction I had come from and said, "Ganga yahan se bahut dur hain (The Ganges is very far from here)!" I said I know, but pointed at the water and said it was from the Ganges. Guns lowered, lightly bobbing their heads from side to side in agreement, they sternly gestured for me to go. With little hesitation, I walked brusquely back in the direction from where I came. I never turned around. The walk felt eternal. I had strayed more than a kilometer from the car.

I recalled people warning me about Naxalites (Maoist "insurgents" as they're often referred to in the media), but I was always told they were more prevalent in states to the east of Uttar Pradesh

(U.P.): Bihar and Jharkhand. Unbeknownst to me, not far from Chandraprabha Wildlife Sanctuary, the district of Chandauli in U.P. was experiencing an increase in Naxal fighters. I came to find out, after doing some research, that women composed a large percentage of the fighting force. I will never know for certain whether or not the three armed women were Naxals or not. Confident that any mention of this incident to the authorities, friends, or even my family would make my research impossible—I told no one of this incident for years.

My intimate encounter with three machine guns had more to do with the Ganges water flowing in the canal, and with the suspicion I met when asking for maps of this territory in Varanasi, than I thought. I began to grasp that in the context of India, the hinterlands were the most politically charged and volatile area of civic space, not the traditional consolidated city—a cliché, but an important one. In an article published a few months later in *The New York Times*, "Governments Tremble at Google's Bird's-Eye View," V. S. Ramamurthy, secretary in India's federal Department of Science and Technology responded to Google Earth with the comment, "It could severely compromise a country's security." No doubt satellite images did make it easier to see the entrances and barriers to government institutions, but they also exposed vast tracks of unguarded infrastructure running between cities and factories, farms and transportation hubs.

Obtaining maps of cities like Allahabad and Varanasi was never a problem. It was everything outside these areas that proved difficult. As I slowly made my way out of Allahabad, traveling to Varanasi, again by car, but eventually by boat, I began to see the lines of a highly engineered surface. When I spoke to people in Delhi, Patna, or Allahabad, more often than not they would ask why I was so interested in a rural part of the country. Yes, the Ganges has great religious significance for Hindus they would say, but its cities always puzzled people. Besides the great architectural display at Varanasi, few saw why anyone would want to venture out into a "rural jungle", as one person indelicately put it. It seemed anything but rural. I felt as though I stumbled upon an endless city, camouflaged by thousands of hectares of agriculture and millions of cows and water buffalo. Between the clusters of villages and the mosaic of fields watered by canals and lift irrigation, there was not a single space not shaped by infrastructure. I knew that the only way to discern this endless city, camouflaged by thousands of hectares of agriculture and remade every year after the southwest monsoon, was to draw the subtle and monumental changes taking place here. As my familiarity with the basin grew, my photographs and drawings framed a volatile meeting of religious and cultural heritage, agricultural cultivation, and diffuse urban settlements—the whole supported by an elaborate system of hydrological and transportation infrastructures. This infrastructural excess rivaled that of the most densely populated blocks of Manhattan or Tokyo.

Enthralled by this hidden complexity, I set out to see the extent of the Ganges from its source at Gangotri Glacier in the Himalayas to the city of Patna. With great hubris, I thought I could explore, map, and write about all of this in one year. It took me nine years to complete. Traveling by boat for hundreds of kilometers, trekking through the Himalayas, and driving up and down national highways was nothing short of extraordinary. The site of a glacier lodged in the Himalayas, especially for someone like me raised in the plains, made me weep before succumbing to altitude sickness. Seeing *Susus* (Ganges River dolphins) jumping out of the water while boats ready to sink from the weight of sand mined from the river's banks, were pointed reminders of what made the Ganges so dynamic. Because my intent was to document this dynamic landscape, I revisited many of these sites multiple times. In total, I snapped over 25,000 photographs, filled fifteen sketchbooks, scrawled over 1,000 journal entries, and made some 350 maps. A fraction of those recordings and interpretations gloss the pages that follow.

INTRODUCTION

Machine...Monuments...Miniatures...Layers...Atlas...Almanac...
Crisis...Transects

Pump House in Allahabad, 2013 (AA)

Beyond the urban density of Mumbai (Bombay) and the technology centers of Bangalore and Hyderabad lies the Ganges River Basin: a fertile alluvial plain of 1.1 million square kilometers in area, which is today home to over one-quarter of India's billion-plus population. Most of the basin sits within India, but it also extends into present day Bangladesh, Nepal, and Tibet. The area is one of the most densely populated river basins in the world and undergoes radical physical changes every year. With the arrival of the southwest monsoon between late-June and late-August, over one meter of rainfall drenches northern India. Remarkably, despite these drastic seasonal changes and high population density, the basin remains agriculturally productive.

This book focuses on the overlaps and juxtapositions of these three conditions: Population Density—Monsoon— Agriculture. It is an atlas of built and unbuilt projects designed to transform the Ganges River Basin.

Since the middle of the nineteenth century, this watercourse has functioned as a laboratory to test and build a new civilization around the culture of water management. Jointly authored by human actors and their shifting natural heritage, the Ganges River Basin today is a machine in which the entire basin functions as a highly engineered hydrological super-surface. This surface has been constructed from innumerable interventions operating at vastly different scales, from massive state-sponsored canals to individually drilled tubewells. Because of the mixture of actors, the scale of inhabitation, and the widely varying techniques of interventions, visualizing this landscape of infrastructure requires a different kind of atlas.

INTRODUCTION

MONUMENTS AND MINIATURES

Since 1854, when water from the Ganges River was first redirected into the Ganges Canal, irrigation has reshaped the built environment of the Ganges River Basin. Over the past 160 years, the main trunks and branches of colossal canals have been cut through cities and hamlets; brick-lined water tanks were embedded deep in the ground to capture rainwater; and countless small-diameter shallow tubewells have been connected to miniature engine-driven water pumps. These interventions have served, in turn, to reorient people's thinking about the environment. The irrigation infrastructures transformed spaces, from the home to the region, often in unforeseen ways. They fostered agriculture, factories, hydroelectric dams, transportation, and new patterns of settlement while further blurring the distinction between rural and urban. By 1950, India was not only independent from British imperial rule, but her leaders encouraged the proliferation of irrigation infrastructure far beyond the ambitions of the imperial regime. Forged in the crucible of colonialism and nurtured by the new republic, these canals, tubewells, pumps, and adjoining buildings, roadways, and railways had become something entirely new.

Of all the ways one might qualify this new space—*rural, urban, suburban, landscape, drosscape, edge city*, and *megalopolis*–none of these accurately defines such elaborately engineered spaces and infrastructures. A sociological term, such as rururban, first used by social scientists in the 1960s to describe continuities between cities and villages, does not adequately identify the conditions of the Ganges River Basin.[1] From the foothills of the Himalayas to the Bay of Bengal, the Ganges Machine cuts across agricultural fields, cities, and hamlets, inscribing in its monumental re-organization of space and infrastructure a new way of life. Throughout this transformed river basin flowed the forces of tradition and innovation, dotted by diffuse urban projects (regional urban capitals), temporary tent cities (*Magh* and *Kumbh Melas*), miniature infrastructures (tubewells), and colossal public works projects (Ganges Canal). Its spiritual and religious significance inspired reverence in pilgrims; its archaeological and architectural monuments attracted painters in search of the picturesque; its seasonal ebb and flow of water perplexed farmers and engineers alike; and its fast paced urbanization vexed geographers, planners, and architects. In short, the physical and cultural complexity of this territory has challenged traditional terminology.

Despite this remarkable background, the infrastructural and architectural projects of the Ganges River Basin have received comparatively little study—with a few notable exceptions. One in particular is worth mentioning: 'The Ganges Water Machine', which appeared in a special "Food Issue" of *Science* on May 9, 1975. Authored by Roger Revelle, director of the Harvard Center for Population Studies, and V. Lakshminarayana, an assistant professor of civil engineering at the Indian Institute of Technology Kanpur, the article designated what others had yet to claim since the opening of the Ganges Canal 121 years earlier. All of the seemingly disparate and incoherent efforts to transform the basin were finally given a name. But, with the exception of their title, the authors did not once mention the word "machine." Instead, they concerned themselves with low productivity and poor management. "Deeply embedded cultural, social, and economic problems inhibit modernization of agriculture and fuller utilization of the water resources," noted Revelle and Lakshminarayana. No maps or illustrations of how to increase productivity and improve management of the basin accompanied their article. Only equations and diagrams documented how best to utilize and promote conjunctive irrigation of "one of earth's great natural resources."[2] 'The Ganges Water Machine' was a technical paper for scientists and international development specialists.[3]

This book is a different interpretation of that machine: the river basin that was transformed into one of the most engineered spaces on the planet. It draws together the moving parts of the machinery outlined by Revelle and Lakshminarayana, but it also looks at the individuals and stories that contributed to the construction of this robust matrix of population density, monsoon, and agriculture. Part of this involves what novelist Joseph Conrad called "map-gazing," which "brings the problems of the great spaces of the earth into stimulating and directive contact with sane curiosity and gives an honest precision to one's imaginative faculty."[4] Part of this map-gazing entails closely inspecting infrastructure.

LAYERS

This book is an atlas—a dynamic atlas—of the Ganges Machine: a collection of transects that expose the juxtaposing layers of infrastructure and adjoining built forms. The book is subdivided into four sections, starting in the Himalayas and ending at the historic city of Varanasi, that illustrate how the super-surface is a continuous infrastructure of both water management and risk management.

The development of these infrastructures is largely a story of linear succession, from monumental public works projects like the Ganges Canal (still growing in length) to the miniature scale of the tubewell (whose exact number and distribution remains purely speculative). And while at first blush the development of these technologies is linear—going from large, top-down approaches (canals), to small, bottom-up approaches (tubewells)—the story is in fact anything but a linear succession of technological innovation. Instead, it is a story of how these technologies were layered over one another, often in unexpected ways, and how their built forms shaped the land. Canals of the nineteenth century were not abandoned once tubewells became popular in the early twentieth century—on the contrary, as tubewells proliferated across the basin, new canals and mega-pumping stations were built across the Ganges Basin.

The Ganges River Basin is best viewed as five distinct modes of irrigation: rivers, canals, tanks and lakes, pumps, and tubewells. At different times of the year these infrastructures compete, coexist, and occasionally feed off of one another. A cut through these five layers reveals a lamination of infrastructural ideologies borne out of the discontinuities of politics, bureaucracies, and cultural practices. Even though these ubiquitous infrastructures affect millions in their daily lives, there is no map that legibly renders their built reality or the relationships they produce.

It might be worth asking: what use are detailed cartographic maps of a territory transformed into a machine? Geographers, surveyors, and hydrographers are not trained to draw plans and sections of machines. This is the job of engineers and architects. Is this perhaps why the Survey of India (established in 1767) has not mapped this hydrological super-surface in over 45 years? Did each Survey General and each generation of bureaucrats agree that static, stale cartographic conventions of representation could not compete with infrastructure embedded in the ground? To make matters more complicated, over the last 45 years the basin's exploding population has produced new clusters of settlement along existing roads and railways, although the basin has nevertheless remained largely agrarian. Who can blame the Survey General for not wanting to trace the movements of a machine of such heterodoxy? Not to mention, how does one possibly map the territory where a meter of rainfall transforms parched earth into gelatinous silt, flooding and submerging thousands of kilometers of land every year? A dynamic basin such as this presents many challenges to cartographers and physical geographers. Even in an age of GPS (Global Positioning System) and GIS (Geographic Information System), the Ganges River Basin remains largely uncharted.

ALMANAC, ATLAS: TRACKING CRISIS

A far older surveying tradition, and one that is more attuned to change (and risk), is that of the *patra* or *jantri*, known in English as an almanac. An annual publication, the almanac plotted changes in the seasons, the positions of the celestial bodies, and noted the high and low tides for those aboard ships at sea. One can find almanac makers today in nearly every town and city in northern India, as they have been for hundreds of years. Their job is to anticipate change. If an almanac is a collection of calendars and astrological positions, and an atlas is literally a collection of maps, then this book draws the dynamic conditions of people and weather (monsoon) along with agriculture and infrastructure. It draws both cyclical/seasonal time (almanac), and historical time (atlas), since the interaction with the river is necessarily time-based. The goal is to manifest time visually—to be able to "see" both the history and the cycles of the river—to perceive it dynamically.

If this super-surface is still in need of an atlas-almanac hybrid—a dynamic atlas— then this dynamic atlas, once drawn, should yield criteria for design that is at once dynamic and stable.

There are any number of ways the dynamism of the Ganges River might have been interpreted. Since its beginnings in 1854, the forces that drove the development of the Ganges Machine exploited these dynamic changes as signs of "crises." Floods, droughts, and lackluster agricultural productivity were all used to justify "technological intervention"—without ever once losing faith in technology to deliver a better managed modern future. After all, both dynamism and crisis are related. They both deal with time and probability.

In a pre-Ganges Water Machine era, officer James Johnson observed in 1808 that due to "the flatness of the country... and its being every where intersected, and a great part annually overflowed by the Ganges, will for ever be the cause of sickness, as well as fertility!"[5] By the 1840s, officers and administrators in the East India Company decided the best way to diminish sickness (famine) and cultivate fertility (agricultural production) was to *re-engineer* the environment through irrigation. Since then, imperialists and nationalists alike have embraced the continual redesigning of this vast alluvial plain.

If there is a way to understand the drive to construct the Ganges Machine, it is through understanding the protagonists' sense that they should live not *with* machines but *in* the machine. They sought, quite literally, to reconstruct their environment into a machine the size of the basin itself.

Without exception, the development of the Ganges Machine was undertaken through the exploitation of crisis and of trauma. Although it was built to avert economic and agricultural crisis, this machine produces its own set of crises. Most crises, such as hurricanes, tornados or earthquakes, are instant or happen in a few days or weeks, but the transformation of the built environment into a water machine over the last two centuries has also produced a crisis of water exploitation that is occurring in slow-motion. The scale and ubiquity of this exploitation is not perceptible in everyday life. Given the complexity of the different phenomena, from urbanization and agriculture to weather and infrastructure, its visibility, registration, and representation is complex. A slow-motion crisis of this scale and speed must be rendered through measured drawings to be fully understood.

The difficulty of bringing the yearly cycles of the monsoon and expanding human settlement into the same picture plane as the hydrological systems of ground and surface water, presents serious challenges to rendering the mixture of water, soil, and culture. Moving beyond the modes of linear, static, and mono-scale image making common to engineering, a dynamic approach to design is perhaps best poised to synthesize across these diverse scales and materials of the built environment. In other words, if engineers traditionally deal in statics, designers deal with variable dynamics. This book, then, provides a designer's survey of the Ganges Machine.

TRANSECTS

Finally, while most maps draw the rivers and canals of the Ganges River Basin as a web of misshaped plumbing lines, this book suggests using the *transect* to expose how water exists as much above the surface of the land as it does below. The transect is a visual technique, devised by Alexander von Humboldt and further developed by Patrick Geddes, that draws a swath through the landscape—a sectional sampling of the characteristic land uses. Von Humboldt used it as a tool to better understand microclimates. Geddes, and later Ian McHarg, used it is as both an analytical tool and an instrument of design. More recently, it has been used by New Urbanists to create a consistent gradient between urban and rural (at least in theory). In the case of the Ganges River Basin, cutting transects exposes the many layers of the Ganges Machine. Given the expansive nature of water and transportation infrastructures crisscrossing the basin, it offers a promising method of surveying the layered super-surface described above.

This book emerges from these transects and layering, representing the first attempt at visually recording the dynamic rhythms and experiences of this territory in over 45 years. It is an introduction to a highly dynamic machine, celebrated in its infancy, but largely overlooked today in its prime.

PART 1

Gaumukh...Ganga...Ganges...Gaumukh to Rishikesh...Theater...Outwash Fans...Traverse Survey...Military Force...Topographic Monumentality... Guidebooks...Global River Basins...Ganges...Monsoon... Pictures... Behemoth... Vapors...Tools

Bhagirathi River at Uttarkashi, 2006.

GAUMUKH

Satellite Photo of Gangotri Glacier at Gaumukh, 2008 (Google Earth).

GAUMUKH

GAUMUKH

Gangotri Glacier at Gaumukh (Source of the Bhagirathi River) 3,892m above sea level, circa 1905 (AA).

Gangotri Glacier at Gaumukh, 2006.

GAUMUKH

Ganges at Rishikesh, 2006.

It is not strange that the Ganga should have such an extraordinary hold on the imagination of the people of India. For millennia she has watered and nurtured an entire civilisation, and has become a symbol of eternity and a theme of art, legend and literature. The moods of rivers are fascinating to watch but even more so are the faith and reverence they evoke in the hearts of millions. To one born in Allahabad this is imprinted on the mind by the constant stream of pilgrims who journey long distances to take a quick dip in the waters of the Ganga.

... what my father [Jawaharlal Nehru] called "a symbol of India's age-long culture and civilisation, ever-changing, ever-flowing, and yet the same Ganga."[1]

Indira Gandhi, 1976

From the Himalayas to the Malabar Coast, the word Ganga designates, in sacred texts and everyday lives, a terrestrial, earth-bound river and a celestial flow of purifying waters. Ganga is a general term for what everyone would agree constitutes a river. For there is no denying the fact that it is indeed a large flow of water, streaming across the land and emptying into the ocean. But the nature of that flow of water (is it natural or artificial if it is dammed?) and its properties (is it pure or defiled if it is full of sewerage?) defines how people name it and how they use it.

Early on, long before explorers mapped the river and modern engineers redirected its flow, Ganga was venerated in numerous sacred texts. In the *Mahabharata*, goddess Ganga is commonly viewed in her anthropomorphic form as a daughter or wife. In the *Puranas*, she is the ultimate water (*jal*) purifier and a central element to myriad rituals. Stone sculptures, from the caves at Ellora to the island of Java, depict Ganga as a great purifier, mother, daughter of Himalaya, and as a stream or wife of Lord Shiva's. Descriptions of the sanctity and importance of the Ganges spread as far as Ancient Greece. Megasthenes, a Greek emissary to the Mauryan Empire (circa 300 BCE) and stationed in Pataliputra (near modern day Patna), spoke of the Ganges as the greatest river of the earth.[2] Strabo, in his 17-volume *Geographica*, written during the reign of the Roman Emperor Augustus (63 BCE–14 CE), described India as a land of rivers, with the Ganges as the largest of them all.[3] Those describing the country, from Dante Alighieri 's *The Divine Comedy* (1308–21) to photographer Raghubir Singh's *The Ganges* (1992), have defined India in terms of the Ganges—its Axis Mundi. *India intra Gangem*—India within the Ganges—and *India extra Gangem*—India beyond the Ganges—defined India and most of what today we call Southeast Asia for centuries.[4] It is no wonder that the entire Indian subcontinent is profoundly inscribed with the geography, imagery, and sanctity of this unequalled watercourse. All other river basins and streams are extremities—given attention in their own right, yet always more significant in relation to Ganges' stream.

The quotation above—written as a foreword to the book *Ganga* in 1976 by Prime Minister Indira Gandhi—describes Ganges' symbolic and cultural importance to modern India: the Ganges as the theater of Indian civilization. Its protagonists are at once Ganga, most notably in her capacity to water and nurture; the farmers and herdsmen who benefit from her nourishment; and the millions of pilgrims, streaming from great distances to take a quick dip.

In the word Ganga we can glimpse the scenes of ancient India or Bharat as depicted in the *Mahabharata*, traditions sustained through the colonial and imperial periods all the way to modern, independent India. Mrs. Gandhi, drawing upon the words of her father, gives us a sense of the scenes of the Ganges, for the river serves at once a cyclical, ritualistic, and religious function in her description, even as it performs in "modern" ways as a domestic and industrial resource.

Its scenes are "ever-changing, ever-flowing, and yet the same Ganga." In other words, its stasis and dynamism, just as its past and future, can be rehearsed *ad infinitum*: for if something is constantly changing and staying the same, it follows that what was created in one moment will give way to destruction in the next. The actions of the river, most notably the seasonal ebb and flow of water during the southwest monsoon, obscure and deform whatever is being created or destroyed. Perhaps this is why Raghubir Singh, having photographed the Ganges for decades, observed: "The mountain, the plain and river are the real protagonists and metaphor's of India's life."[5]

But does it end there? Is it purely a question of geography? Perceptions of the Ganges are constantly mediated, and these mediations produce artifacts like canals and water pumps that regulate the river's flow, alongside the temples and *ghats* (steps) that make up the ritual infrastructure of the river.

GANGA

View En Route From Gangotri (Village) to Gangotri Glacier at Gaumukh, 2006.

View Upstream From Gangotri (Village), 2006.

Neither geography nor tradition, neither engineering nor ceremony, these sites together create a dynamic interplay of a changing river and its changing culture. Is the river shaping society and society reconfiguring the river? Are cause and effect distinguishable? The drama that unfolds does not progress in a linear narrative towards some over-arching conclusion. It is, rather, the cyclical replication of a single spectacle: destruction always at the heels of creation, forever interlinked coevals.

This means factoring in time as one interprets the river; it means understanding the rhythms of mechanical pumping alongside the recurring sanctity of a site, and observing how these layers interact with one another. What people call Ganga often depends on how they see these layers, each with its own story, each contributing to the shared story that explains how all of them came together into a single, thickened, "ever-changing" landscape. And it is this layering that accounts for the disorderly, contested usage of the term *Ganga*, which names the celestial stream at the same time as it identifies an agro-industrial resource. Drawing and writing about these layers may give some sense, but it is only ever momentary, fragmentary, the separated strata of the continuous, unified Ganga. To 'see' the Ganga, we have to visualize these shifting layers—only then might we be able to see how they interact to define this robust and varied landscape.

How does one write and talk—or draw—about such a restless place?

THEATER

If there is a map of this theater, then perhaps the *Ramayana* is its most enduring blueprint. An epic Sanskrit poem spun by Valmiki some twenty centuries ago, the *Ramayana* has a cyclical logic and drama that has sustained, like Jawaharlal Nehru said of the Ganges, "an extraordinary hold on the imagination of the people of India" for millennia. Recited and rewritten over countless generations, this unmatched narrative presents the Ganges as a goddess, a river, and a setting—a landscape of sorts.

How far up in the Himalayas does the Ganges begin? Where and how does this water become purified? Even if they have never seen the Ganges, itself, nearly every Indian has read or heard how Ganga descended to earth: King Sagar's great grandson, Bhagiratha, spent one thousand years in the Himalayas praying ceaselessly and living on fruits, roots, and water. After this millennium of supplication, the boon-granting goddess Ganga appeared. Bhagiratha recounted to Ganga the story of the death of his ancestors, when a Hindu sage burned 60,000 of King Sagar's sons to death. Bhagiratha asked for Ganga's stream to purify their burnt remains. Ganga, daughter of Himalaya, was greatly moved by this tale. She agreed to his request, but cautioned that the impact of her fall from heaven would be so great that she would flood the earth. The only god capable of sustaining her fall was the "Blue-throated Great Lord" Shiva, who would catch and secure Ganga by entangling her in his hair (*jatas*). As she descended through Shiva's locks of hair, she spread across the vast plains to the Bay of Bengal, at long last washing the burning remains of Bhagiratha's ancestors. Other stories of Ganga's descent involve Vishnu cracking the vault of heaven with his foot and Ganga's waters pouring through the crack. Ganga always comes into contact with three of the most significant Hindu deities: Shiva, Vishnu and Brahma, spreading their divinity across the terrestrial plane. Thus, Ganga is also known as *tripathaga* (the union of three aspects).

"The stream of Ganga, with lovely tortuous waves meandering on Siva's head, create[s] doubts in the minds of the celestials as to whether they are lily garlands, digits of the moon, the sprouts of deeds of merits, the sloughs of snakes, or ashes falling from Siva." The Ganges, in this stream of metaphors, is identified as a creation almost beyond comprehension.

For millennia the *Ramayana* has been materialized in the world through the enactment of rituals and the construction of landmarks. From Gangotri Glacier in the upper reaches of the Himalayas to Sagar Island in the delta of the Hooghly River, the Ganges forms a constellation of terrestrial sites imbued with celestial significance. These sites are marked by *Tirthas*—sacred crossings such as a river ford or mountain—where devout Hindus perform ritual purification, "taking the waters cupped in their hands and pouring them back into the river as offerings to the *pitrs* and the *devas*. They present in the river, as in the sanctum of a temple, offerings of flowers."[6] To commemorate a holiday or an auspicious astrological juncture, pilgrims trek hundreds of miles, often camping along or near the shores of the Ganges to bathe at providential hours. They carry back home jugs of Ganges water (*jal*) along with devotional images. *Tirthas*; *tirthayatras* (pilgrimages to sacred sites); Hindu rituals, festivals, and temples; and devotional images help create and sustain a scenography that is "ever-changing, ever-flowing, and yet the same Ganga."

This procession starts at Gangotri glacier at Gaumukh: 3,892 meters above sea level where the Bhagirathi River begins—the principal tributary of the Ganges River. Gaumukh is where Ganges makes her entrance into the terrestrial plane and descends over 3,000 meters in less than 250 kilometers. But to get to Gaumukh, one must first get to Gangotri, a small town on the banks of the Bhagirathi. There, one sees temples, *kunds* (tanks or basins), along with bridges and *ghats* that mark topographic features. At Surya Kund, half a kilometer from Gangotri Temple, Bhagirathi descends some fifty meters into a box canyon of polished bedrock granite. Ghats carved into the granite mark its ritual uses. But how long have they been there? Or, for that matter, how old are the other kunds in the area? Voices of tradition, such as a local priest or resident, commonly say such sites are as old as time itself.

In 1808, authorized by the Bengal Government, Lieutenant William Spencer Webb led an expedition to survey the Ganges from the foothills of the Himalayas at Haridwar to Gangotri. The first Anglo-Europeans to traverse and map this terrain, the goal of their expedition was straightforward: to ascertain the sources of the Ganges, Yamuna, Alacanda, and Bhagirathi Rivers. Unable to ascend all the way to Gangotri because of threats and protests from the Gorkhas, Webb sent a team of

Indians to see Gangotri instead. On May 19, 1808, the team reached the town. "There are three Cundas or basons [kunds or basins], where the pilgrims bathe, called Brahma-cunda, Vishnu-cunda and Surya-cunda, formed in the bed of the river" noted Captain Felix V. Raper in his report on the expedition to the Asiatic Society in 1812. In the town itself, they noted a single temple, eight to ten feet high, with images of Ganges and Bhagirathi amongst others. They also observed the area's landscape, "The mountains in the vicinity have a very barren appearance; the only tree produced there is the Bhurjapatra."[7] (Bhurjapatra [*Betula utilis*], or Himalayan Birch, is found throughout the Himalayas, usually at altitudes above 3,000 meters.) Beyond three kunds, a temple, and the barren landscape, the team reported little else to Lieutenant Webb. Perhaps this was because their trip was made exceptionally difficult by poor roads, a violent squall, and snow.[8]

Getting to Gangotri today is easier than in the past; the roads are paved although often without guardrails. As the snow begins to recede in the lower Himalayas, usually near the end of April, pilgrims begin their *tirthayatra* by foot or bus in the city of Haridwar. Haridwar (also spelled Hardwar) is where the Ganges enters the plains, some 290 kilometers south of Gangotri.[9] If one is in a hurry, the ascent by bus or car typically takes two days. But by foot, it can take several weeks. Temples and ghats dot the basin, often marking *tirthas* on the route to Gaumukh, and pilgrims stop at many of these sites for hours and sometimes days. Once at Gangotri, far fewer pilgrims or tourists continue to Gaumukh. Twenty kilometers (12.4 miles) further upstream, the narrow, winding path is fit only for walking or donkey riding. Those who do continue onward are rewarded.

With a clear sky in July, a hiker coming from Gangotri village can see, from ten kilometers or so to the southeast, what at first appears as a colossal wall of burnished turquoise in the distance, softly glowing in a hummocky valley. Moraines—debris carried and deposited by a glacier—form a variegated grey wall along the lower valley. Advancing a few more kilometers, one approaches the snout of the Gangotri glacier: a 75-meter (246 ft) tall wall of ice. The terminus of a glacier is often referred to as its snout, and fortuitously so at this place, for Gaumukh roughly translates as "mouth of a cow."

Along the banks of the Bhagirathi River in Uttarkashi, 2006.

THEATER

Panorama of Outwash Fans En Route to Gangotri Glacier at Gaumukh, 2006.

1 of 2: Vertical Panorama Across Bhagirathi River Valley, 2006.

Panorama at Gangotri Glacier at Gaumukh, 2006.

En route to Gaumukh, hikers traverse the compacted gravel path and notice changes in the consistency of the ground underfoot; places where the rock is coarser, squishy and loose, and prone to slippage. Unawares, they are crossing an outwash fan—a deposit of sediment from a melting glacier or other flow of water that, from the air, resembles the shape of a Gingko tree leaf. These bright white gutters of the mountains puncture the otherwise continuous moraines of the valley. Dry most of the year, they turn into torrents filled with rocky debris during the summer southwest monsoon. Before the monsoon, they may have a trickle of water passing through them, or they may be bone dry. Made of loose gravel and pebbles, they are easy enough to identify on the opposite side of the valley because they appear as white stems; they are harder to see along a narrow and winding path like the one to Gaumukh. In the months after the snow melts and before the monsoon, hikers must tread carefully. Not only might a loose boulder roll down, but the loose gravel outwashes are weak underfoot. A careless step may lead to a fall.

2 of 2: Vertical Panorama Across Bhagirathi River Valley, 2006.

THEATER

Top and Bottom: Terrace Farming Along the Bhagirathi River, South of Gangotri (Village), 2006.

Bhagirathi River at Gangotri (Village), 2006

Maneri Dam Along the Bhagirathi River, 2006.

With fewer trees, taller moraines, and wider outwash fans, the torrential water flowing through a rocky bed produces a distinct background noise in the final few kilometers to Gaumukh, punctured by rocks falling and chunks of ice breaking off the glacier snout. Those chunks of ice that break off and don't settle into a bed of ice-crystals at its base float further downstream, creating an archipelago of miniature icebergs. Glacial flour—fine sediment produced by the glacier grinding its rocky bed—gushes from the snout. Ice cold and thick with sediment, this grey, milky water grows in volume and velocity as other brooks and outwashes join it. By the time the stream reaches Gangotri, it has attained a perilous volume and velocity. It gushes through mountainous valleys—passing Uttarkashi, Devprayag, Rishikesh—all along the way gaining in speed and energy.

Before arriving at Haridwar, the water's descent is transformed into electricity first at the Maneri Dam, producing 42MW, and later at the Tehri Dam, which produces 1000MW of electricity and diverts water to urban dwellers in New Delhi. Along the way, terraced fields subdivide steep slopes. Villages and hamlets hug the contours of paved roads active with pilgrims and tourists in the months between winter and the southwest monsoon. Temples mark *tirthas* between these infrastructures. One can date the age of these dams and barrages, but it is difficult to ascertain the age of temples and ghats. Pilgrims rarely ask how old a structure is and, when asked, they will speak to you about why this site is revered. Never mind its age.

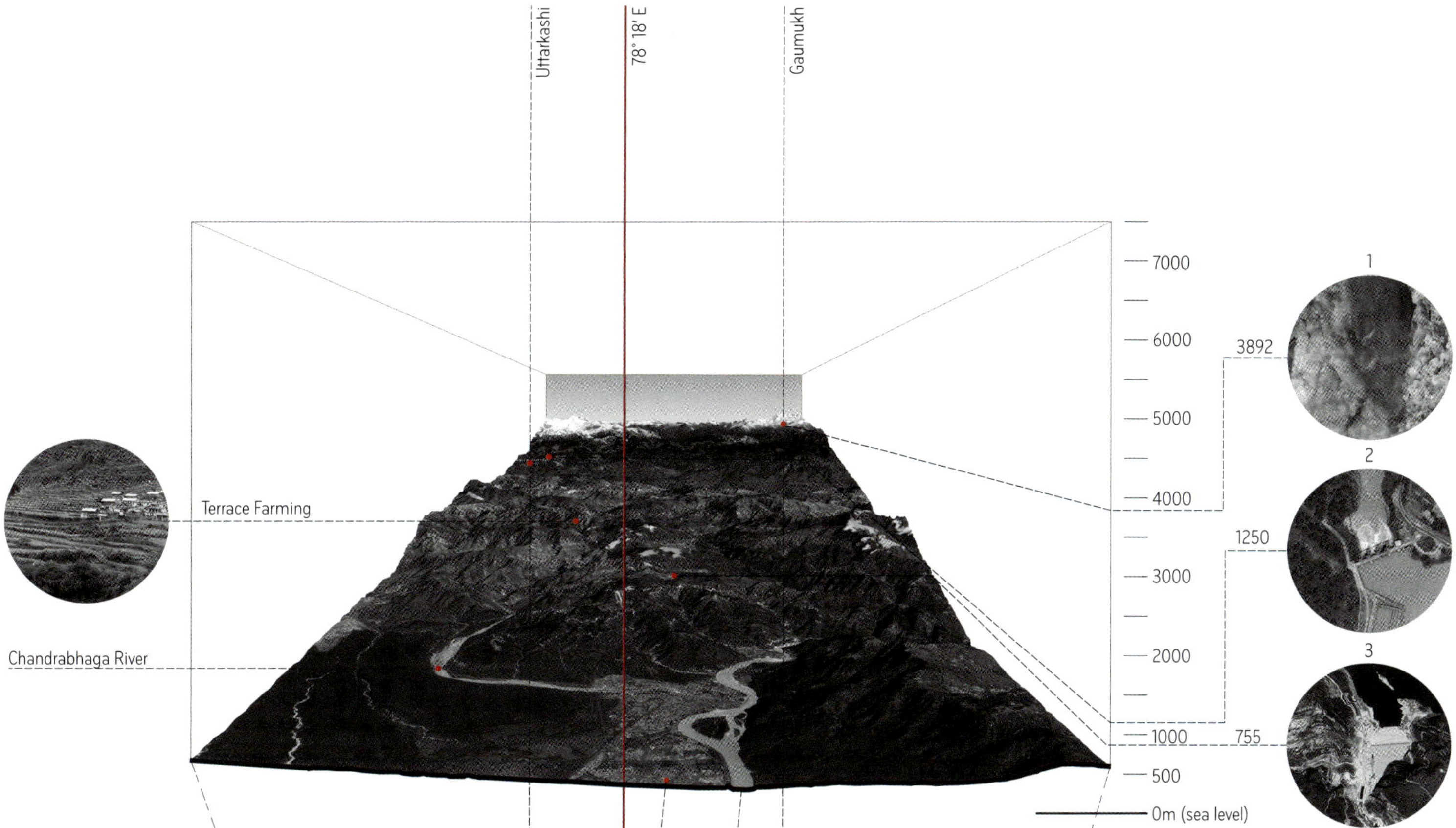

Hydrological Regimes
in the Himalayas, 2009 (AA).

(1) Gaumukh (Gangotri Glacier)

(2) Maneri Dam
- Catchment area: 4,024km²
- 39m tall Concrete Dam
- Gross Storage Capacity: 600 X 10³m³
- Produces 42MW of power
- Completed in 1984

(3) Tehri Dam
- Catchment area: 7,511 km²
- 260.5 m tall Rock and Fill Dam (largest in Asia)
- Gross Storage Capacity: 3,540,000 X 10³m³
- Produces 1000 MW of power
- 300 Cusecs (162 million gallons per day) of drinking water for Delhi (600,000 people)
- Completed in 2005

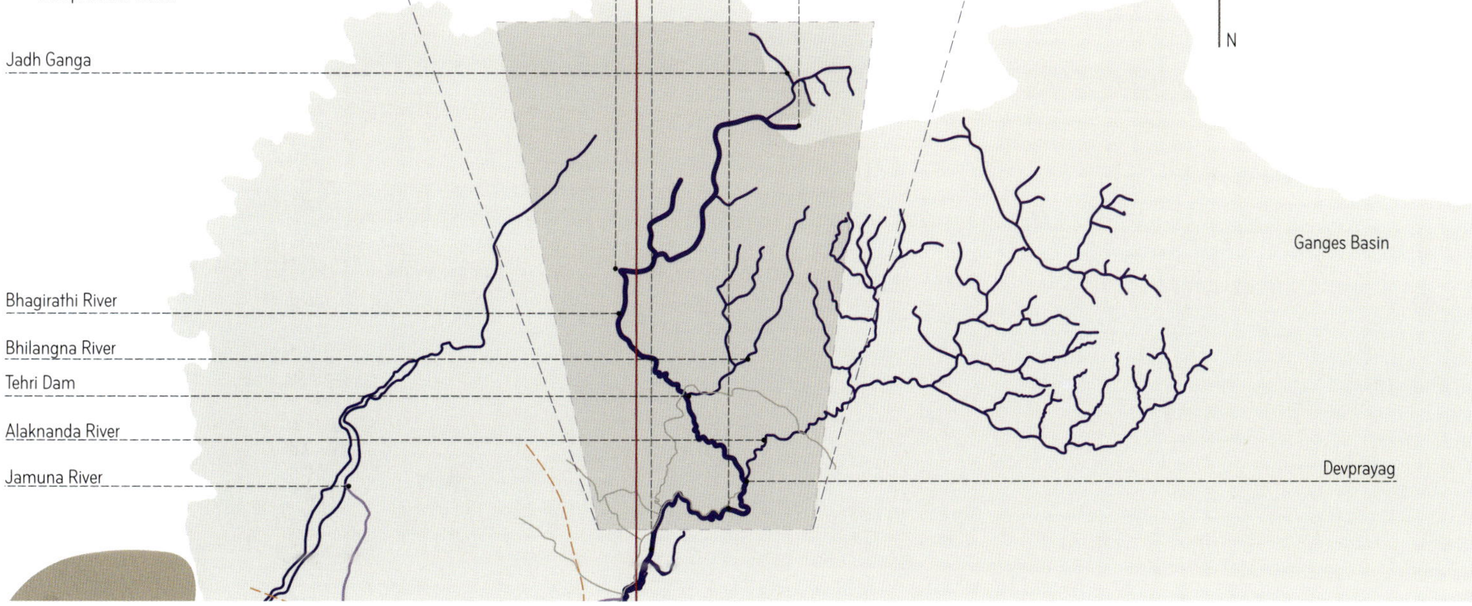

THEATER

TWO KINDS OF MAP: SURVEYING THE GANGES

In the late-eighteenth and early-nineteenth century, tourists, primarily Anglo-European travelers and artists, half-mesmerized and half-horrified by devotees ascending to the upper reaches of the Himalayas, began writing and painting scenes of the area. More generally, India, particularly outside of urban centers like Calcutta (Kolkata) and Madras (Chennai), "provided rich and varied subject matter for some of the finest illustrated books ever published in England."[10] Their visual and written descriptions foregrounded topography, with an eye to "paint their lonely desertness, or describe the undefinable sensation of reverence and dread that steals over the mind while contemplating the deathlike ghastly calm that is over them." For "it is not so simple, to many surely not very possible, to convey an adequate idea of the stern and rugged majesty of some scenes...Surely such a scene is Gungotree!"[11] So exclaimed James Fraser, the first modern European painter to set eyes on Gangotri in the summer months of 1815. He was impressed not only by the great numbers of pilgrims he observed hiking to Gangotri and other sacred sites, but also by the sites themselves, noting that "A great portion of these places, like the acts they record, and of which they were the theatre, are fabulous..."[12] At this time, the East India Company was expanding its territory. After Lord Lake captured Delhi from the Marathas in 1803, the Company moved further into Upper India and the Himalayas. By 1814 they were enmeshed in the Anglo-Nepalese (Gorkha) War.

Traveling with his younger brother William, a Civil Servant in the Company, James kept a journal and sketched throughout his trips, which he hoped would not only add "mite to the general stock of geographical knowledge," but also "exhibit a picture of its inhabitants."[13] Following in the picturesque travelogue tradition of artist William Hodges' *Travels in India, 1780, 1781, 1782 & 1783* (1793)—which, for Hodges, was to counteract the ways in which "reasoning assumes the place of observation, and the traveler is lost in the philosopher" —Fraser's journal, too, placed a premium on observation.[14] And whereas Thomas and William Daniell, in their six part *Oriental Scenery* published from 1795 to 1805, only spent a month in the Himalayas before moving to Sri Nagar, James Fraser had the opportunity to spend far more time in the mountains. This afforded him the chance to not only watercolor and sketch his experiences, but also map what was "hitherto a blank in the Geography of Hindostan."[15] This was no easy task. Though he was armed with a theodolite (a surveying instrument), small pocket compass, and a perambulator (similar to an odometer), the uneven ground and "dark[,] foggy days robbed" him of fixing his bearings. In such circumstances, he made use of his small pocket compass carried in hand to note the general direction of the land.

To call the Fraser brothers militaristic trek a picturesque tour would be misleading. In May of 1815, almost seven years to the day that Webb's team reached Gangotri, and just days after the Gorkha chiefs ceded the Garhwal range to the Company—William Fraser had just been appointed Commissioner—the brothers traversed the lower Himalayas for three months. In a show of force, William corralled some six hundred troops to escort them on what must have appeared a triumphal march to local villagers. This was no small caravan of amateur artists in search of the picturesque. Part I of his travel journal, titled "Historical Sketches of Nepal, and of the Ghoorka Conquest" suggests as much to the reader, as does his first sentence, "In the end of 1814 it was deemed expedient by the British government to declare war against that of Nepal," that this is as much a travelogue as a military survey.[16] Fraser's Appendix, replete with references to the Population of Nepal and the Military Force of Nepal, bookends what was clearly as much a military survey of newly acquired lands *and* people as it was an artistic adventure.

Where not sylvan or mountainous, the land was chiefly agricultural or pastoral. James paid careful attention to the ubiquity of terraced farming. It gave a "ridgy appearance," which both "chiefly attracts the traveler's eye" and "produces a strange regularity, which frequently takes from the dignity of the landscape." Every mountainside in the lower areas, regardless of slope, bears the imposition of this regularity and shows "the mark of former tillage." Laced within this regularity, James observed an elaborate "irrigating system" that "is chiefly necessary for the rice crop, which, though not put into the ground till the rains have set in, frequently requires the assistance of artificial flooding." The singularity of this somber, stunning flood was impressive—and informative for Fraser, as was the neatness and skill with which these ridges were leveled, so as not to collect water nor erode the thin layer of soil. When this artificial flooding was imposed on the landscape, "the singularity is heightened by the aspect of a hilly country partly under water."[17] Such an observation shows Fraser at his most imaginative and observant. And while the regularity of the ridges followed the contours of the land, the deeper he went into the hills and mountains, it is clear, both to Fraser and any reader at the time: this topography was shaped by people. It was many things, but it was not pure wilderness.

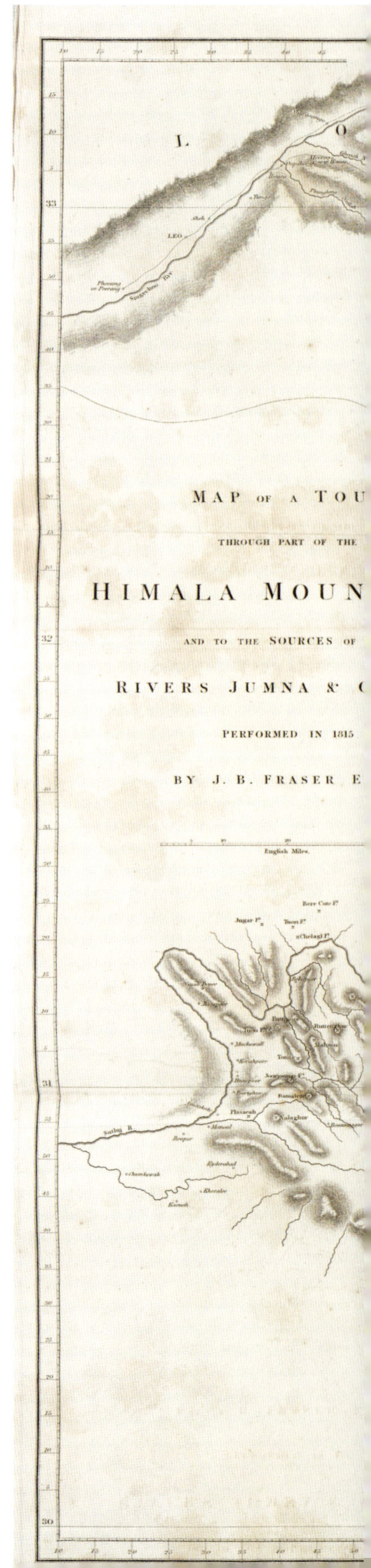

James Baillie Fraser, "Map of a Tour Through Part of the Himala Mountains and to the Sources of the Rivers Jumna & Ganges Performed in 1815 by J.B. Fraser, ESQ." **Journal of a Tour Through Part of the Snowy Range of the Himālā Mountains, and to the Sources of the Rivers Jumna and Ganges**. London: Rodwell and Martin, 1820. (Columbia University Library).

James Baillie Fraser, "Views of the Himala Mountains by J. B. Fraser, Esq." **Views in the Himala Mountains**. London: Rodwell and Martin, 1820. (Princeton University).

By mid-July, William was called to Sri Nagar to help settle the newly acquired lands; James continued to the sources of the Ganges and Jamuna. From time to time, the topography at certain points reminded him of his home in the Highlands of Scotland. After an eleven mile hike in early May, fatigued though he was, the scenery was such that: "Asia was almost lost in our imagination: a native of any part of the British Isles might here have believed himself wandering among the lovely and romantic scenes of his own country." He continued, "The delight of such association of feeling can only be understood by those, who have lingered out a long term of expatriation, and who anxiously desire the moment of reunion with their native land."[18] Not only did some hills resemble "those of the Scotch highland sheepwalks, totally bare, save of the brown dry grass," but the architecture of certain villages "exactly resemble the ruins of old castles frowning from their heights over the valley."[19]

James's vision of the topography was situated within the reassuring landscape of the Scottish highlands and glens, for the foothills and Himalayas presented scenes with "so highland an air, that it was difficult to believe it to be Indian."[20] Though quick to draw parallels for the reader between the scenes of Scottish highland sheepwalks and those of the hills leading to the Himalayas, Fraser pointed out the differences in topography. He underlined the difference by recounting that a typical mountain hike in Scotland can cover twenty to fifty miles, often with little effort, while a thirteen-mile march in the Himalayas may produce moderate fatigue. He contended that whereas no hill in Scotland exceeds 4,200 feet (1,280m) above sea level and the paths of the slope generally rise slowly, hills in the Himalayas ascend 5,000 to 6,000 feet (1,524 to 1,829m) and "the great source of labour is in the excessive and universal steepness and abruptness of their sides, always rising suddenly at a great angle from the streams, that fill the chasms at their feet."[21]

James Baillie Fraser, "Jumnotree The Source of the River Jumna." **Views in the Himala Mountains**. London: Rodwell and Martin, 1820. (Princeton University).

After months of trekking up and down and across these steep hills, by August, William and James returned to Delhi. James continued to sketch and edit his journals before returning to Calcutta, where he would spend the next four years studying with artist George Chinnery and eventually working closely with William Havell on aquatint engravings to improve his skills as a painter. In 1820, he published *Journal of a Tour Through Part of the Snowy Range of the Himālā Mountains, and the Sources of the Rivers Jumna and Ganges* in England. William Havell, whom he met in Calcutta, engraved the drawings in Fraser's *Views in the Himala Mountains*, published in 1820 as well, a book of twenty scenes meant to add detail to his journal—particularly his key map of the tour.

By linking the geographical space represented in the map with the perspectival space of the picturesque paintings, Fraser situated these views within a larger set of coordinates. A viewer was not limited to just one mode of viewing this newly acquired territory. The map, by virtue of its coordinates relating to a picturesque perspective of a topographic landmark, was not just a field of points plotted with a compass and level—it was a field of specific spaces. Each type of representation reinforced the other.

The topographic monument balances these different types of representation. In the context of the Himalayas, the topography provided key landmarks to define boundaries and hierarchies of space. James's sketches of Gangotri attest to such a designation. Consider 'Gungotree The Shrine of Mahadeo.' Nine men appear in the foreground, all in different poses, while a tenth man wearing a red turban appears to be walking due north, poised to either round the corner and enter the religious compound or enter some cavernous grotto. The man's turban matches the tattered flag hanging over a single temple, no more than four meters tall, surrounded by a stone wall. Using the same pigments and line work in the surface of the temple as he does the bark of the tree, Fraser and

James Baillie Fraser, "Gungotree The Holy Shrine of Mahadeo." **Views in the Himala Mountains**. London: Rodwell and Martin, 1820. (Princeton University).

Joseph Meyer, "Die Ganges—Quelle in Indien." **Meyer's Universum**, circa 1830s. (AA).

Havell conflate the works of nature with those of the inhabitants. If it were not for the hard edges and tattered flag flying above, the temple might melt into the ground like the roof of the stable to its left does. Such conflation between "native" architecture and "nature" starts with the opening cartouche of the illustrated volume. Here, vegetation, hills, and mountain peaks dominate the scene while two men sit in front of the freshly carved title. To the right is the dome of the temple found at Gangotri. And, like the aquatint engraving, the color of the flag matches the red of the men's turbans.

James's painting depicted a landscape dominated by mountains and trees, but it went further than earlier descriptions in its precision—the painting demarcated thresholds between the forest, the grey-brown rock (barren landscape), and snow-capped mountains in the background. For instance, in the central valley, where the shadow from the mountains on the left falls, it not only creates a sharp shadow, but it also forms a boundary for the four parallel furrows of vegetation.

Fraser's vision characterizes the picturesque perception of the late-eighteenth and early-nineteenth centuries depicted in so many travelogues, guidebooks, and printed images. In such representations, human figures, always representations of "natives," define scale and inhabitation. Their footprint is not minimal and they are not inconsequential. Prints, like "Assemblage of Hillmen," from *Views in the Himala Mountains*, show careful attention to groups and their individual attire. They reflect Fraser's interest in topography and ethnography. "With the advent of works like Fraser's the tendency to look simultaneously at both the physical and human components of the Indian world seems more marked," observed English art historian Mildred Archer.[22] Nonetheless, like his predecessors and contemporaries, Fraser shared the period's preoccupation with the picturesque furniture of nature—rocks, mountains, and lakes. Fraser's notoriety and ingenuity was in how he represented both the inhabitants and topographic conditions.

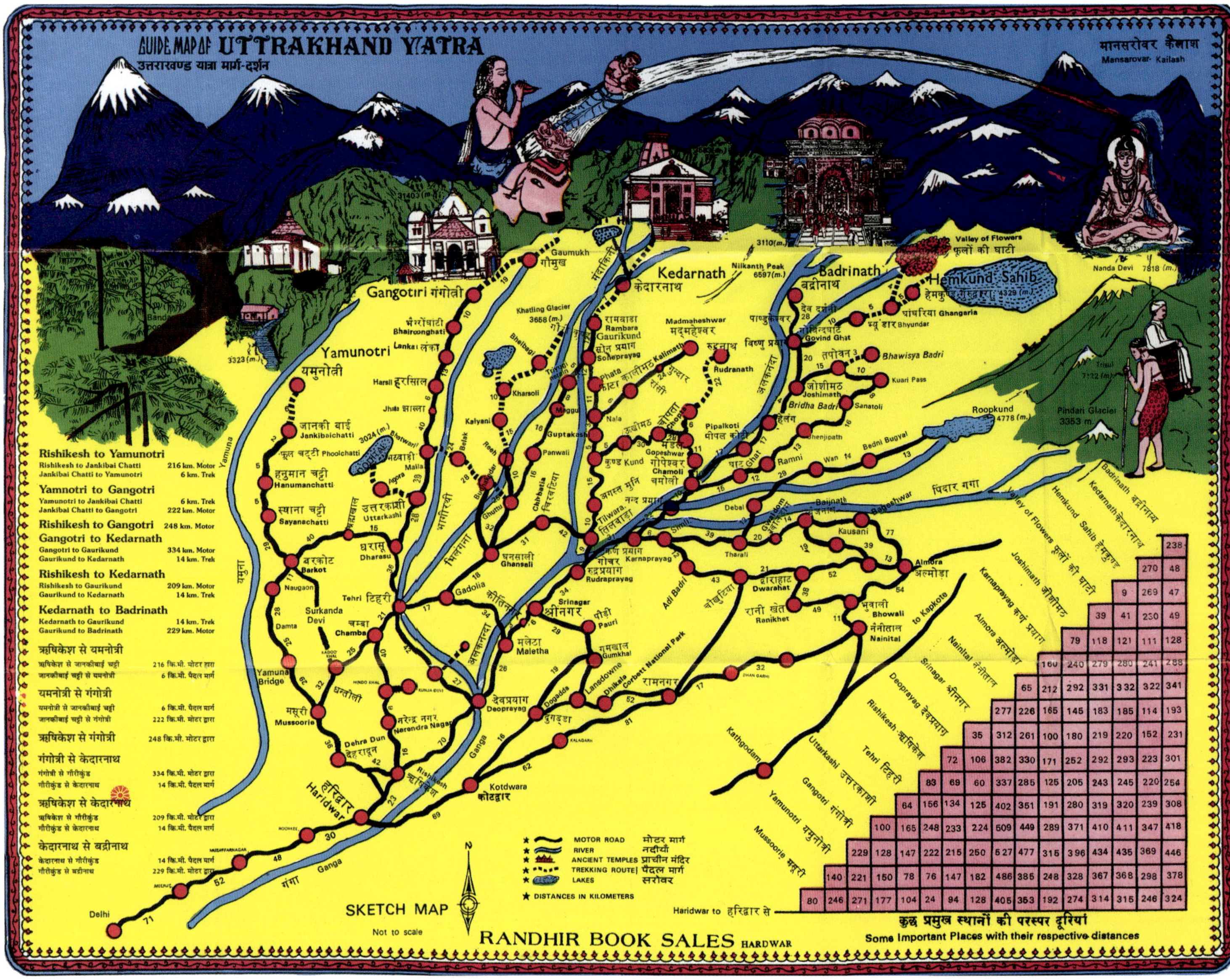

"Guide Map of Uttarakhand Yatra." **Tapotumi Uttarakhand**. Haridwar: Randhir Book Sales, circa 2006.

These conditions are not made evident in the map per se, but the reader is able to relate the painting to the circle and label on the map, both of which are reinforced by the thick, topographic description he gives. This back and forth between text, picturesque perspective, and map—a pictorial triad—stitches spaces together. Through this layered depiction—dramatizing the theater of the Nepal War or comparing the mountains to familiar Scottish highlands—Fraser "brought to the notice of the British public scenes much nearer to the snows."[23]

Travel and written description shaped the popular imagination of the picturesque qualities of the Himalayas, but images corroborated it. By the 1830s, monthly publications like Joseph Meyer's *Meyer's Universum*, printed in Hildburghausen in Thuringa, printed scenes of nature and monuments from around the globe. Published over a decade after Fraser's views of Gangotri, 'Die Ganges—Quelle in Indien' presented a similar, if less sophisticated, picturesque sensibility: just before a dark recess in a craggy cliff stands a lone temple. Bathers, like their fellow brethren near the temple, give a sense of the scale of the mountains and river—of the topographic monumentality. The water of the Ganges does not so much reflect the blue patches of the sky as it dimly glows a pale cerulean. The grey sediment flowing downstream is nowhere to be seen. (Most likely the artist etched this scene by copying another artist's work.) Nonetheless, the circulation of images like 'Die Ganges—Quelle in Indien' imbued a sense of topographic monumentality.

Entering the Himalayas meant experiencing this monumental topography, from Haridwar to Gangotri. It does so to this day. But what has changed are the authors and audiences for guidebooks to the Himalayas. If, during the reign of the Company and later the British Raj, Anglo-Europe-

"A Bird's Eyeview of Uttarakhand Chardham Yatra."
Haridwar: Randhir Book Sales, circa 2006.

ans emphasized the sublimity of the landscape and timeless habits of pilgrims, the majority of guidebooks today are produced for Hindu pilgrims on *yatra* and foreground the sacred geography of the Himalayas. Booklets for pilgrims, like *Tapotumi Uttarakhand*, a guidebook for pilgrims, or 'A Bird's Eyeview of Uttarakhand Chardham Yatra,' a map for pilgrims en route to sacred sites in the Himalayas, present a way of traversing this vast terrain. Like any guidebook or map, they key in important information about how to see, move through, and describe this region, emphasizing the monumental topography and the religious markers that define this territory. Yet the dams, barrages, and hydroelectric facilities of the region are absent from these maps and guidebooks, unlike the named temples, mountains, and tributaries. Ritual and topography come together in a way that dams and electricity do not. These guidebooks define a specific set of boundaries and objects in the landscape, and what they edit out is as important as what they choose to represent.

These rituals, seen today in religious maps and devotional images that are somewhat reminiscent of Fraser's picturesque paintings, do more than just create a powerful scenography: they provide a set of ethical guidelines for the management of the Ganges, one which comes into conflict with modern planning forces, most notably the central Ministry of Water Resources and the different State Governments. Because they do not always align, state and religious authorities articulate their legitimacy to manage the flow and uses of the Ganges on very different grounds. Religious authorities, representing the views of large segments of the Indian population, point to the aforementioned sacred texts and longstanding provenance of the Ganges' divine purification powers. Political parties—both religious and secular—mobilize religious beliefs and practices to contest the authority of government planning and infrastructure. In parallel, National and State water authorities discuss the Ganges River Basin in physiographic, empirical, engineering, and legal terms. Although the government points to a long heritage of water management that dates back to the Mughal Empire, or even to the Mauryan Empire, contemporary day-to-day practices and institutions are largely the vestiges of British colonial administration. As a consequence, the motives and mechanics of how best to manage this shared resource have produced political and material conflicts dating back to the mid-nineteenth century, as the different responsibilities of government agencies overlap with inherited British traditions.

Lord Shiv Statue at Haridwar, 2006.

COMPARATIVE WATERSHEDS: EIGHT RIVER BASINS

Even if one disregards the religious and national significance of the Ganges, its physical form remains remarkable. Though it has considerably less surface area than other major river basins (when divorced from the Brahmaputra and Meghna basins, it is not even half the size of the Mississippi or Amazon) and it is not one of the longest rivers in the world (there are more than forty rivers that are longer), the Ganges contains the largest water management bureaucracy in the world. It is a highly engineered and calibrated landscape—perhaps more designed than the densest city blocks of New York or Tokyo. Comparative maps of eight major river basins—Amazon, Danube, Mississippi, Murray-Darling, Nile, Yellow, and Indus—reveal just how intensely the Ganges is artificially irrigated.

AMAZON BASIN

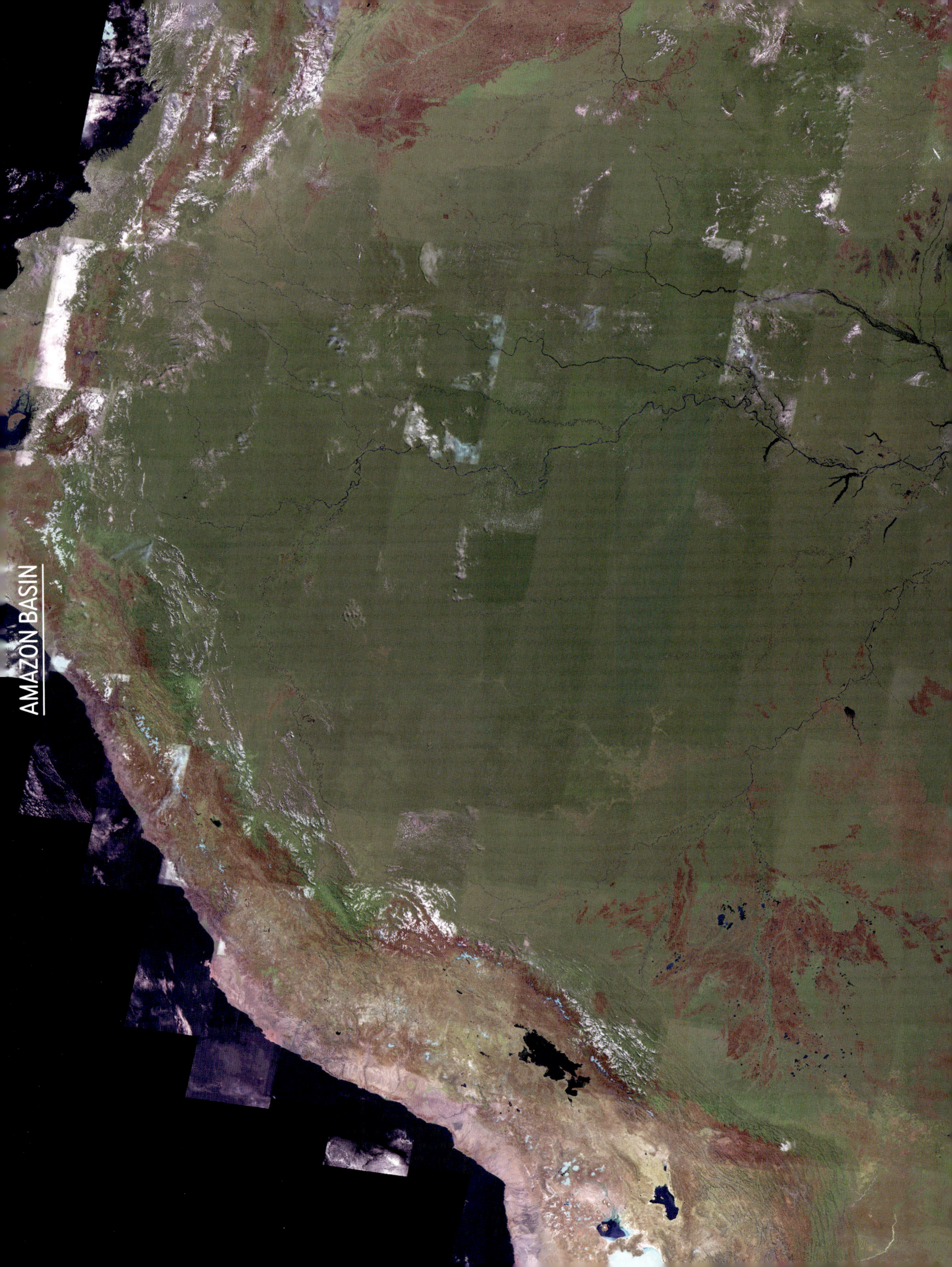
AMAZON BASIN

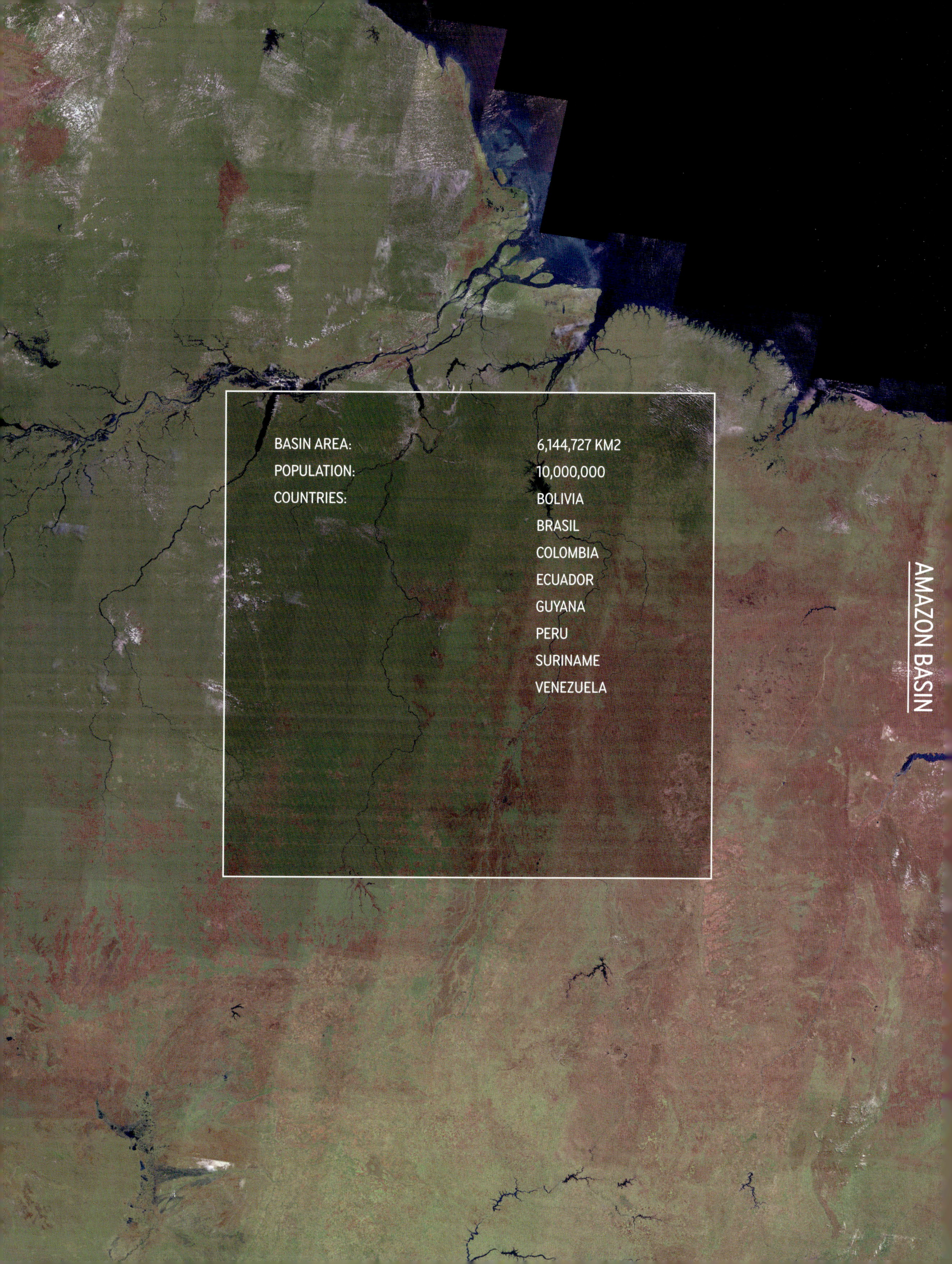
BASIN AREA: 6,144,727 KM2
POPULATION: 10,000,000
COUNTRIES: BOLIVIA
BRASIL
COLOMBIA
ECUADOR
GUYANA
PERU
SURINAME
VENEZUELA
AMAZON BASIN

A4_ 0.853584° N, 52.0135° W

B2_ 2.6438° S, 54.2628° W

C3_ 3.37037° S, 60.7383° W

D2_ 3.82447° S, 73.3405° W

AMAZON BASIN

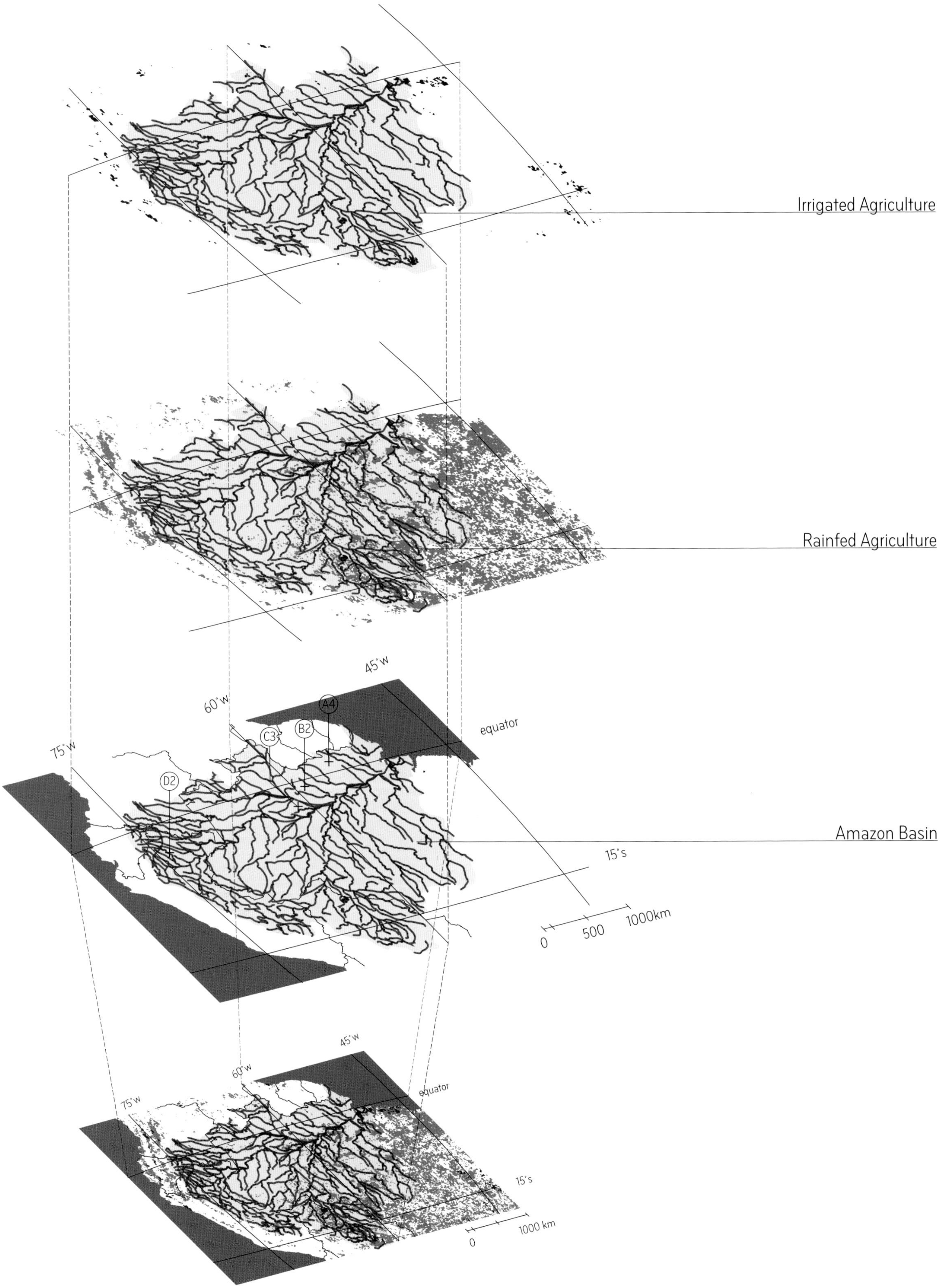

DANUBE BASIN

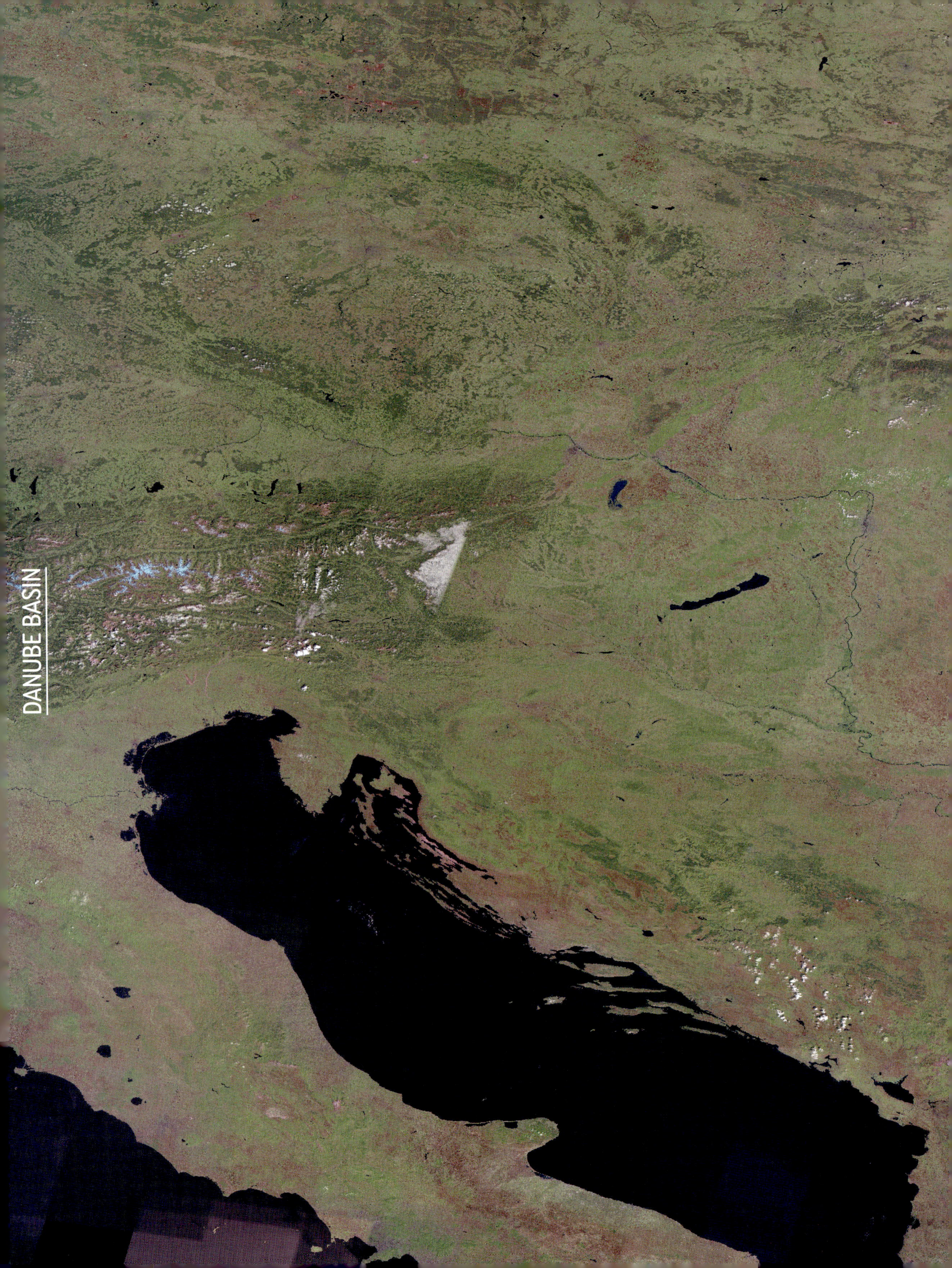
DANUBE BASIN

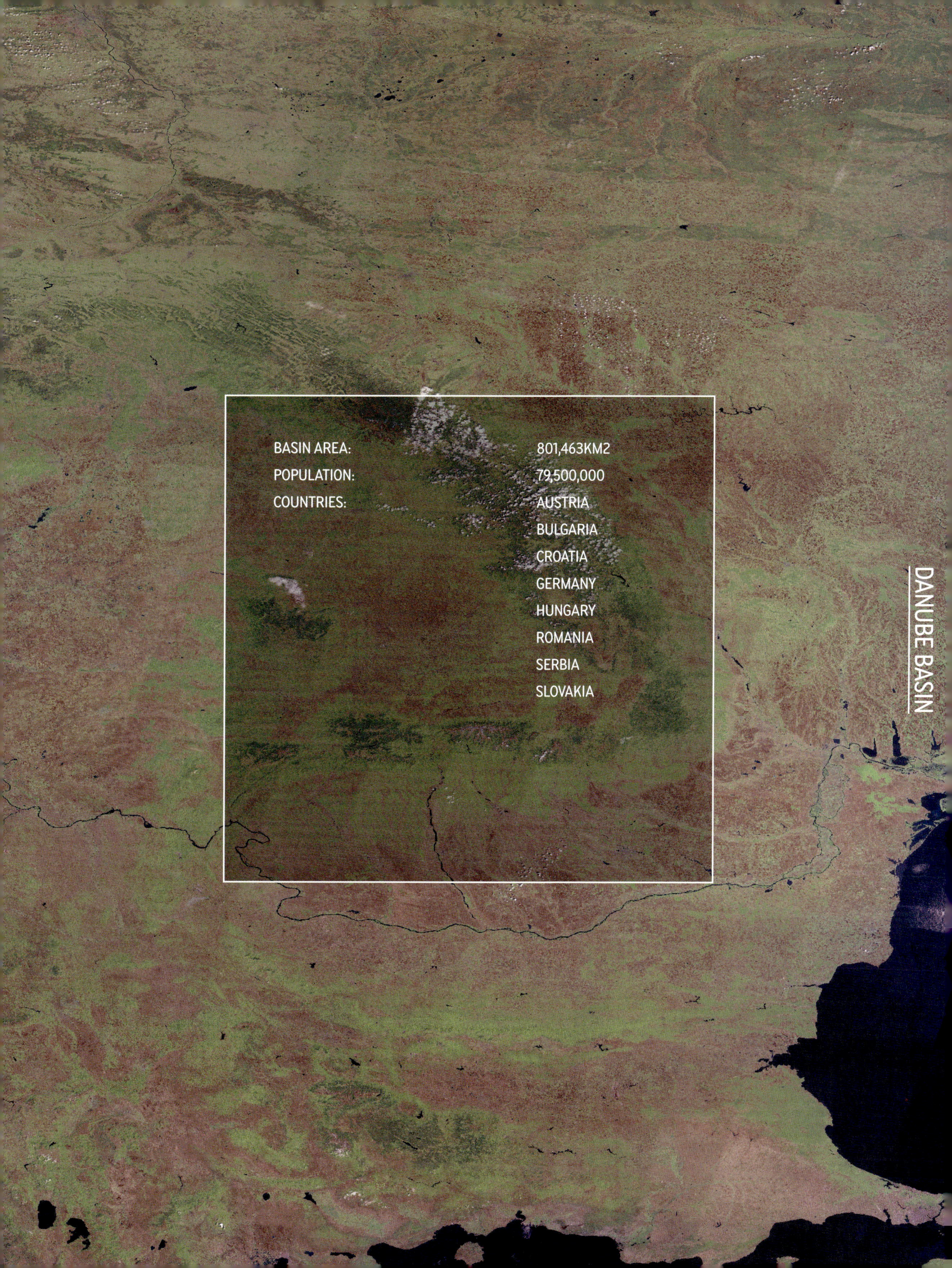
DANUBE BASIN
BASIN AREA: 801,463KM2
POPULATION: 79,500,000
COUNTRIES: AUSTRIA
BULGARIA
CROATIA
GERMANY
HUNGARY
ROMANIA
SERBIA
SLOVAKIA

A3_ 45.4733° N , 29.4692° E

B2_ 43.9565° N, 24.1052° E

DANUBE BASIN

DANUBE BASIN

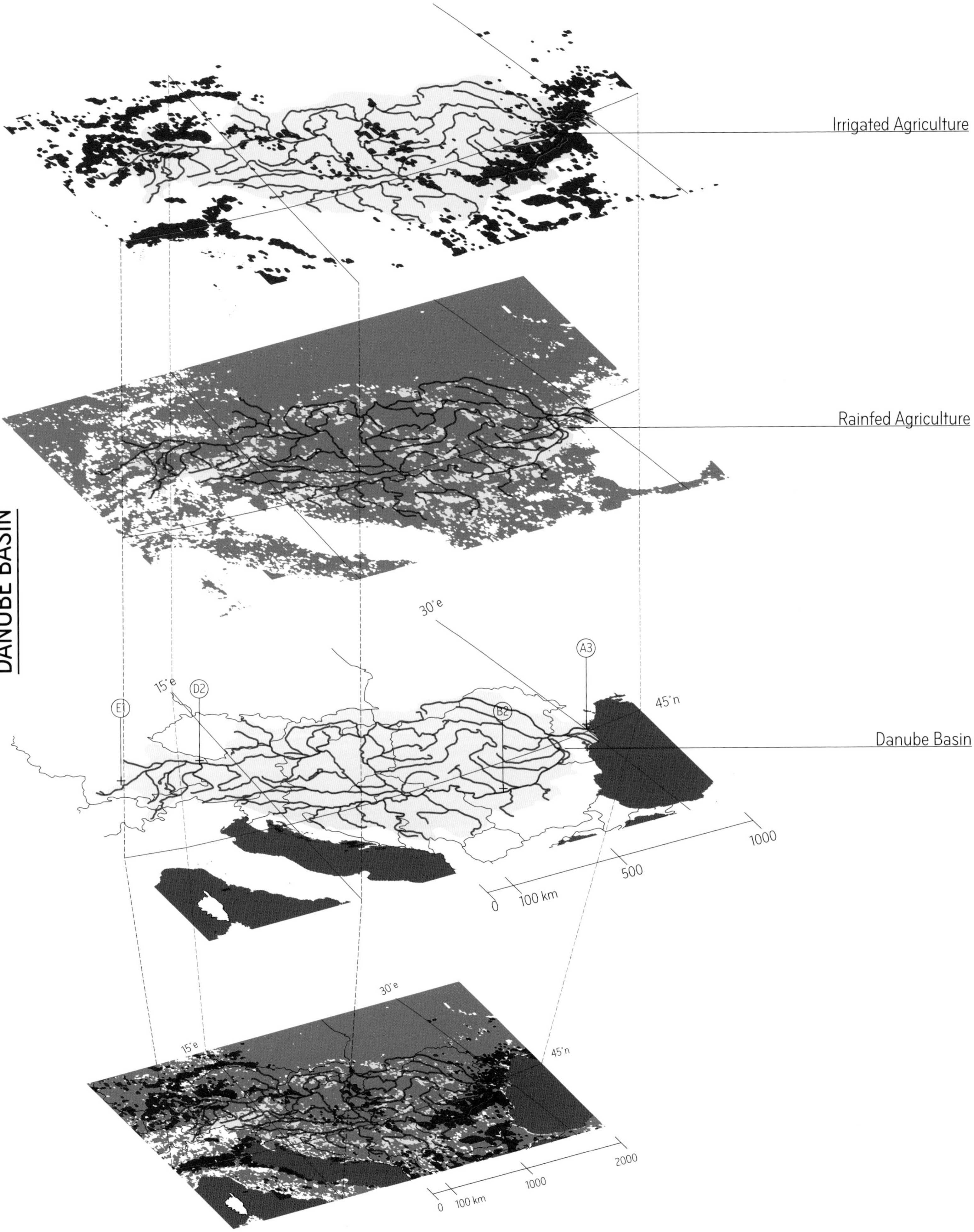

MISSISSIPPI BASIN

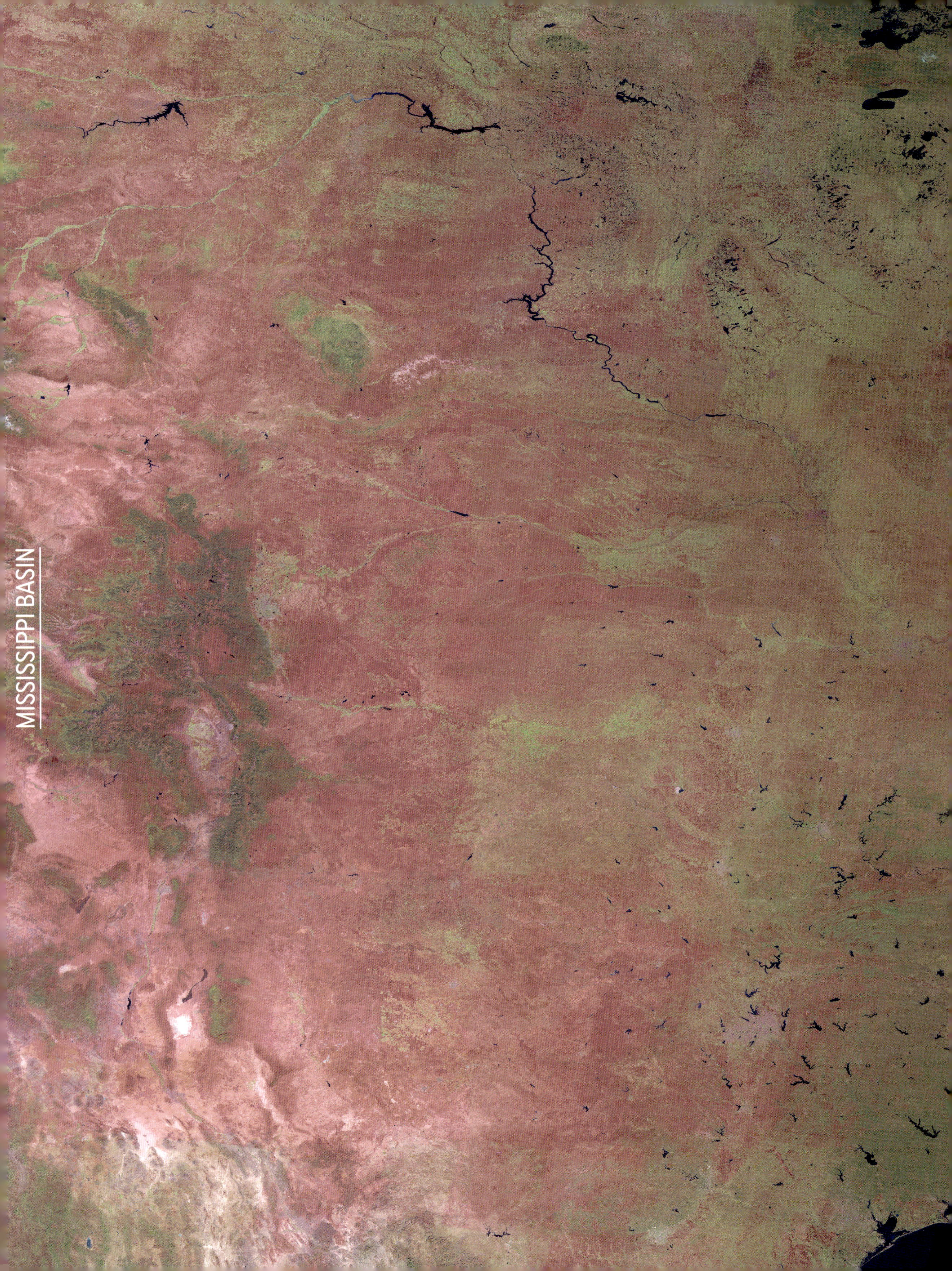
MISSISSIPPI BASIN

MISSISSIPPI BASIN
BASIN AREA: 3,202,230 KM2
POPULATION: 72,000,000
COUNTRIES: UNITED STATES

A2_ 29.4393° N, 90.3114° W

B2_ 33.6245° N, 90.315° W

D5_ 42.1609° N, 88.469° W

E2_ 45.6145° N, 96.3548° W

MISSISSIPPI BASIN

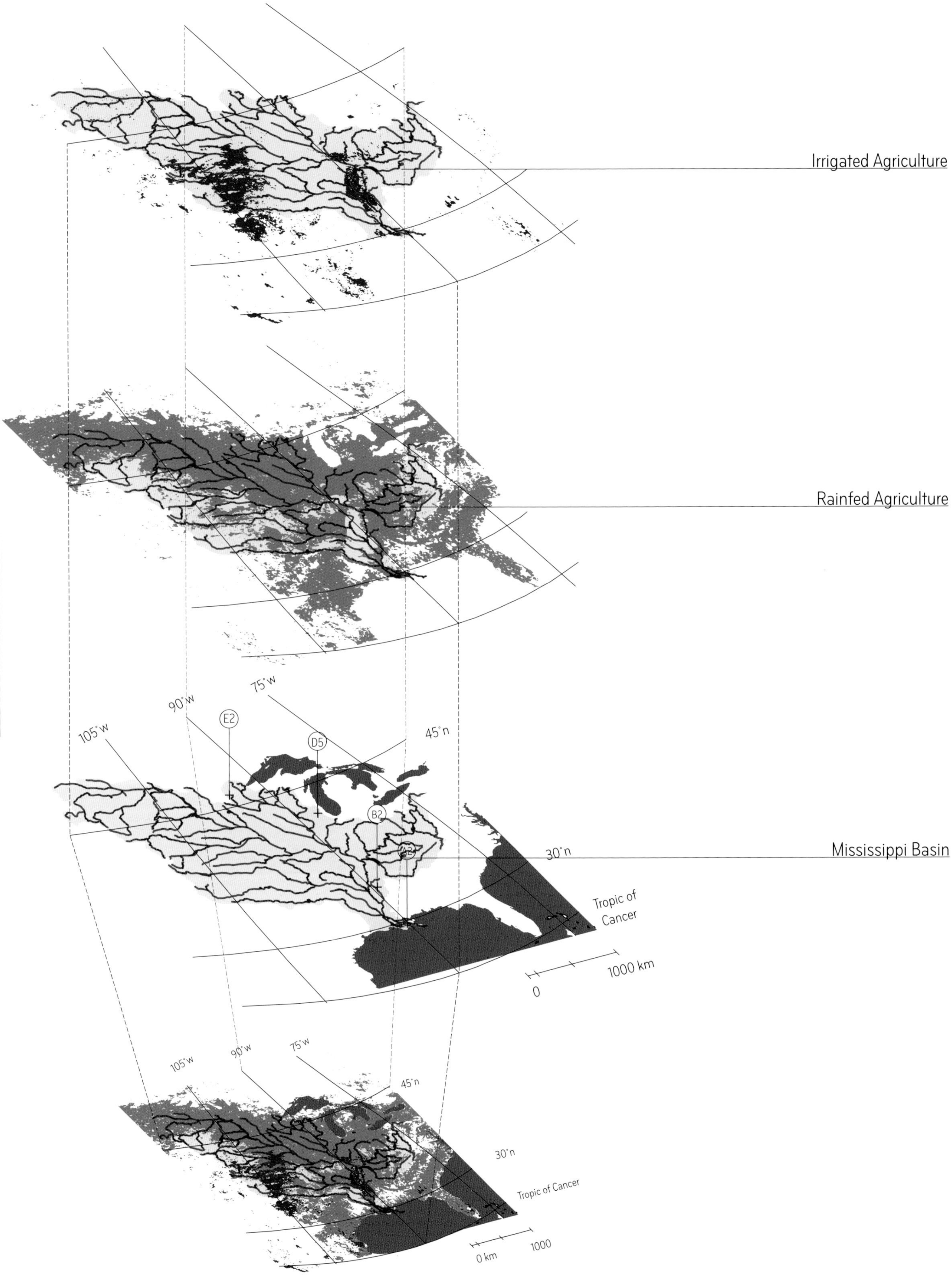

MURRAY-DARLING BASIN

MURRAY-DARLING BASIN

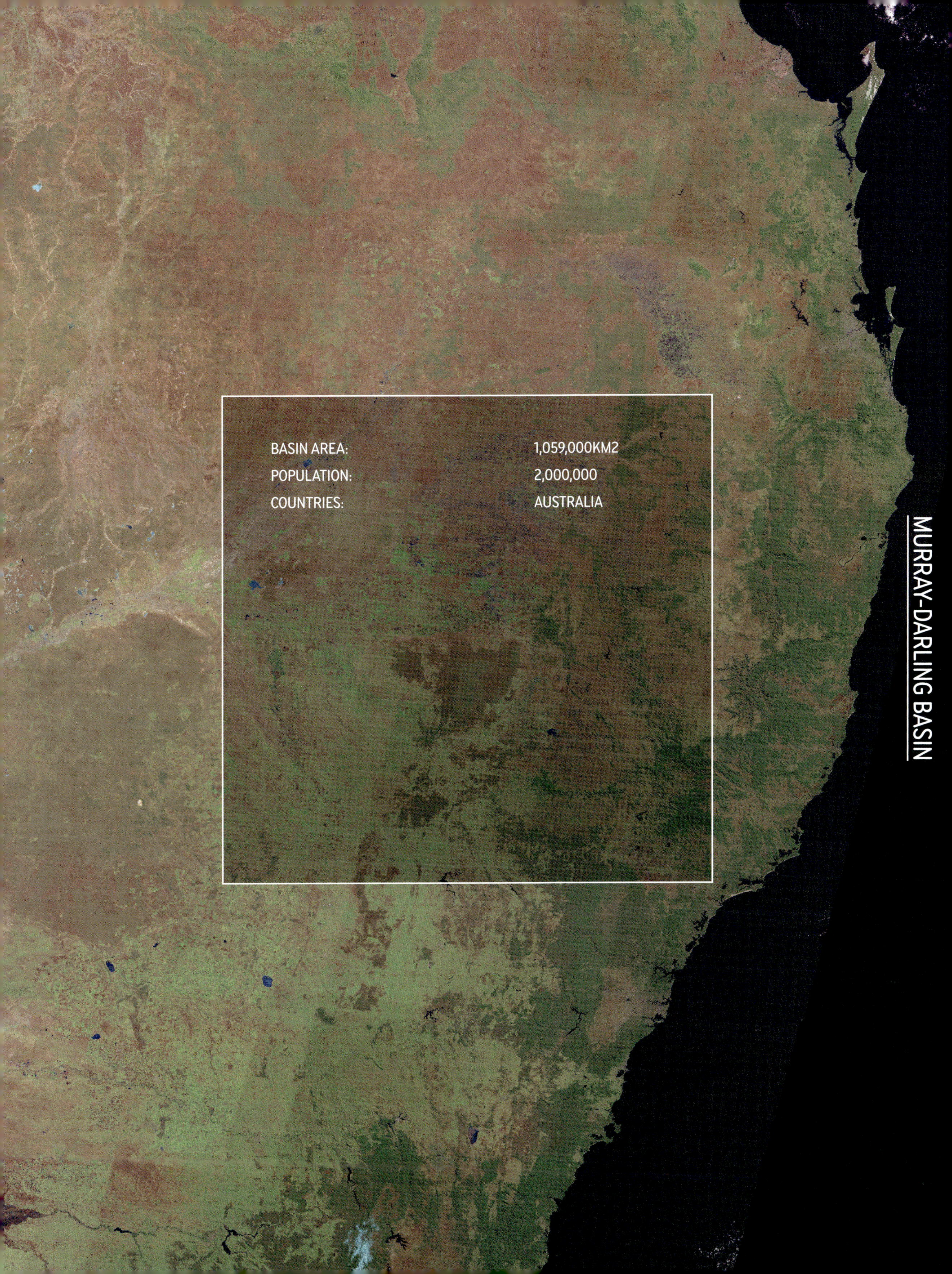
MURRAY-DARLING BASIN
BASIN AREA: 1,059,000KM2
POPULATION: 2,000,000
COUNTRIES: AUSTRALIA

A2_ 35.1872° S, 139.327° E

B2_ 34.2298° S, 142.145° E

C2_ 32.4097° S, 142.395° E

E1_ 29.1019° S, 149.833° E

MURRAY-DARLING BASIN

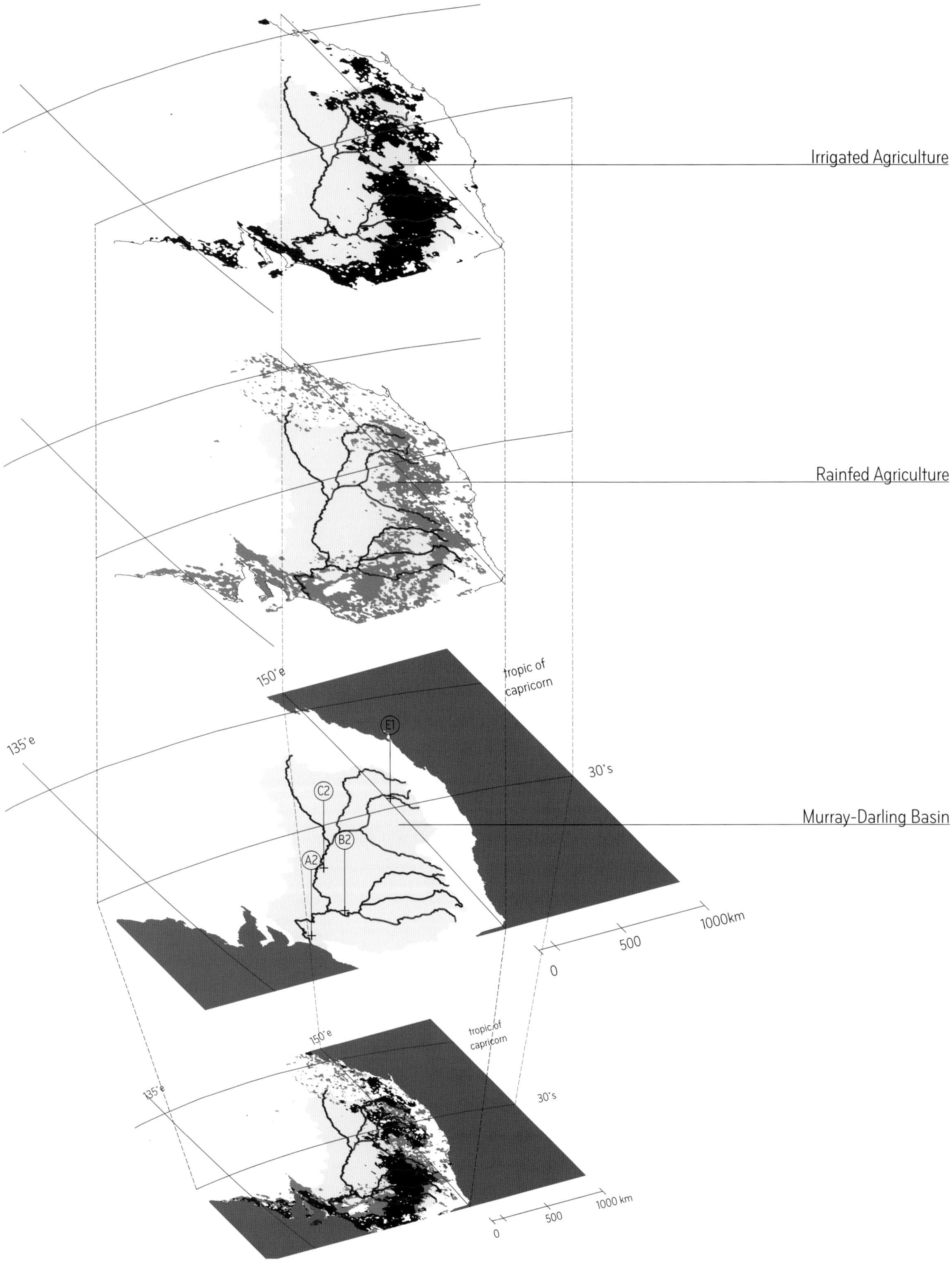

NILE BASIN

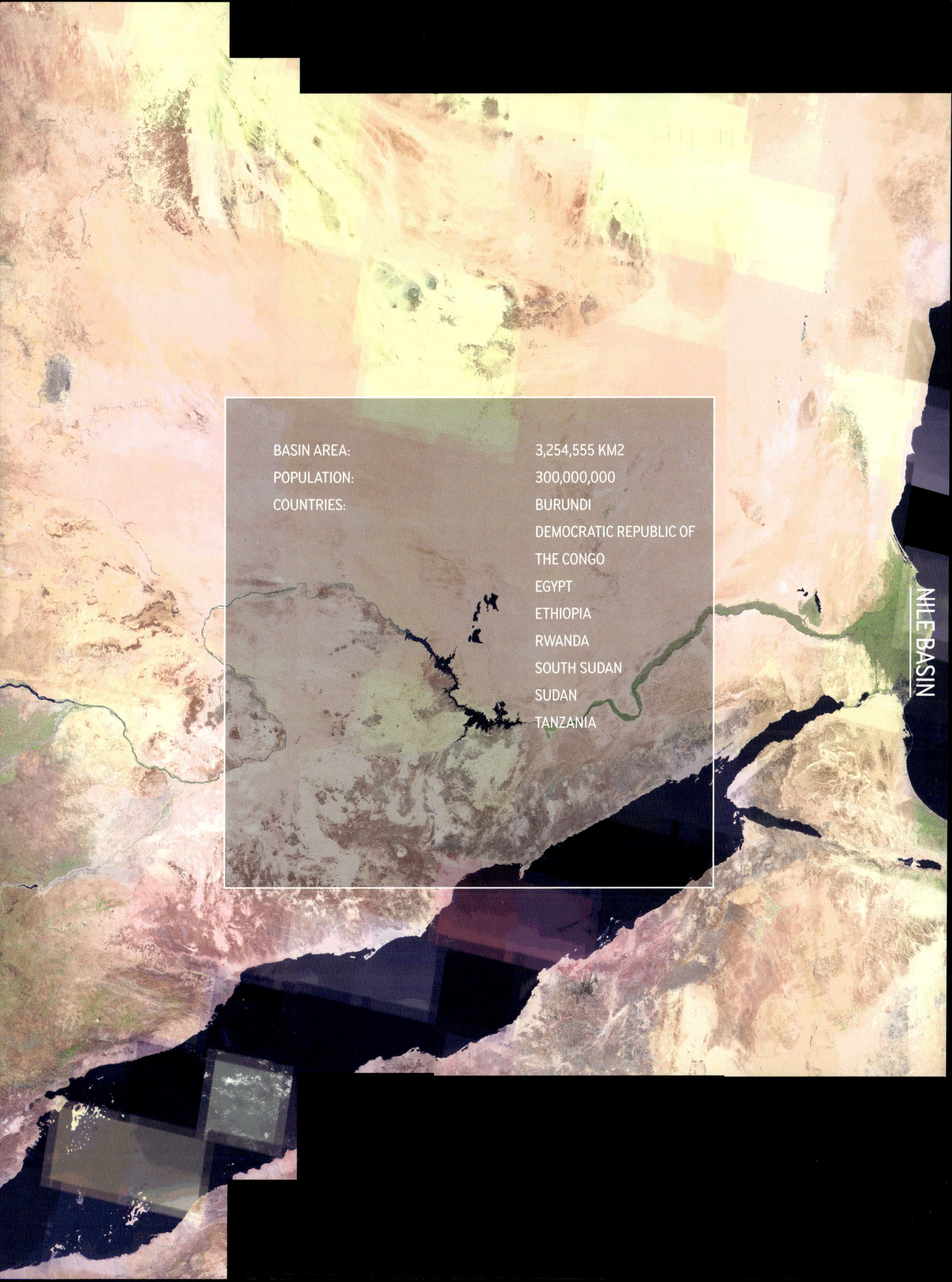
NILE BASIN
BASIN AREA: 3,254,555 KM2
POPULATION: 300,000,000
COUNTRIES:
BURUNDI
DEMOCRATIC REPUBLIC OF THE CONGO
EGYPT
ETHIOPIA
RWANDA
SOUTH SUDAN
SUDAN
TANZANIA

A4_ 30.9888° N, 32.384° E

B1_ 20.0839° N, 30.5578° E

C2_ 16.5167° N, 33.0815° E

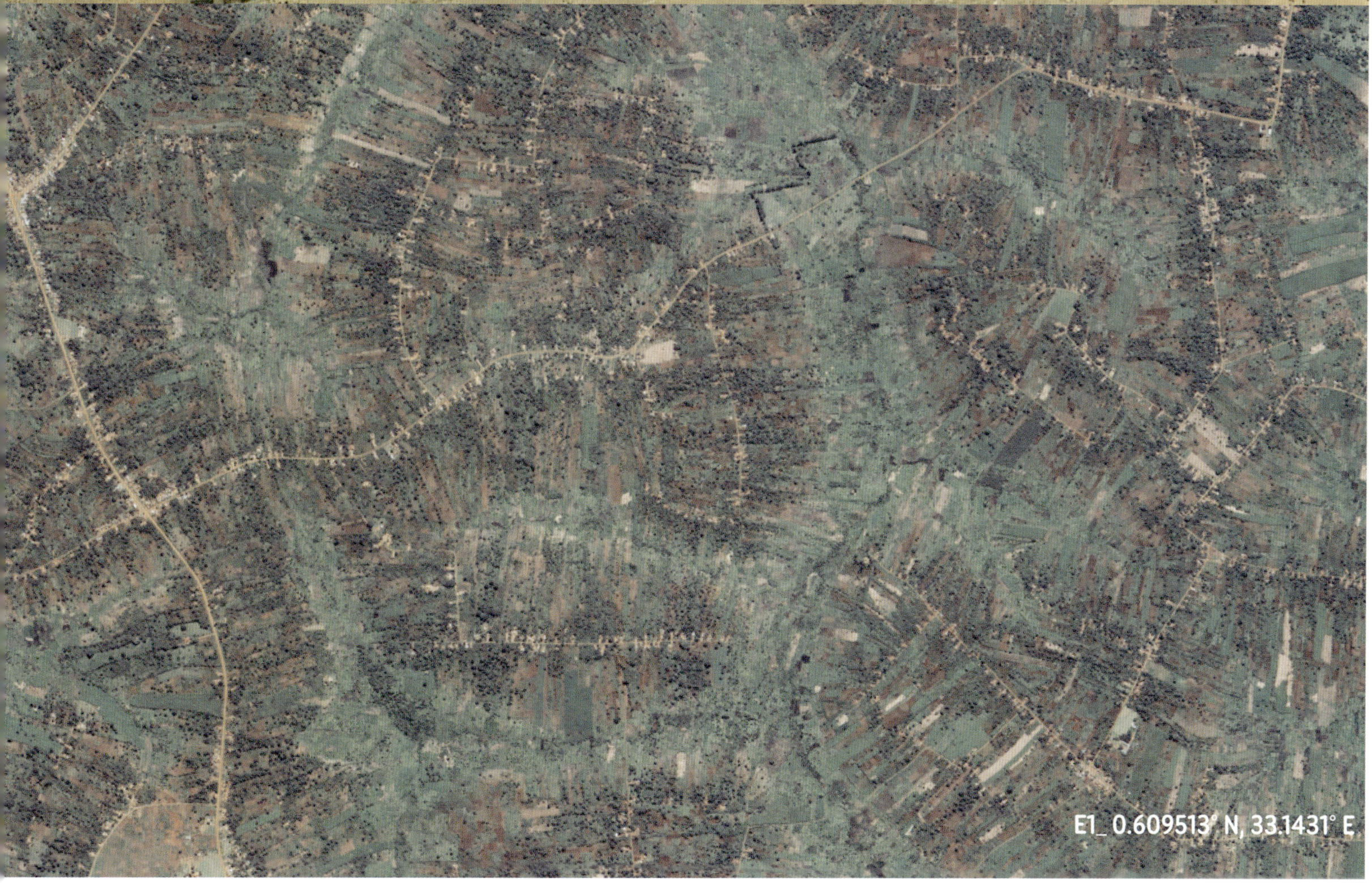
E1_ 0.609513° N, 33.1431° E

NILE BASIN

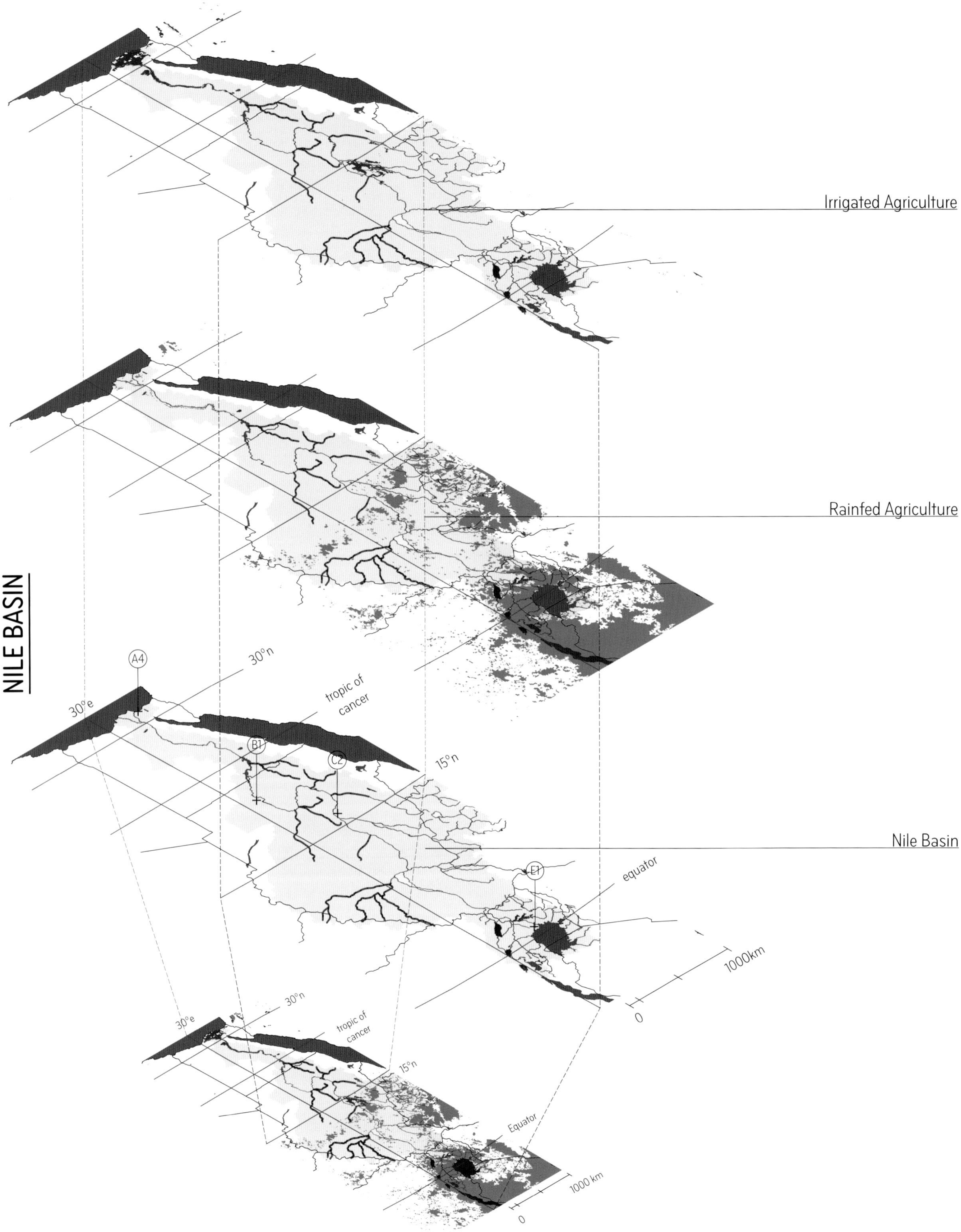
Irrigated Agriculture
Rainfed Agriculture
Nile Basin
A4
30°n
30°e
tropic of cancer
B1
C2
15°n
E1
equator
1000km
0
30°n
30°e
tropic of cancer
15°n
Equator
1000 km
0

YELLOW BASIN

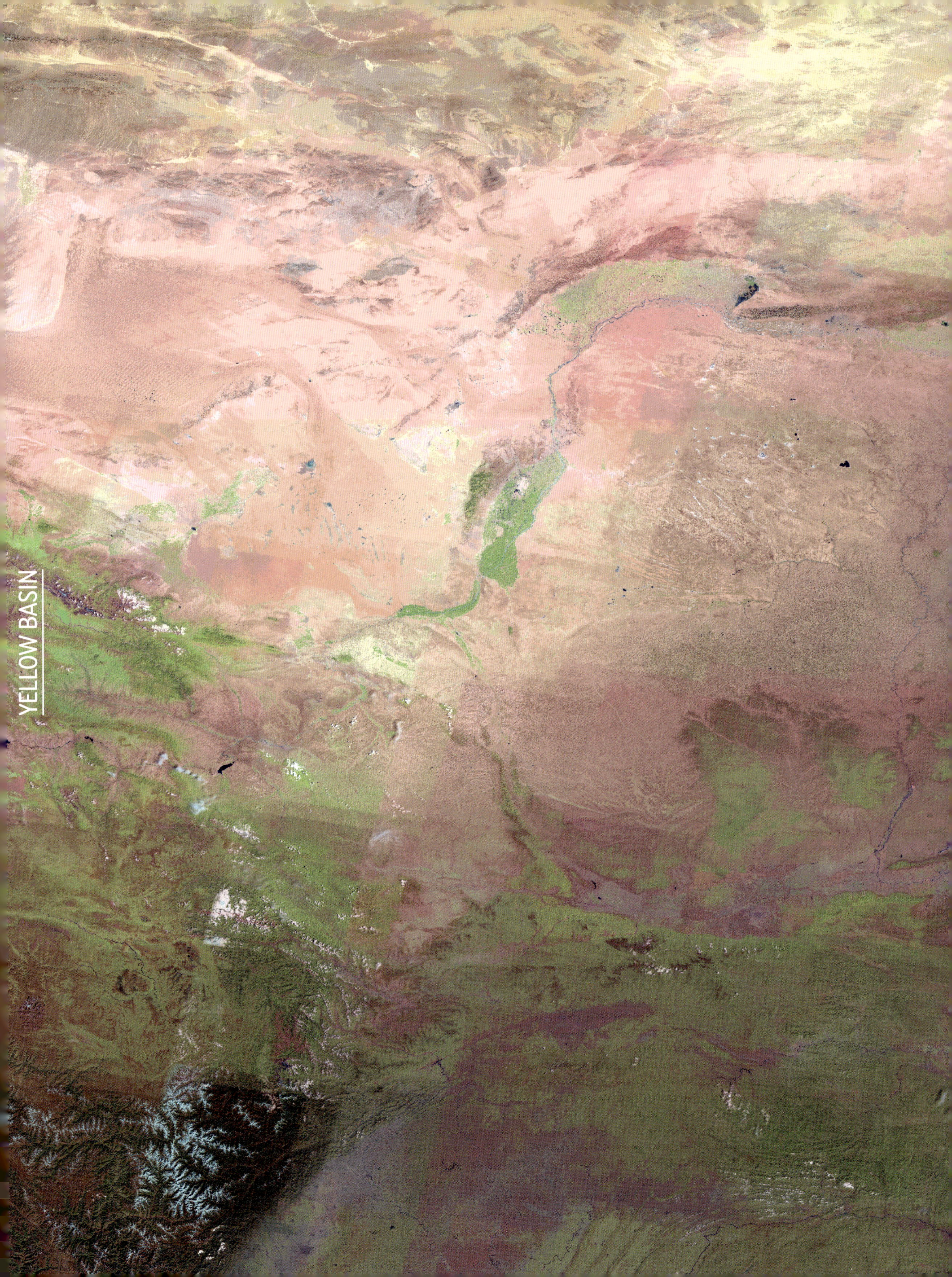
YELLOW BASIN

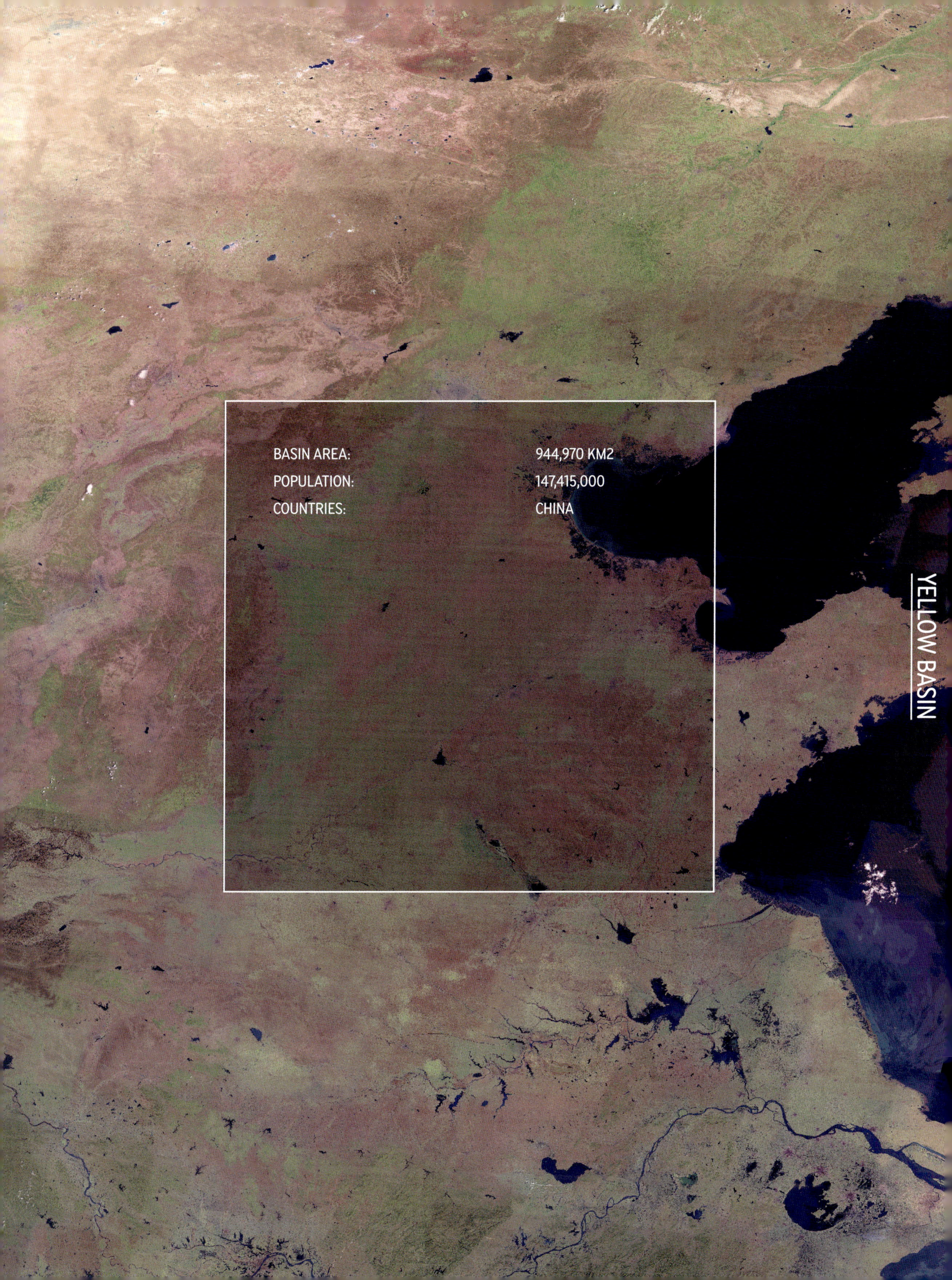
BASIN AREA: 944,970 KM2
POPULATION: 147,415,000
COUNTRIES: CHINA
YELLOW BASIN

A1_ 38.3133° N, 117.684° E

C5_ 40.4183° N, 109.903° E

D3_ 37.5192° N, 105.325° E

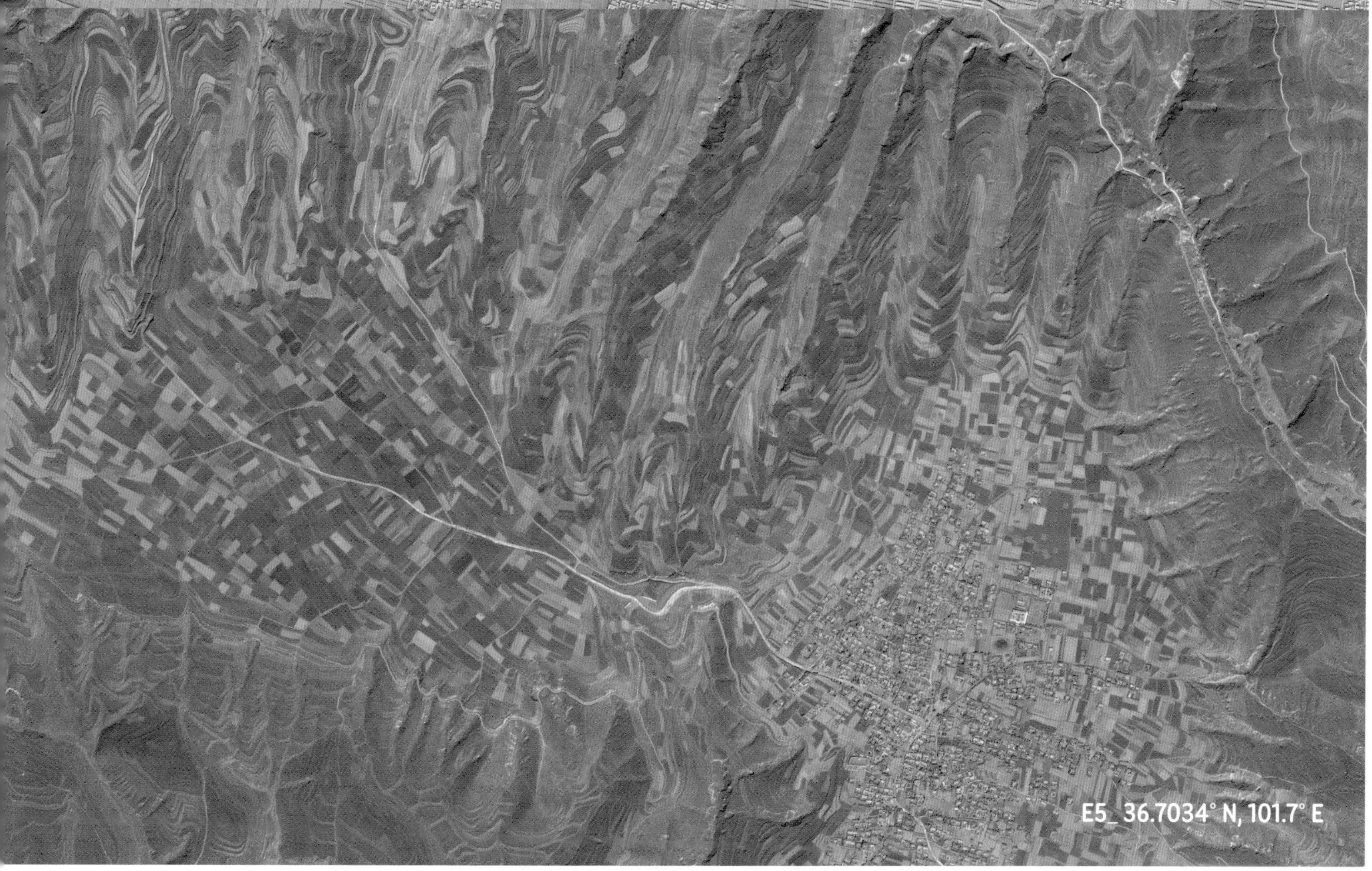
E5_ 36.7034° N, 101.7° E

YELLOW BASIN

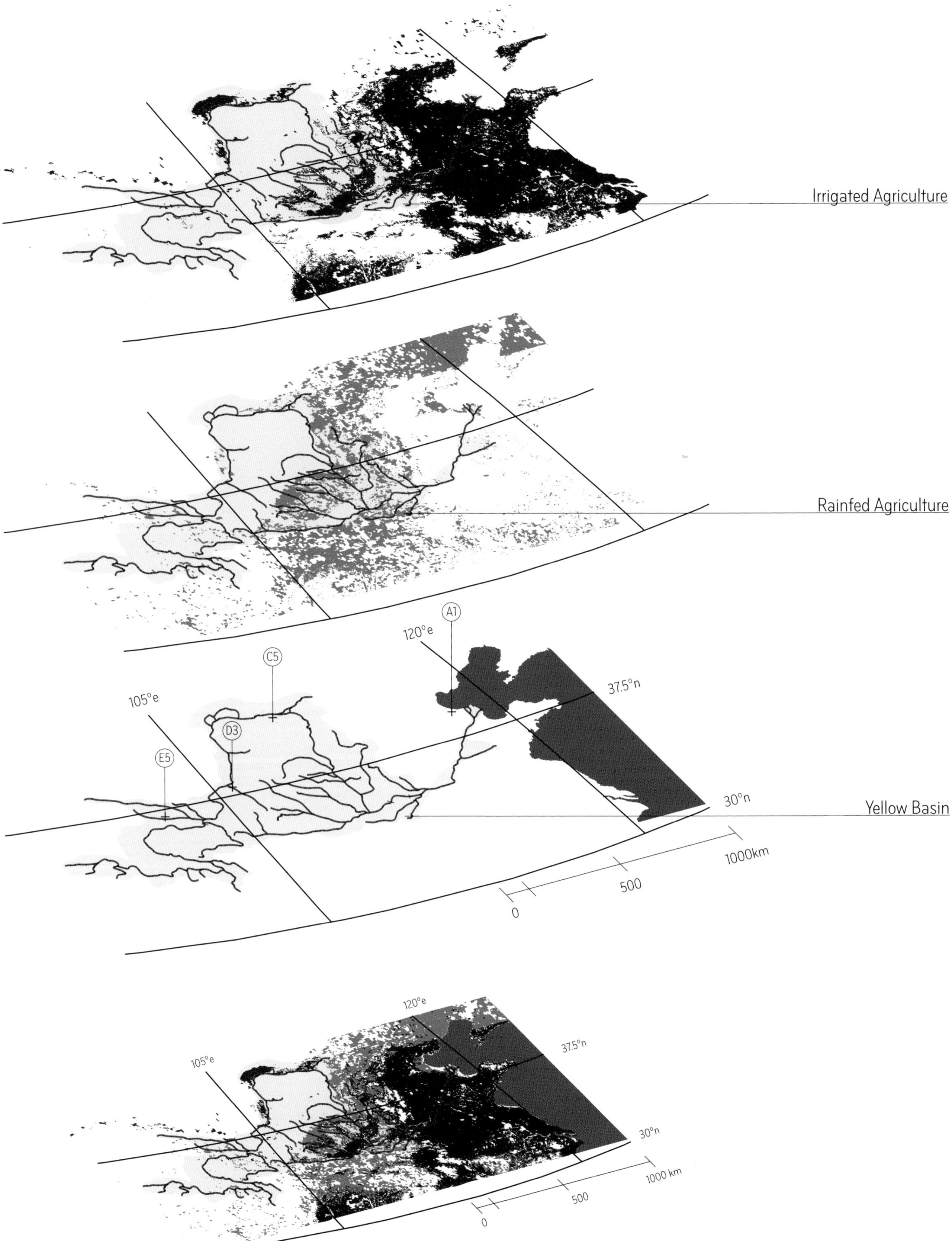

INDUS BASIN

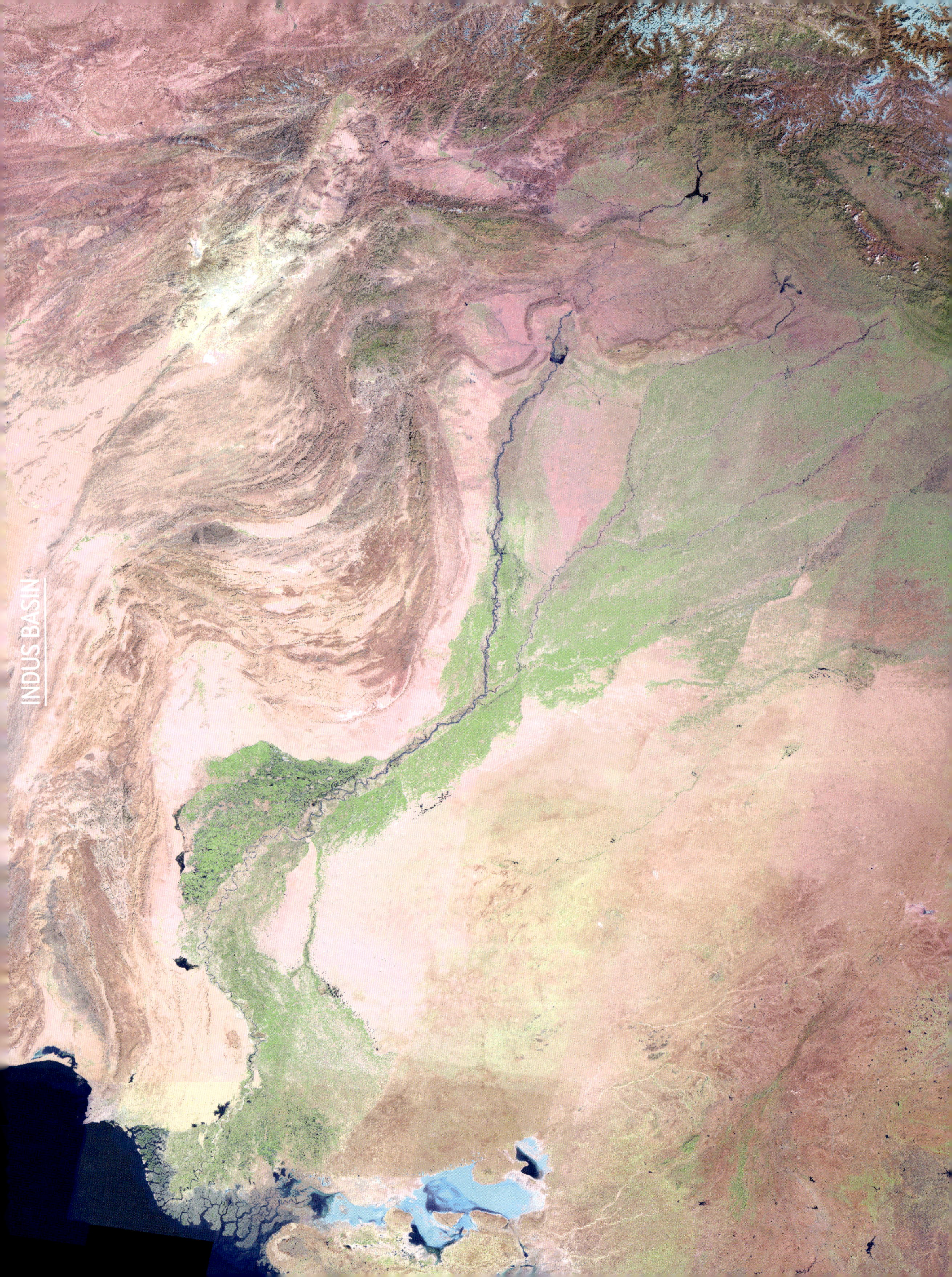
INDUS BASIN

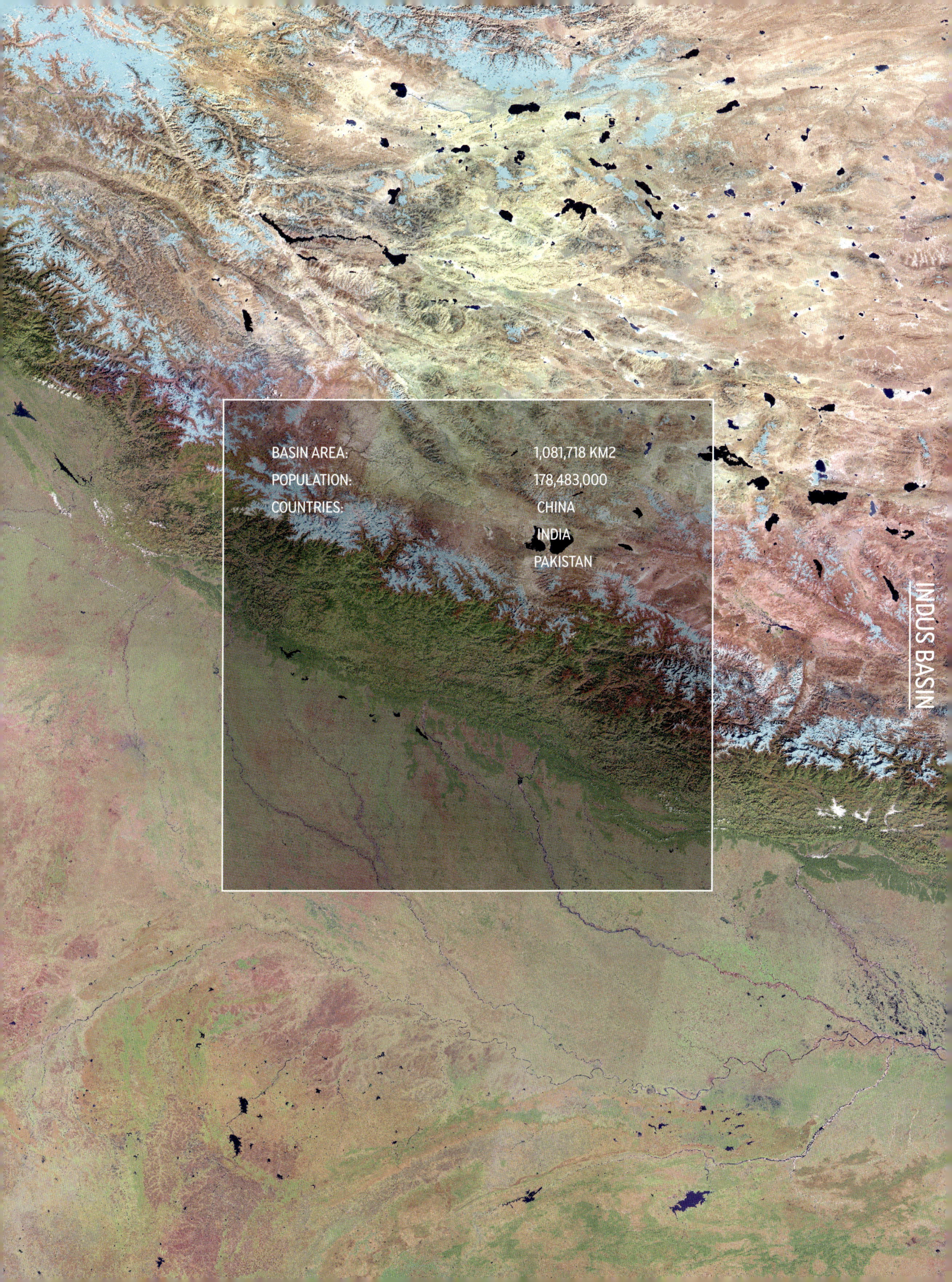
BASIN AREA: 1,081,718 KM2
POPULATION: 178,483,000
COUNTRIES: CHINA
INDIA
PAKISTAN
INDUS BASIN

A4_ 25.3411° N, 68.3174° E

B3_ 27.4606° N, 68.1893° E

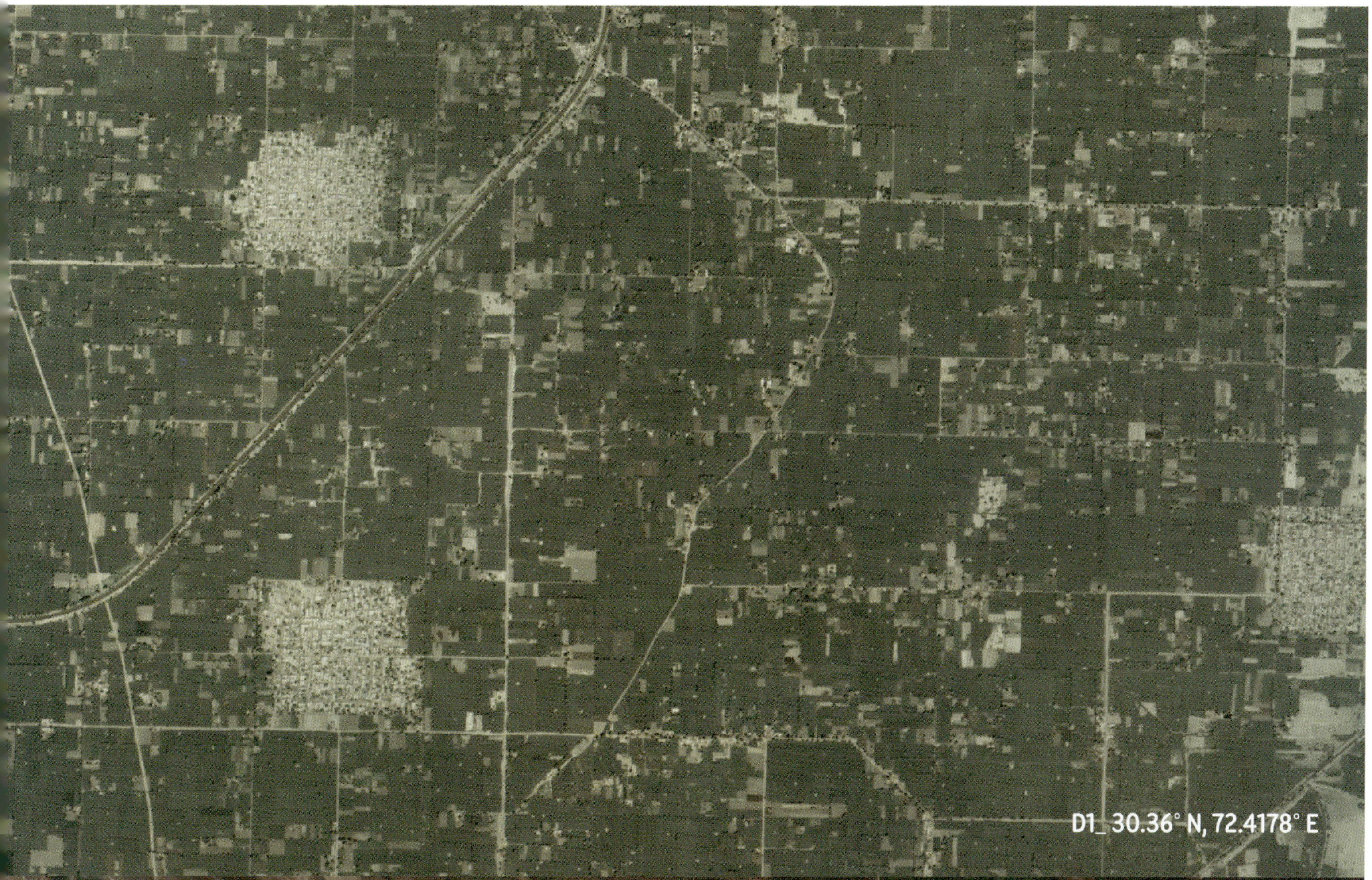
D1_ 30.36° N, 72.4178° E

E1_ 32.6623° N, 74.5163° E

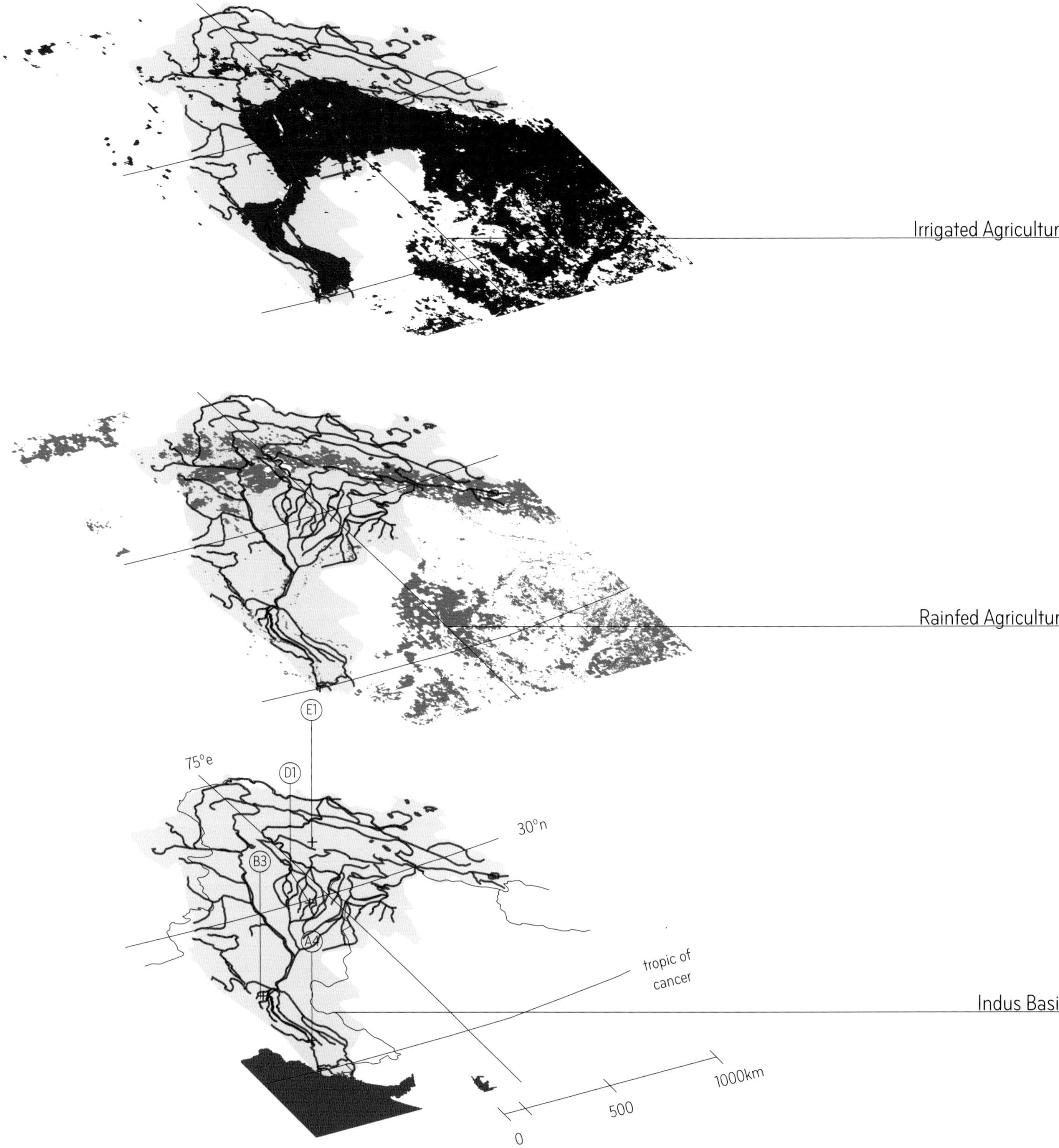
Irrigated Agriculture
Rainfed Agriculture
E1
D1
75°e
30°n
B3
A4
tropic of cancer
Indus Basin
0
500
1000km

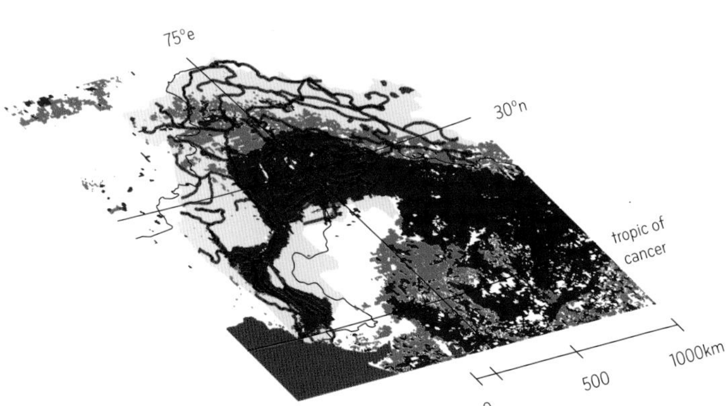
75°e
30°n
tropic of cancer
0
500
1000km

GANGES BASIN

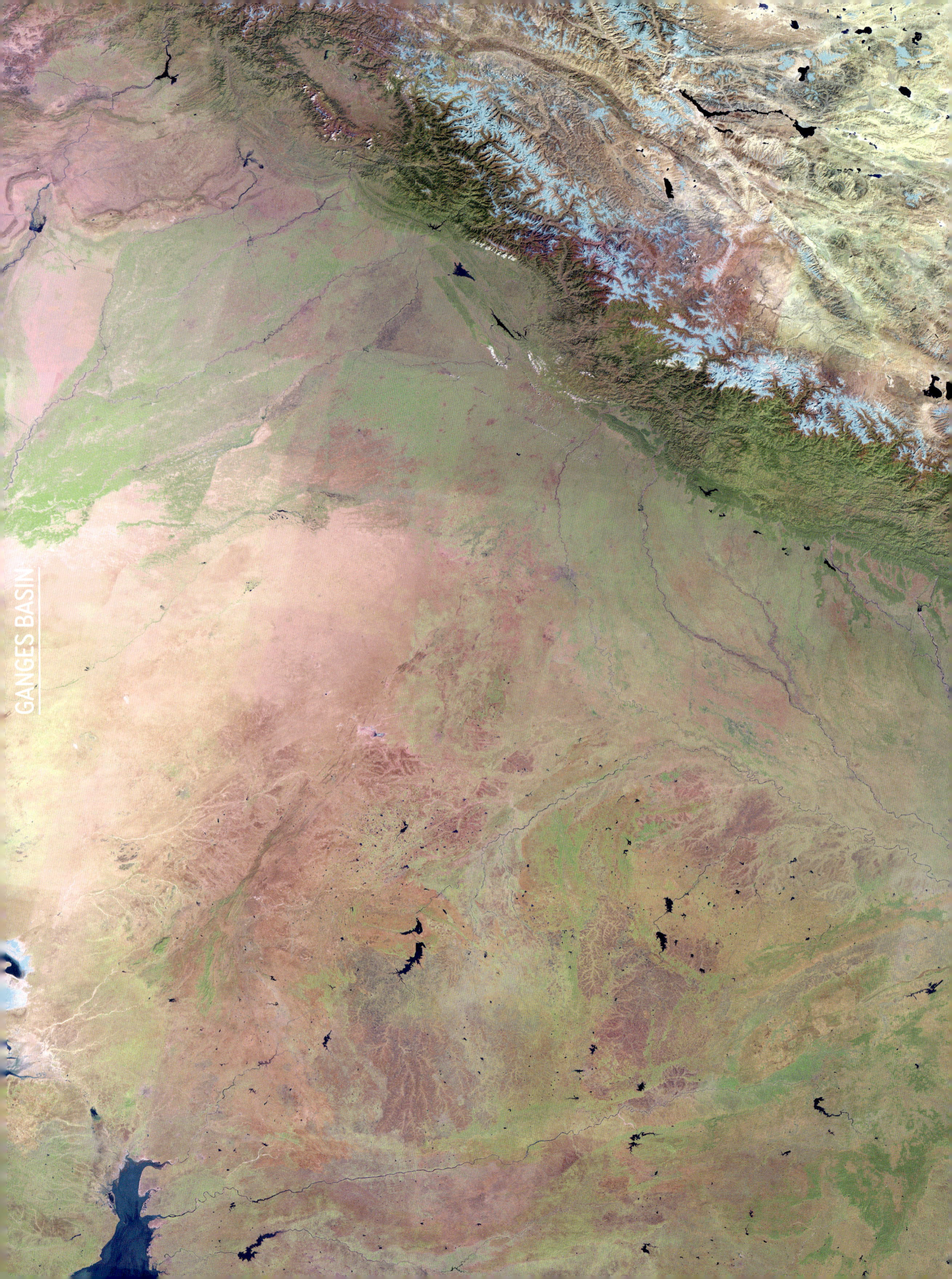
GANGES BASIN

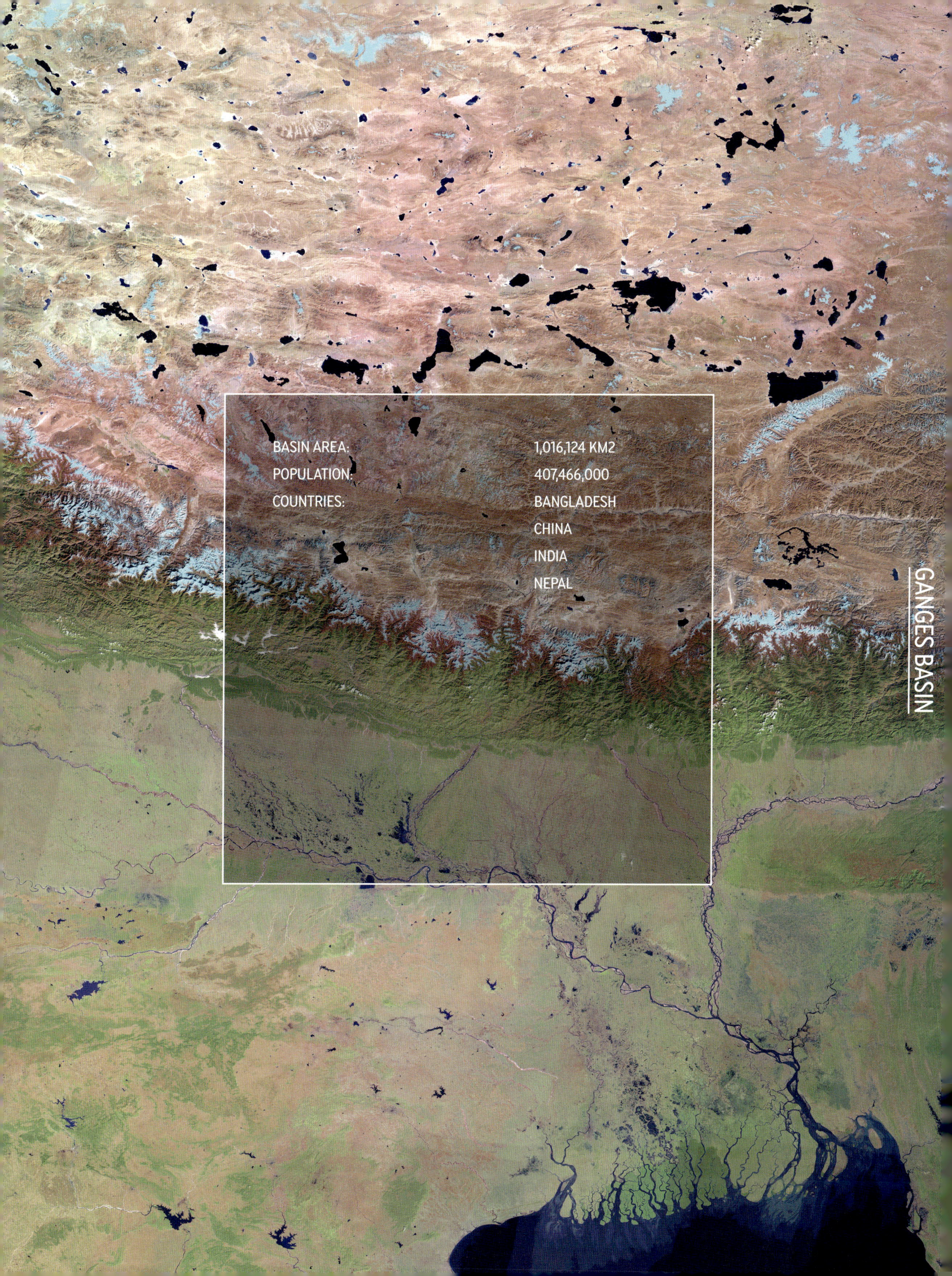
GANGES BASIN
BASIN AREA:
1,016,124 KM2
POPULATION:
407,466,000
COUNTRIES:
BANGLADESH
CHINA
INDIA
NEPAL

A1_ 21.945° N, 90.2159° E

C4_ 25.5726° N, 83.7133° E

D1_ 26.5297° N, 79.5759° E

E4_ 28.5138° N, 78.1946

GANGES BASIN

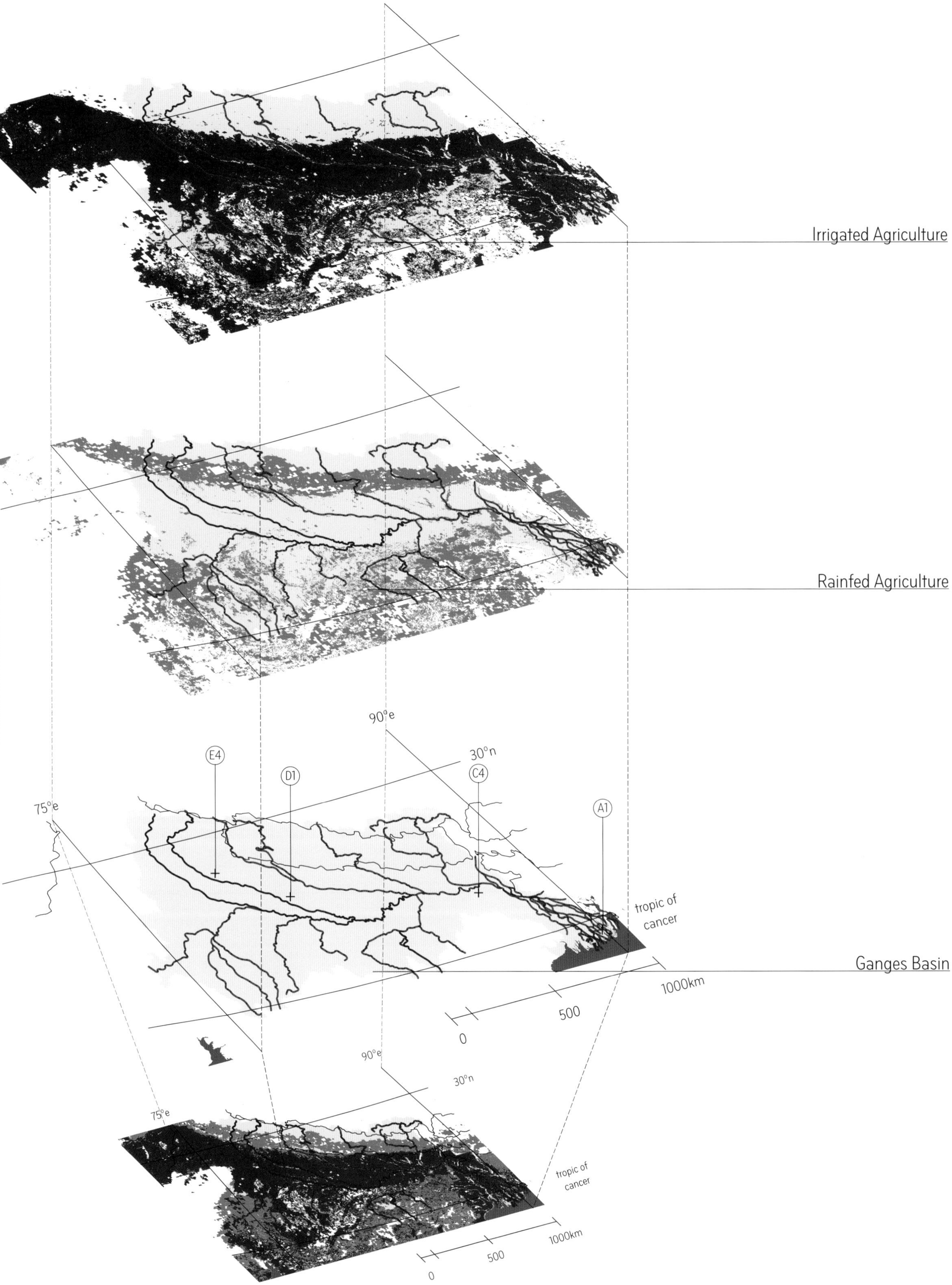

Instead of the passive surrender to an irrational monsoon, the Ganges River Basin shows a culture obsessed with water management to the point of paranoia—an unequalled exploitation of irrigation technologies to regulate capricious rainfall patterns.

MONSOON

The landmass of the Indian subcontinent, separating the Bay of Bengal from the Arabian Sea, shapes seasonal weather patterns each year, producing two periods of tempestuous rain: the southwest monsoon and the northeast monsoon. From June through to late-September or early-October, the southwest monsoon delivers extreme weather changes—dispersing anywhere from twenty-six meters of rain in Cherrapunji (1,400 meters above sea level in the state of Meghalaya) to less than one meter of rain in Allahabad (98 meters above sea level in the state of Uttar Pradesh). The northeast monsoon delivers sporadic rain between mid-October and mid-December, but its effects are primarily felt on the eastern side of the Indian subcontinent.

What produces the monsoon seasons is as much a question of geometry as it is geography: the lopsided triangle of the Indian Peninsula and its relationship to the Indian Ocean. The Eastern and Western Ghats create a thin coastal margin between the Arabian Sea on the western coast and the Bay of Bengal on the eastern edge. Crowning the mountainous rim of the Ghats is the Deccan Plateau. Taller in the west than in the east and still tallest at its southern tip, the great plateau gradually slopes downwards to the north into a set of rolling, rocky hills—the Vindhyas. These hills make up an integral part of the Ganges River Basin, forming an extended threshold between the Great Deccan Plateau and the Gangetic Plain. These plains stretch for several hundred kilometers to the north of the Vindhyas. At the far eastern and western edges, the Plains burst into the great north-south river basins of the Indian subcontinent—the Indus and the Brahmaputra. During the southwest monsoon, the winds from the Bay of Bengal and the Arabian Sea converge on the Vindhya Range. The rivers on the southern side run back to the sea: the Wardha and Wainganga flow south to join the Godavari and the Narmada and the Tapti flow west to the Arabian Sea. The northern slope rivers flow into the tributaries of the Ganges: the Chambal and the Betwa to the Jamuna before it converges with the Ganges at Allahabad and the Son River forms a confluence with the Ganges above Patna. The flow of these rivers, both in the volume of water and alluvium, pales in comparison to the flow of the Himalayan fed rivers. The Ganges and Jamuna form the axial rivers of northern India. Glacier fed most of the year, these rivers transform during the southwest monsoon—often into torrents prone to flooding. Because the winds from the Arabian Sea scarcely penetrate the Gangetic Plain, the majority of rainfall during the monsoon enters from the Bay of Bengal. As it moves northwest from the delta of the Ganges and Brahmaputra rivers towards the Punjab, it dumps less and less water. Managing the different volumes of water and alluvium of these rivers in the months before, during, and after the southwest monsoon has been a concern for hundreds if not thousands of years. Examining how water and soil move during the southwest monsoon offers a significant vision of the hydrology and physiography of the Ganges basin; examining how people negotiate such a flow offers another, equally important, viewpoint.

A corruption of the Arabic word *mausim*, which means seasonal wind, monsoon came to mean seasonal rains in Anglo-European descriptions from the late-sixteenth century onwards. These winds and their rains were a central signification, for European travelers, of a harsh difference between the supposedly temperate climates of Europe and the extreme conditions of the tropics. What made the Indian subcontinent seem particularly harsh was not only the time of year when rainfall occurred, but also the character of the rain and how it transformed the topography. "June 29.—Soon after we had commenced our journey this morning, it began to rain in the most pitiless manner I had ever experienced" bemoaned Captain Thomas Skinner on his descent from Gangotri back to the plains in *Excursion in India* (1832). He continued, "The paths were so slippery, it was nearly impossible to move over them; and we were so enveloped in clouds that we could not seek more widely for a surer footing." Skinner describes watching the storm, after falling into a cave no deeper than six or seven feet, "howling along the glens; the black clouds were curling over the brow of the precipice on which the cavern hung." "Nothing could be wilder, or more magnificent," he concluded, "than the scenery..."[24]

The wild magnificence Skinner spoke of is not limited to the Himalayas. As far as the Ganges basin is concerned, nearly eighty-percent of all precipitation occurs during the summer's southwest monsoon. It begins in the south and migrates northward: Colombo (Sri Lanka) experiences the monsoon before Chennai (formerly Madras), and Chennai before Kolkata (formerly Calcutta). Such a deluge, whether one thousand meters above sea level or one hundred, transforms arid and parched land to gelatinous silt every year. Pre-monsoon, the water flowing through the Ganges River may be less than forty meters wide. But once the southwest monsoon arrives, those same stretches can swell to over a kilometer in breadth. By the time of the vernal equinox (mid-late March), with the sun beating down on northern India, these stretches turn into an arid open-air furnace. In such conditions, droughts are all too common. The riverbed's sun-scorched silt is too heavy and particulate to be transported by what, in late-March, is a trickling rivulet of a river. In the months before the southwest monsoon, the dried silt is swept up by the wind into mild sandstorms and also harvested by men and women into mounds to be used for mixing concrete. This annual landscape of sandstorms and silt shows the Ganges is not just about the dynamic movement of water, it is also about the dynamic movement of soils and debris.

This cycle begins in the Himalayas, the youngest major mountain range in the world, which annually shed millions of tons of alluvium—the particulate clay, sand and silt that rolls down the mountains. Nearly all of the Ganges' major tributaries—Bhagirathi, Ramganga, Jamuna, Ghaghra, Gandak, Bagmati, Kosi, Sone, and Damodar—descend from the mountains in a similar fashion. Geologists estimate that 794 million tons of sediment strips from these peaks each year. Some 729 million tons make their way to Farakka Barrage (completed in 1975) in West Bengal. Of the 729 million tons, only 95 million tons billow out into the Bay of Bengal. The rest is trapped in the muddy water of the delta. At Haridwar along the Ganges River and Allahabad along the Jamuna River, this sediment is distributed at a rate between 150–350 tons per square kilometer per year.[25] Further east, past Patna, sediment budgets increase to as high as 4,600 tons per square kilometer per year at the Kamla-Balan River. These rivers, such as the Kosi and Gandak, shift channels frequently, sometimes moving several kilometers in a single year. Whereas most rivers to the west of Patna are not prone to flooding, rivers to the east flood every year in the months surrounding the southwest monsoon.[26]

This variability in the flow and stability of the Ganges and throughout its tributaries necessitates different ways of incorporating rain for agricultural, industrial, and domestic uses. Building homes, roads, railways, and cultivating the land in a flood plane requires a different way of engaging water and soil. There is a temporal dimension to not only how and when it rains, but also to how water circulates, both across the surface of the ground and vertically into tanks, lakes, and aquifers. These temporal and physical differences also necessitate different modes of visual representation. A river prone to flooding its shallow bed might look like a clearly defined stream of water in the months prior to the monsoon; but once the monsoon arrives, this same river may resemble a splotch or a stain. Likewise, a stable or incised riverbed may produce flooding if drains and canals are unable to redirect run off from rains. Drawing any river as two parallel lines is misleading and the Ganges River is no exception. In fact, given the dearth of drawings of the Ganges and its dynamic nature coupled with a rapidly growing population, there is perhaps no better river to redraw and reimagine.

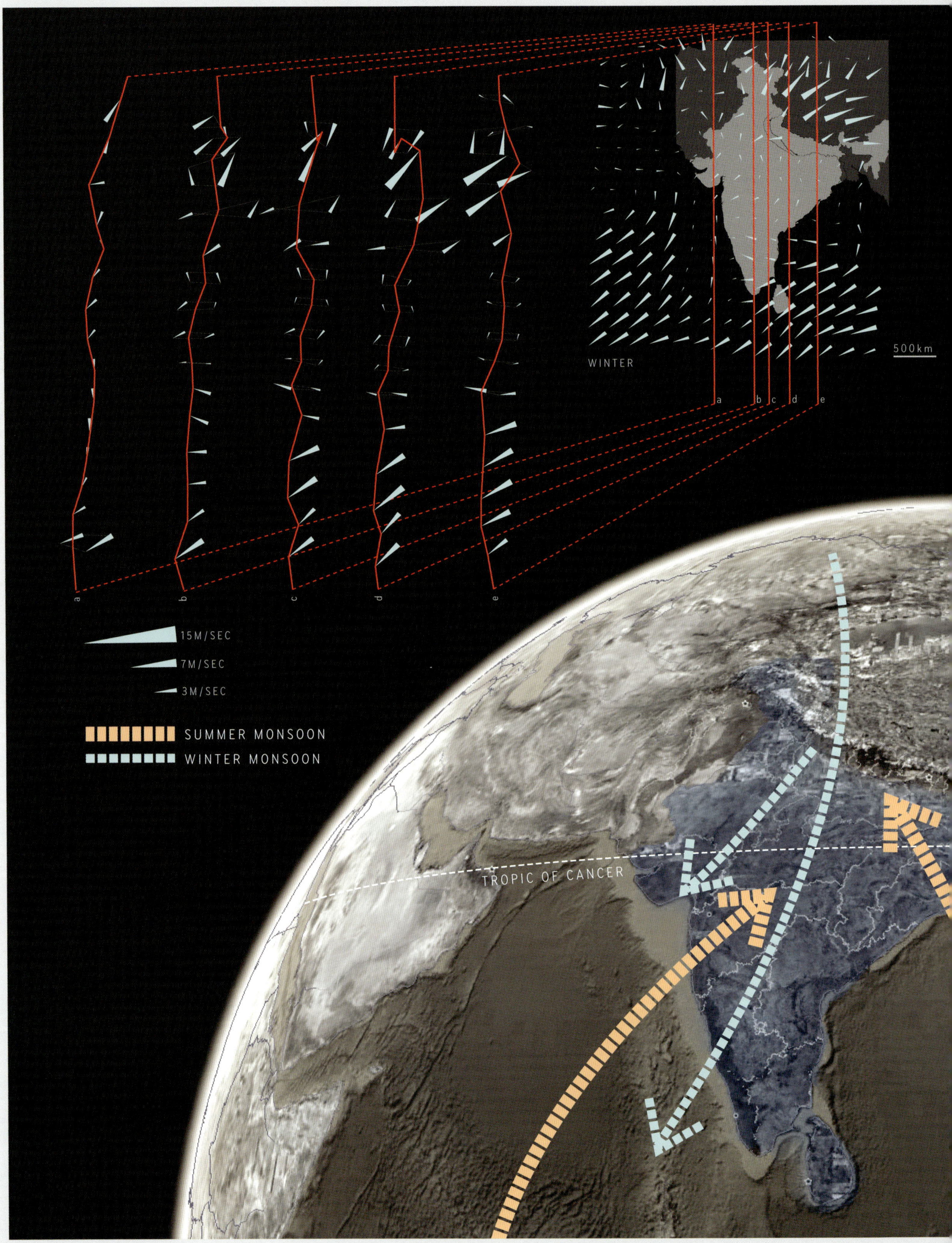
WINTER
500km
a
b
c
d
e
15M/SEC
7M/SEC
3M/SEC
SUMMER MONSOON
WINTER MONSOON
TROPIC OF CANCER

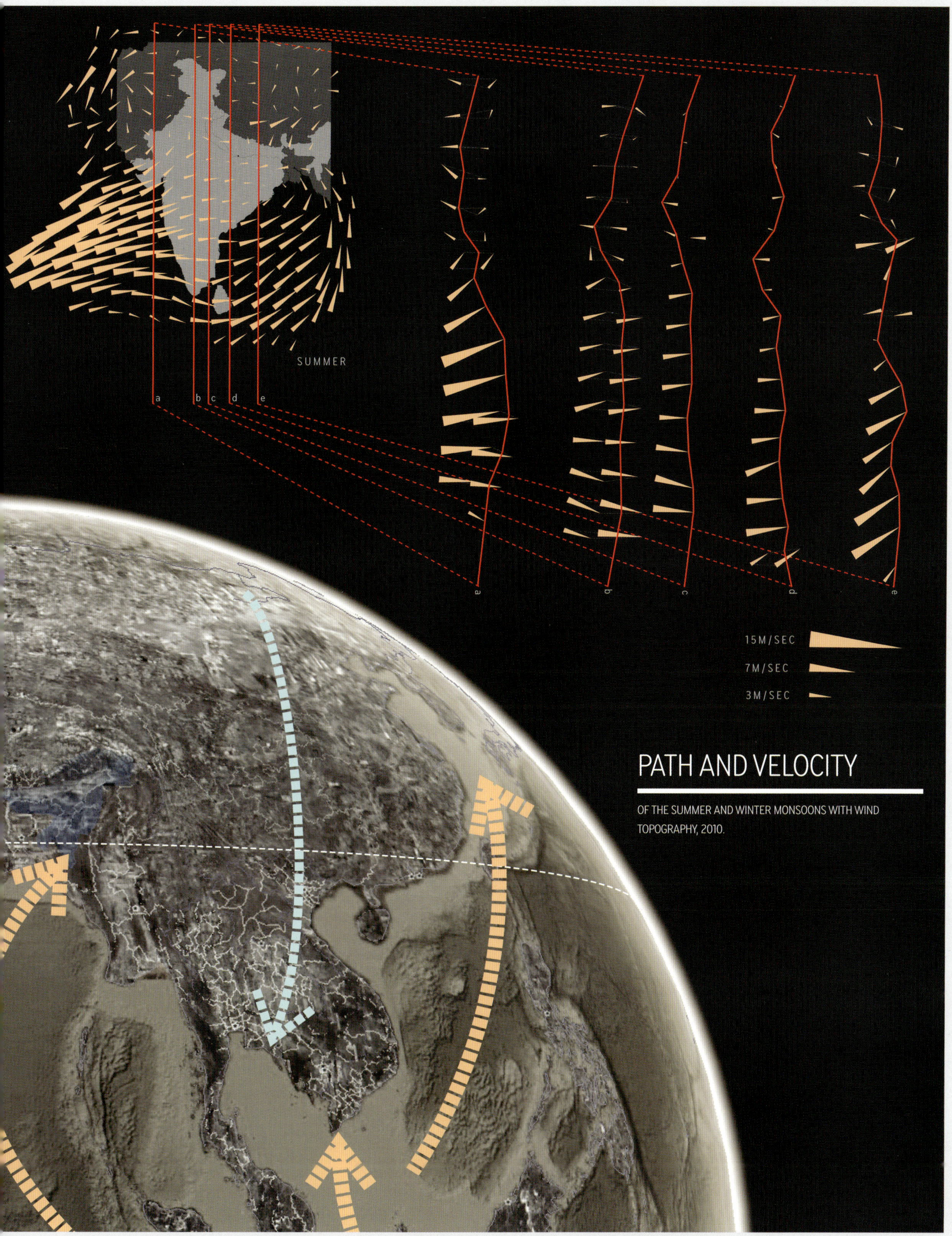

PATH AND VELOCITY

OF THE SUMMER AND WINTER MONSOONS WITH WIND TOPOGRAPHY, 2010.

VERTICAL SECTIONS CUT THROUGH INDIAN SUBCONTINENT

FROM THE HIMALAYAS TO THE INDIAN OCEAN, 2010.

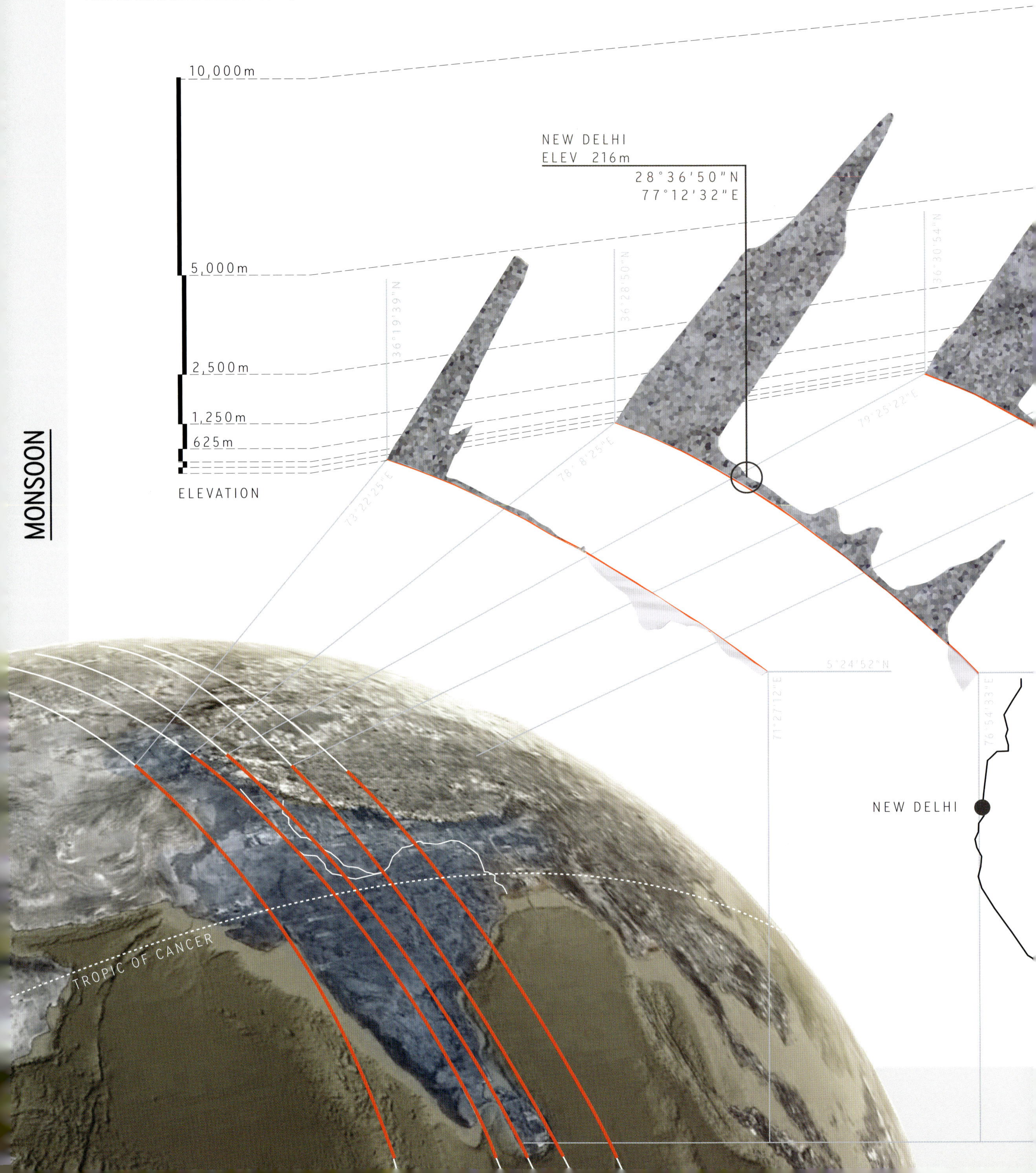

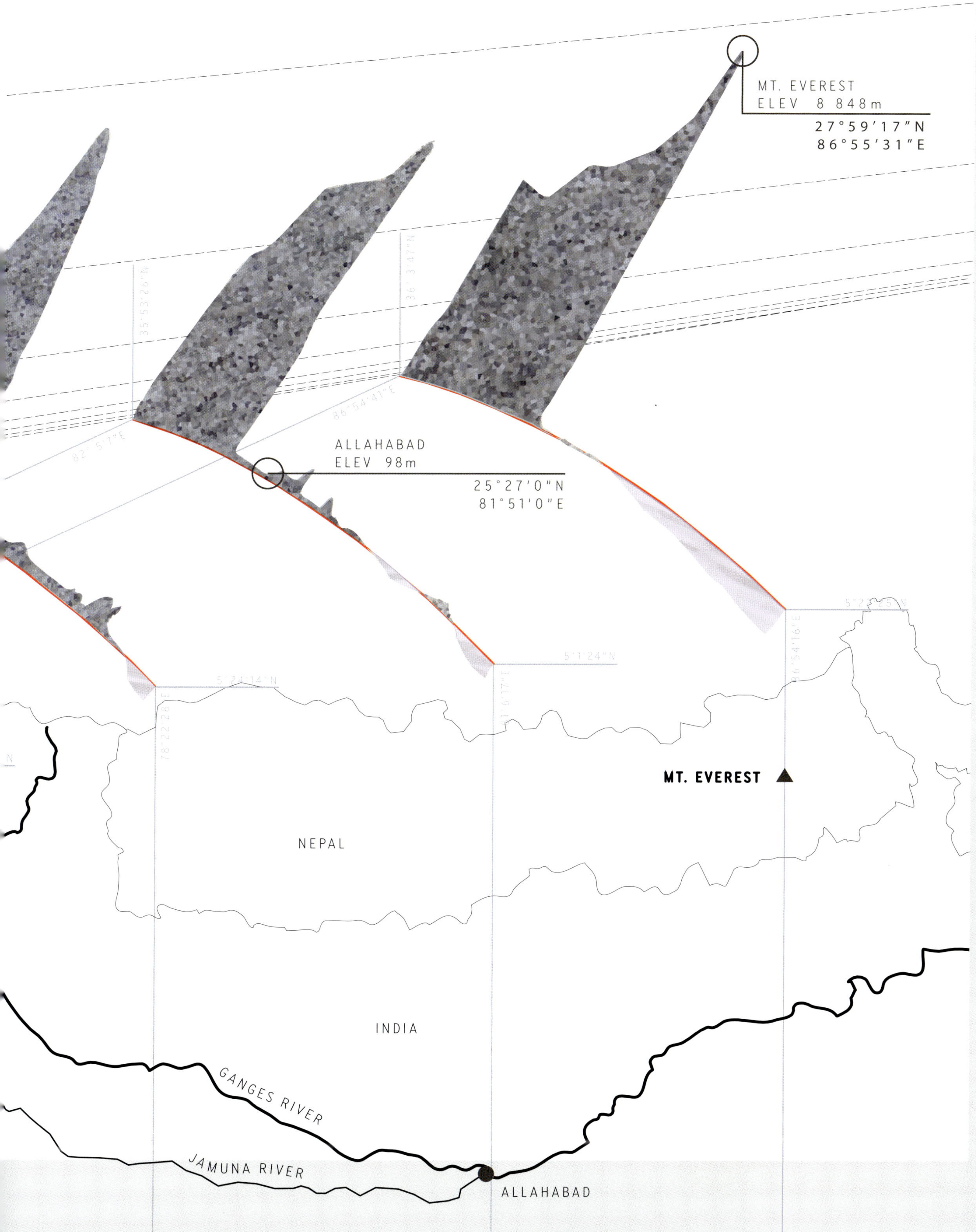
MT. EVEREST
ELEV 8 848m
27°59'17"N
86°55'31"E
ALLAHABAD
ELEV 98m
25°27'0"N
81°51'0"E
MT. EVEREST
NEPAL
INDIA
GANGES RIVER
JAMUNA RIVER
ALLAHABAD

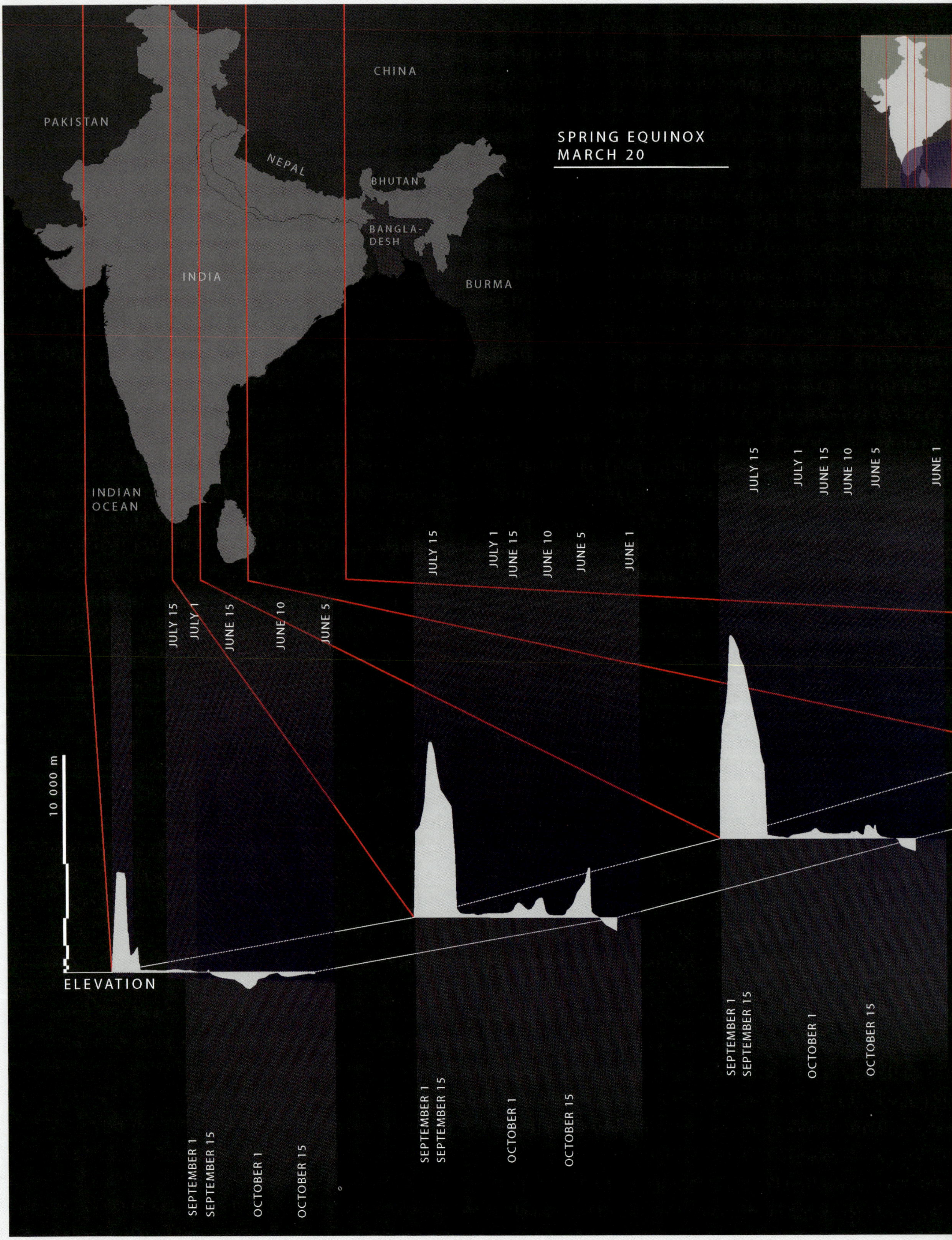
CHINA
PAKISTAN
NEPAL
BHUTAN
BANGLA-
DESH
INDIA
BURMA
INDIAN
OCEAN
SPRING EQUINOX
MARCH 20
10 000 m
ELEVATION
JULY 15
JULY 1
JUNE 15
JUNE 10
JUNE 5
JUNE 1
SEPTEMBER 1
SEPTEMBER 15
OCTOBER 1
OCTOBER 15

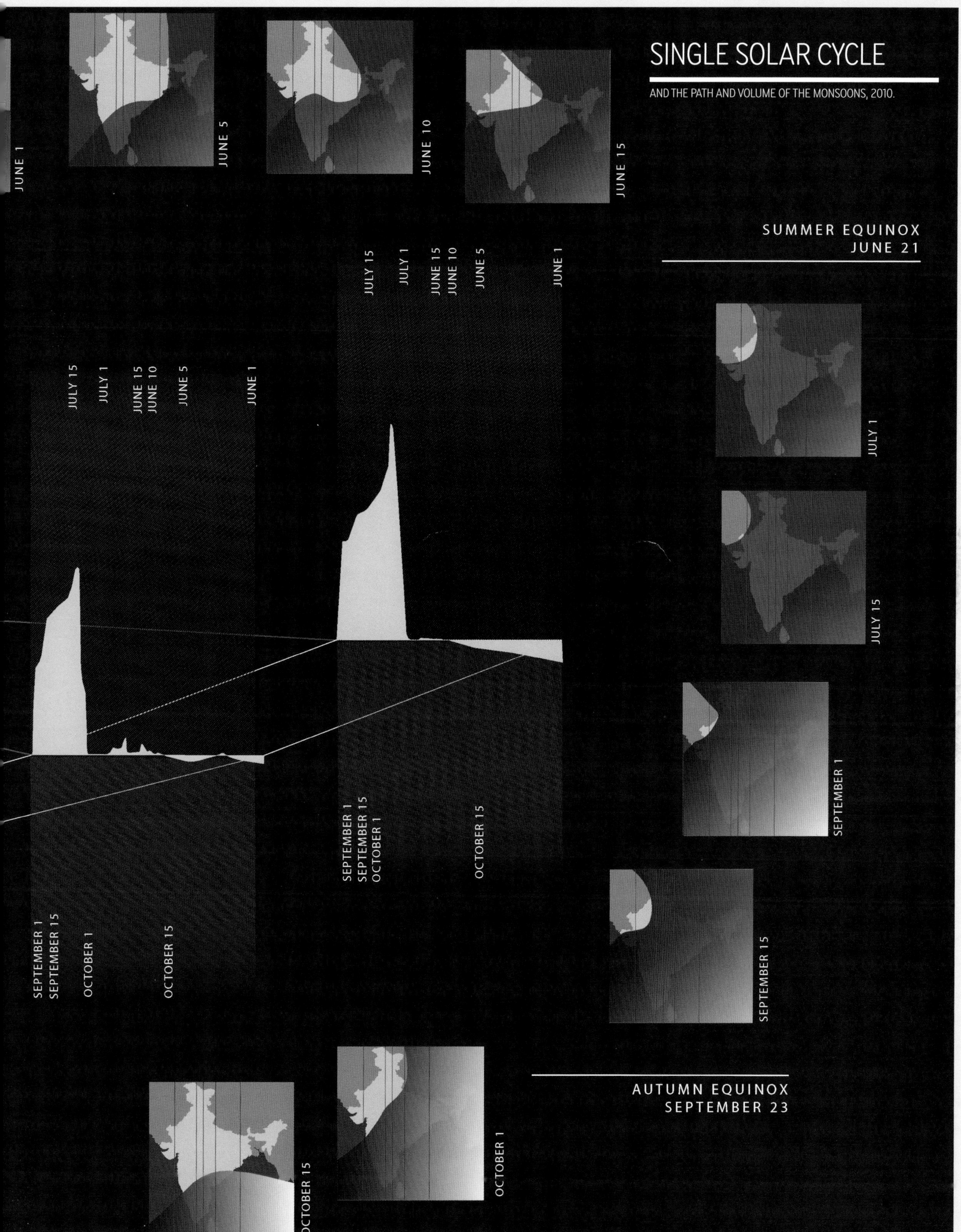
SINGLE SOLAR CYCLE
AND THE PATH AND VOLUME OF THE MONSOONS, 2010.
JUNE 1
JUNE 5
JUNE 10
JUNE 15
SUMMER EQUINOX
JUNE 21
JULY 15
JULY 1
JUNE 15
JUNE 10
JUNE 5
JUNE 1
JULY 1
JULY 15
SEPTEMBER 1
SEPTEMBER 15
OCTOBER 1
OCTOBER 15
SEPTEMBER 1
SEPTEMBER 15
AUTUMN EQUINOX
SEPTEMBER 23
OCTOBER 1
OCTOBER 15

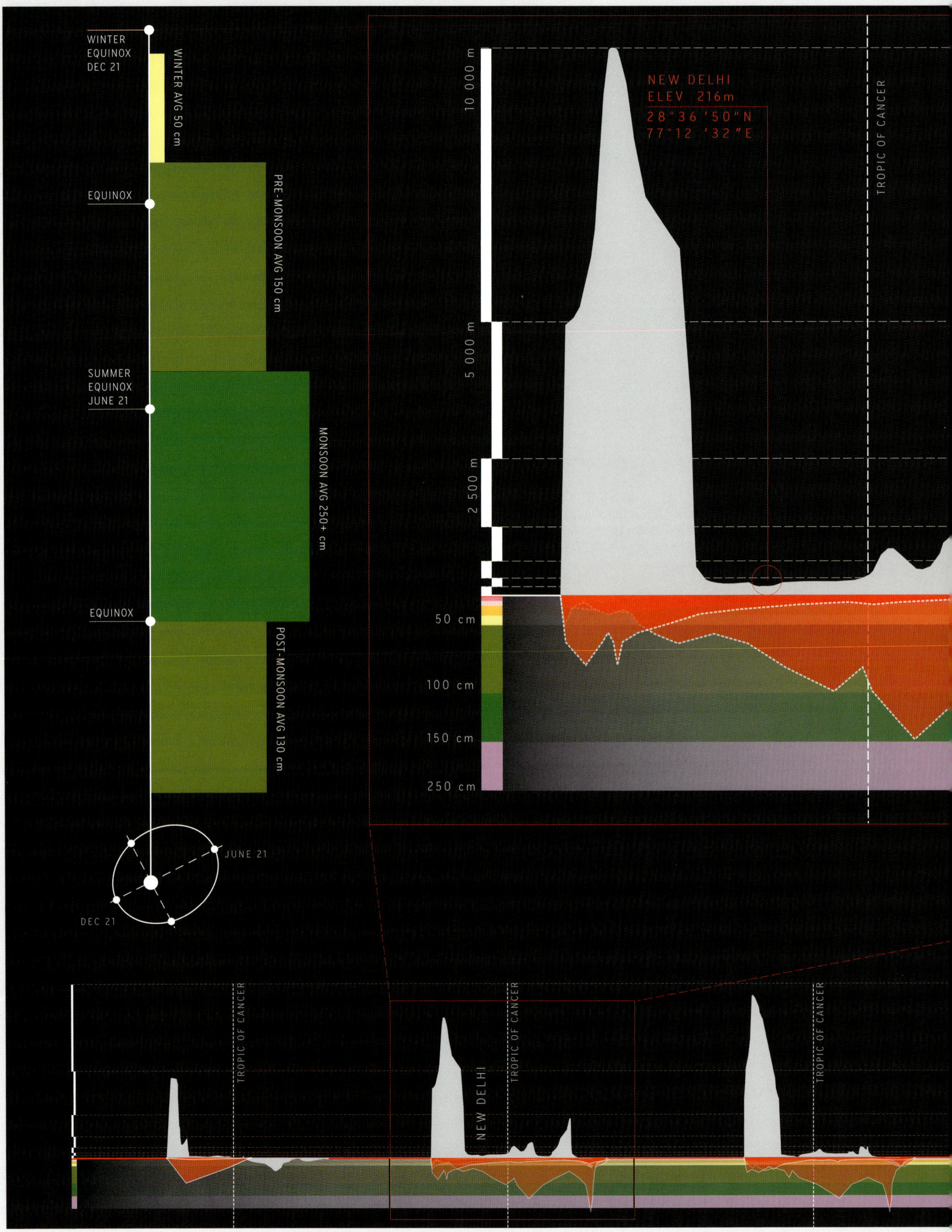
WINTER
EQUINOX
DEC 21
WINTER AVG 50 cm
EQUINOX
PRE-MONSOON AVG 150 cm
SUMMER
EQUINOX
JUNE 21
MONSOON AVG 250+ cm
EQUINOX
POST-MONSOON AVG 130 cm
JUNE 21
DEC 21
10 000 m
5 000 m
2 500 m
NEW DELHI
ELEV 216m
28°36'50"N
77°12'32"E
TROPIC OF CANCER
50 cm
100 cm
150 cm
250 cm
NEW DELHI
TROPIC OF CANCER
TROPIC OF CANCER
TROPIC OF CANCER

SECTIONS OF TOPOGRAPHY AND RAINFALL

COMPARING THE VOLUME OF RAINFALL IN SECTION WITH THE TOPOGRAPHY, 2010.

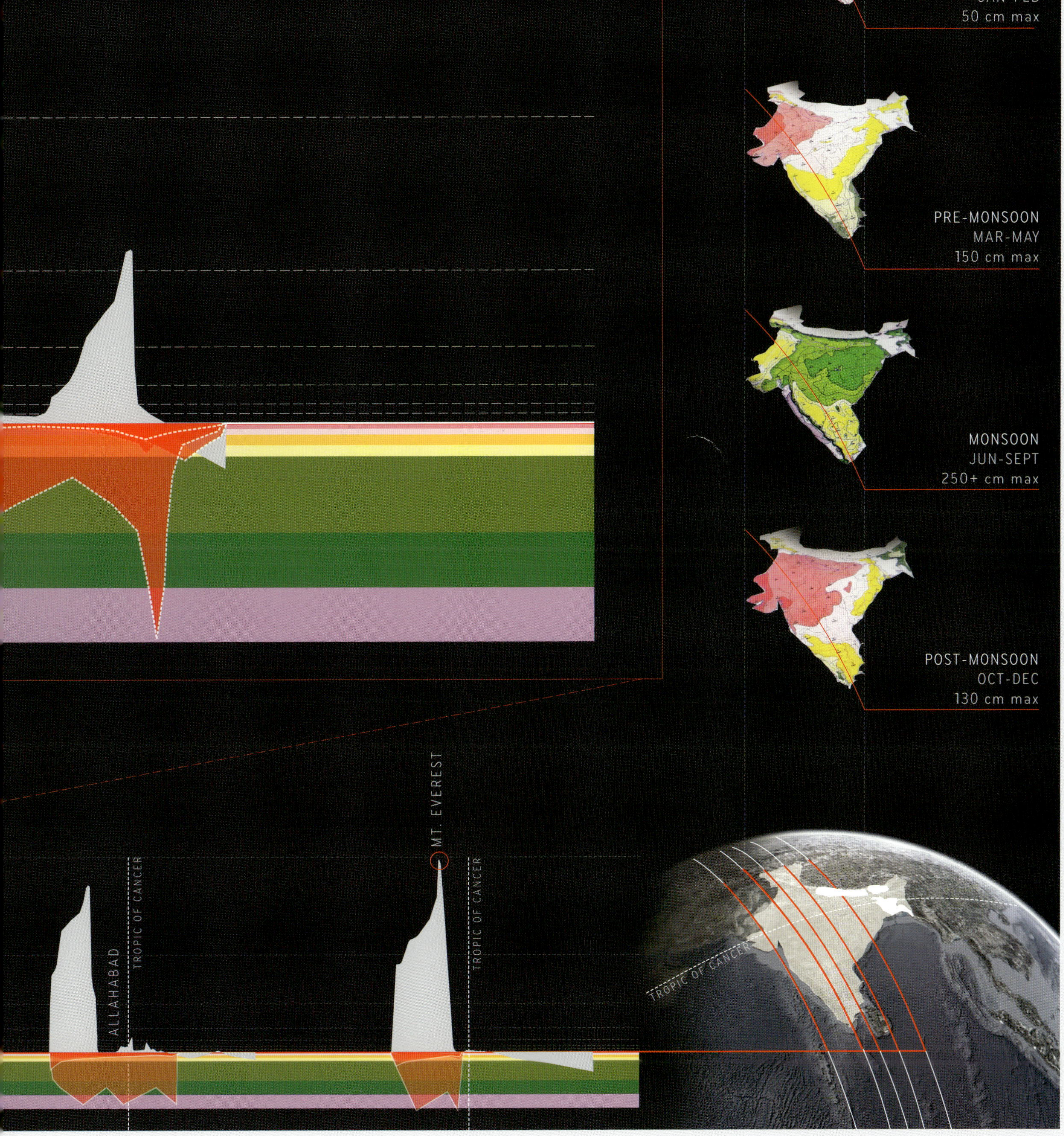

INDIAN SUBCONTINENT
DAILY RAINFALL MONSOON SEASON

SECTIONS OVER TIME

MAPPING THE VOLUME OF RAINFALL OVER THE COURSE OF A SINGLE SOLAR CYCLE, 2010.

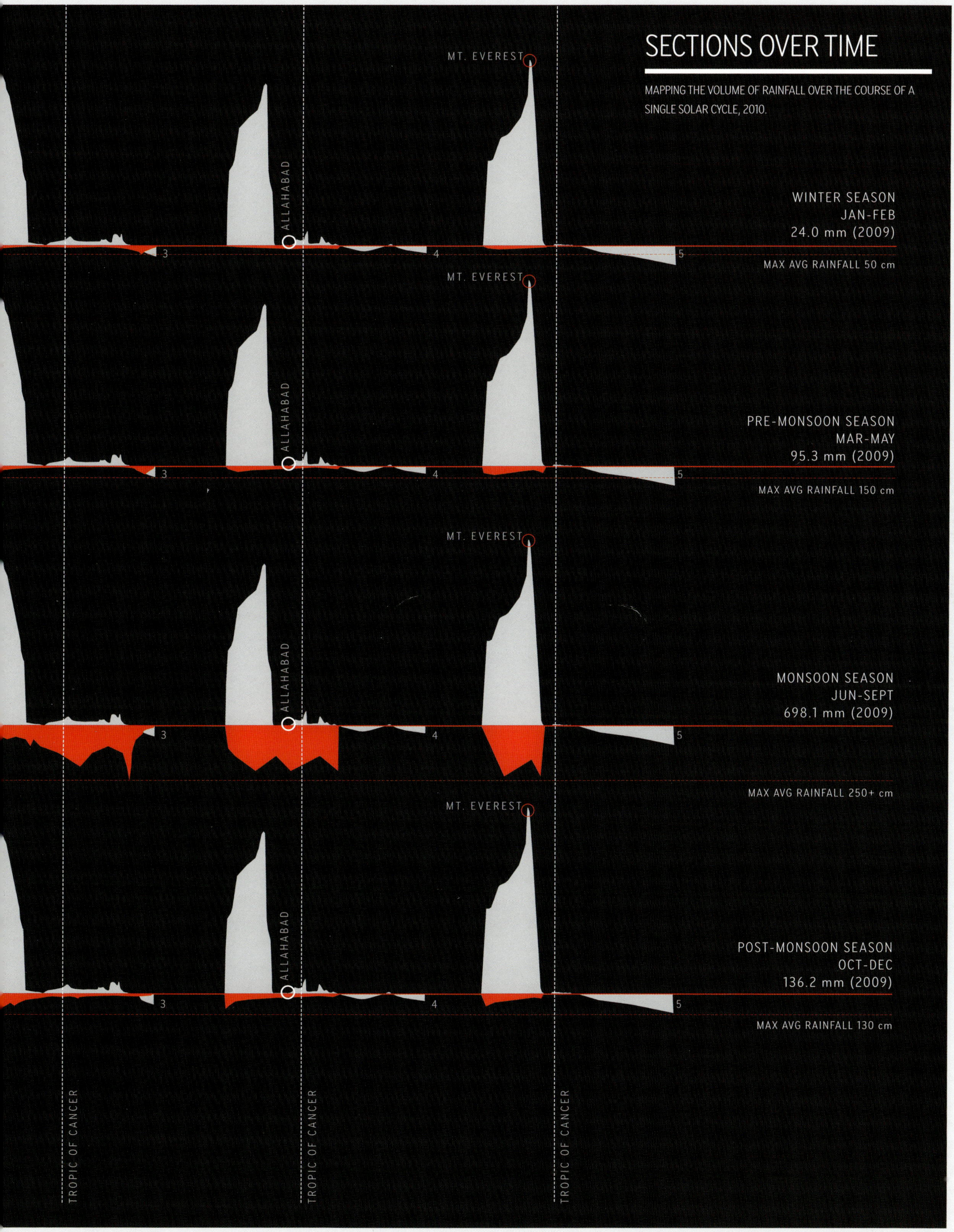

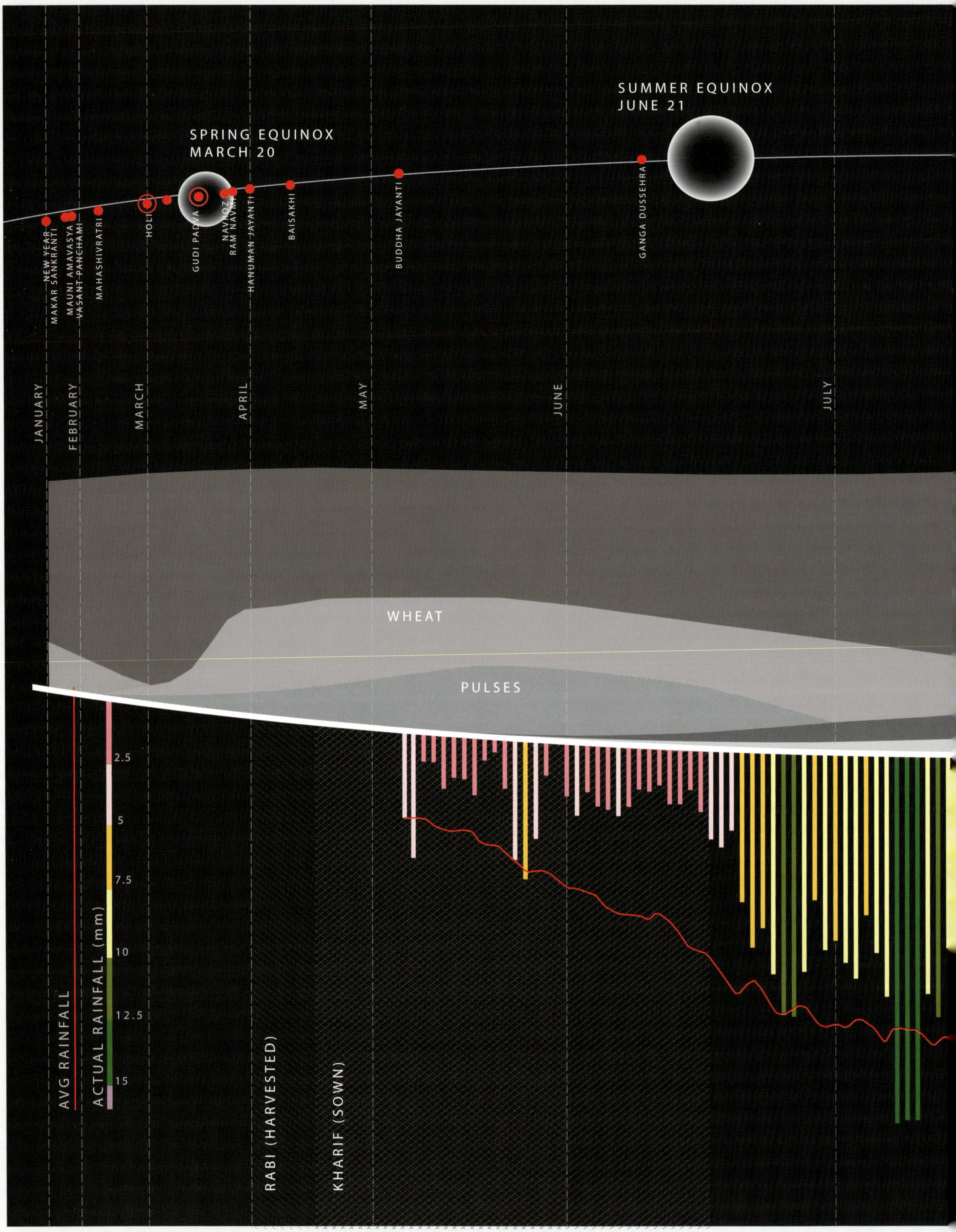
SUMMER EQUINOX
JUNE 21
SPRING EQUINOX
MARCH 20
NEW YEAR
MAKAR SANKRANTI
MAUNI AMAVASYA
VASANT PANCHAMI
MAHASHIVRATRI
HOLI
GUDI PADVA
NAVROZ
RAM NAVMI
HANUMAN JAYANTI
BAISAKHI
BUDDHA JAYANTI
GANGA DUSSEHRA
JANUARY
FEBRUARY
MARCH
APRIL
MAY
JUNE
JULY
WHEAT
PULSES
2.5
5
7.5
10
12.5
15
AVG RAINFALL
ACTUAL RAINFALL (mm)
RABI (HARVESTED)
KHARIF (SOWN)

CROP CYCLES

MAPPING THE DIFFERENT CROP CYCLES OVER THE COURSE OF ONE YEAR, 2010.
*All dates are provisional as most are based on luncar cycles.

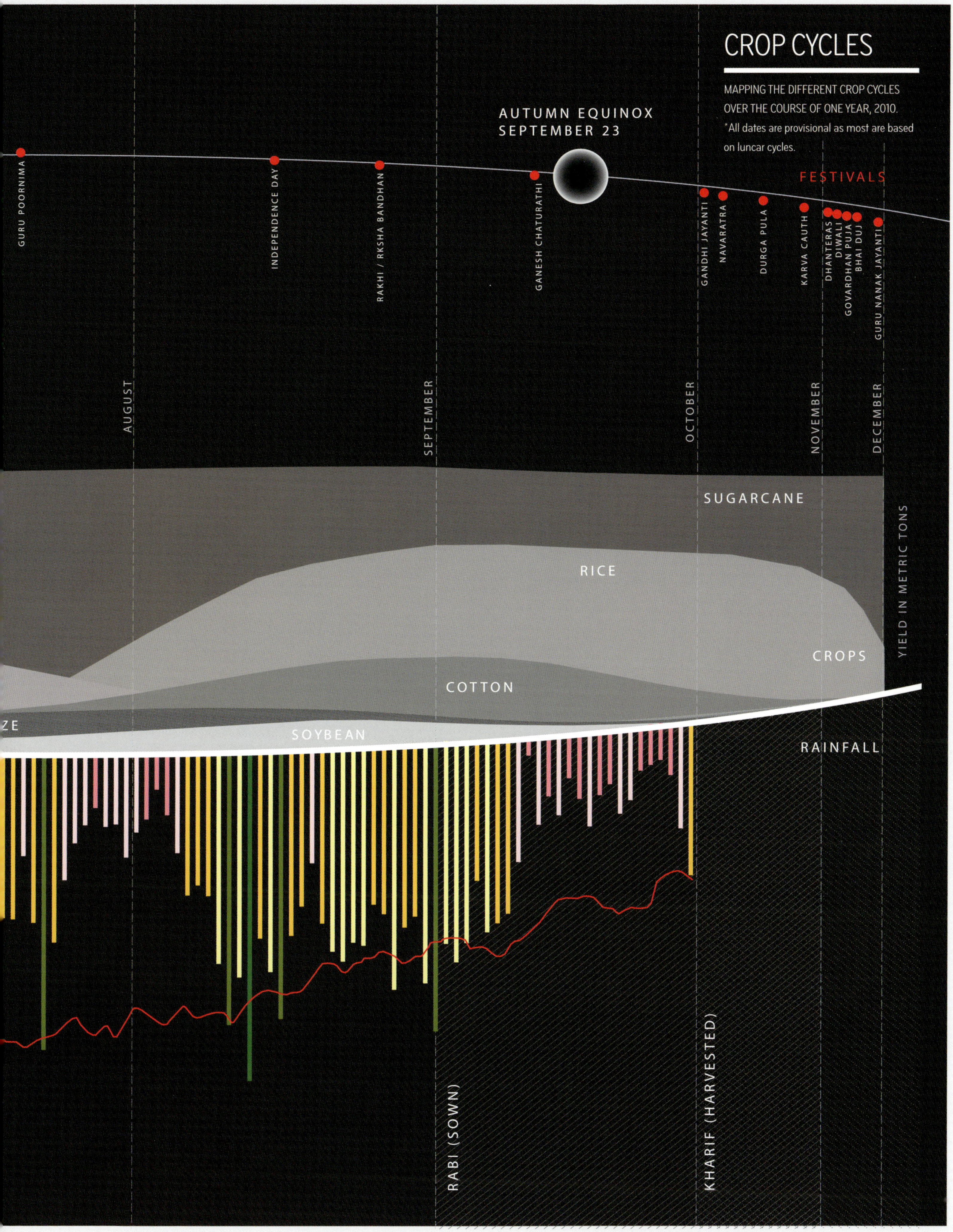

A MORE COMPLETE PICTURE: JULES VERNE AND JAMES RENNELL

Understanding the dynamic interplay between soil and water—vividly *seeing* this process—involves knowing something about descriptions and representations of this vast alluvial basin. Meteorological and geological statistics can only tell us so much; what is needed is a coherent picture of the life of the Ganges—a *tableau vivant*. These fragments of the Ganges River Basin, seen in present-day guidebooks and nineteenth century colonial painting, in *tirthas* and outwash fans, in government planning reports and local farming practices, begin to reveal the layers of the indivisible Ganges. But how might one produce a coherent, satisfying picture from such disparate pieces?

One description, written in 1880 by the novelist Jules Verne, mystified and intrigued readers about the peculiar practices and peoples of the Ganges River Basin:

> *"What you see there, my dear Maucler," said Banks, "is a conquest won by the sacred river Ganges from the not less sacred Bay of Bengal. It has been a mere affair of time. There is probably not an atom of that soil which has not been transported hither, by the mighty current, from the Himalayan heights. Little by little the stream has robbed the mountains in order to form this province, through which it has worked its bed--"*
>
> *"And changes incessantly!" broke in Captain Hood. "There never was such a whimsical capricious lunatic of a river as this same Ganges. People take the trouble to build a town on its banks, and behold, a few centuries later the town is in the midst of a plain, its harbours are dry, the river has changed course! Thus Rajmahal, as well as Gaur, were both formerly situated on this faithless stream, and now there they are dying of thirst amidst the parched rice-fields of the plains."*[27]

This quotation, from Jules Verne's 1880 *La Maison à Vapeur* (translated as the *Demon of Cawnpore*)—portrays how Anglo-Europeans traveling in the plains of northern India during the nineteenth century apprehended the Ganges River Basin. "Capricious," "lunatic," and "whimsical" were not uncommon descriptions in the nineteenth century.[28] In 1864, Sir Proby Cautley, engineer of the great Ganges Canal, described the character of tributaries to the Ganges when they passed through low lands as "the most treacherous...in the highest degree capricious."[29] Official reports on permanent settlement in the northern stretches of the Ganges plain, particularly in the district of Moozuffernugger (Muzaffarnagar) in the 1870s, describe both the Ganges and Jumna (Jamuna) rivers as "so capricious in their actions, so violent in their floods..."[30]

Verne's novel recreates these long held beliefs about the landscapes and rivers of British India, a few decades after the "sanguinary time of the mutiny [of 1857]..."[31] Narrated by Monsieur Maucler, a denizen of France traveling with three British subjects—engineer, captain, and colonel—it describes a fantastic Orientalist voyage up the Ganges. The travelers do not move by boat in "the country of Vishnu and Brahma," nor do they move by train.[32] Instead, they travel across a "perfect network of magnificent roads made by the English all over India," in the comforts of a fantastical wheelhouse drawn by an elephant "encased in steel" with "an actual steam-engine... concealed within its sides."[33] This behemoth, capable of reaching 150 horses without causing an explosion, cruises comfortably at fifteen to twenty-five miles an hour.

Traveling in the birthplace of Rama—the province of Oude (formerly Ayodhya)—before the arrival of the southwest monsoon, the group notes that on the sunrise of June 3rd, the horizon to the west "appeared misty" for the first time. The travellers then witness "a magnificent spectacle — the phenomenon of the mirage, which is called in some parts of India "seekote," or castles in the air, and in others "dessasur," or "illusion." This vision presents "a complete chain of low hills, crowned by castles of the most fantastic form, resembling the rocky heights of some Rhenish valley with their ancient fastnesses of the Margraves."[34] Banks, the engineer, asks the group if they know "what the phenomenon presages?" "'Say on, great engineer!' quoth the captain." Banks tersely replies: "Nothing short of a great change in weather." Later that night, peering out of his moving bungalow, Mau-

Jules Verne, "The fanatics sprang up." **The Demon of Cawnpore**. New York: Charles Scribner's and Sons, 1881. (Princeton University).

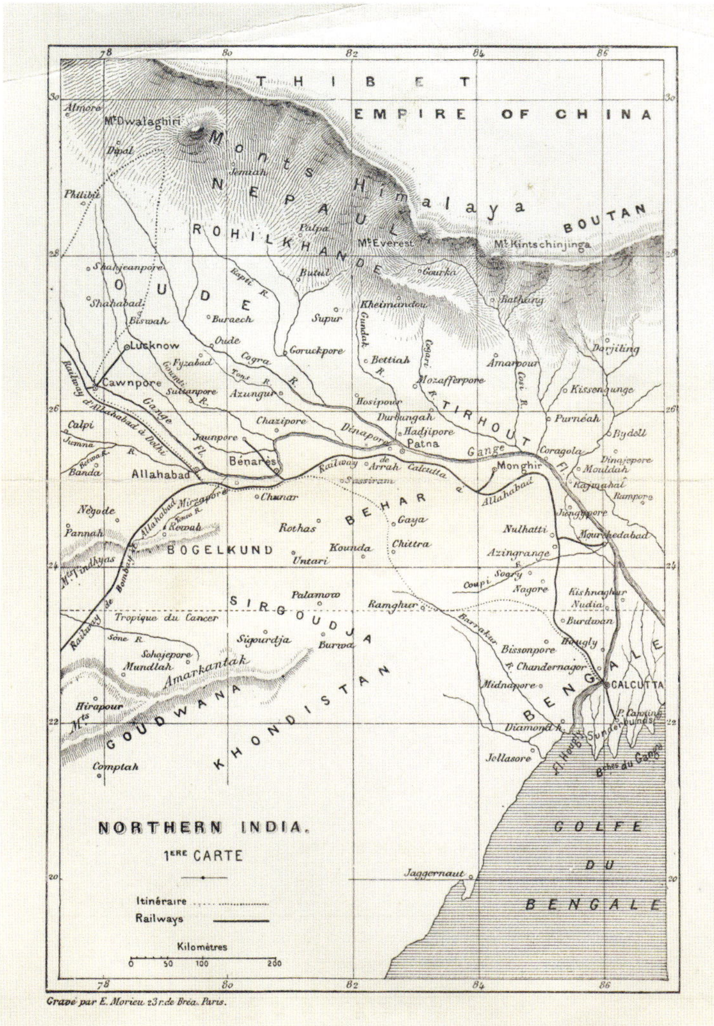

The fanatics sprang up.

Frontispiece.

Jules Verne, "Northern India: 1ere Carte." **The Demon of Cawnpore**. New York: Charles Scribner's and Sons, 1881. (Princeton University).

cler witnesses what Banks had predicted: "Towards evening the western horizon became obscured by vapours, showing that the monsoon, as frequently happens, would commence during the night. These mists, charged with electricity, came across the peninsula from the Indian Ocean, like so many vast leathern bottles out of the cellars of Aeolus, filled full of storm, tempest, and hurricane."[35]

The following day, Maucler finds "Spiral columns of very fine dust whirled along the roads, in a manner quite unlike that which was raised by our heavy wheels." To him, "They resembled a number of those tufts of downy wool which can be set in motion by an electrical machine. The ground might, therefore, be compared to an immense receiver in which for several days electricity had been stored up." He noted how the "dust was strangely tinted with yellow," and "each atom seem[ed] to shine from a little luminous centre. At times we appeared to be travelling through flames, harmless flames, it is true, though neither in colour nor vivacity resembling the *ignis fatuus*."[36]

Maucler's depiction remains one of the most potent of the *changes of state* produced by the onslaught of the southwest monsoon. Such a thick description captures, all at once, an electrical charge, capriciousness, and groundswell leading up to the storm. Verne's tropical steampunk adventure emphasizes nature's dominance over Indian habits. Aboard a mechanical elephant (a cyborg of sorts) his protagonists experience the *accouterments* of a technological wonder while they marvel at the chaotic landscape that confounds their Anglo-European experience. Not unlike many illustrated novels from the second-half of the nineteenth century, readers of *The Demon of Cawnpore* had 49 illustrations and a map to help contextualize and illustrate the exotic places, peoples, and topography of the Ganges basin and Himalayan mountain range. It is but one way to imagine the cultural and geological layers of the landscape. But there are other ways as well, to which Verne's work is most definitely indebted to—most directly to those travelogues and memoirs from the late eighteenth and early nineteenth centuries like those of James Fraser.

As we have seen, nineteenth century accounts can be written in many ways, but they all usually present accounts of monsoons, rivers, and conflict. They observed, as had millions before them, the interplay between water and soil. "The vast collection of vapours, wafted from the sea by the southerly or south-west monsoon, are suddenly stopped by the lofty ridge of mountains that runs from east to west through Thibet," observed Captain James Rennell in a footnote in his "Account of the Ganges and Burrampooter Rivers" in 1781.

Rennell, the first surveyor general in Bengal to the Company and eventually a member of the Royal Society in London, was perhaps the first geographer to conceptualize the profile of the monsoon and the Ganges River in vertical cross-section as opposed to horizontal plan.[37] To do this, he observed the monsoon at the scale of the region (how "the vast collection of vapours" collide with the Himalayas) and how this event produced not only the rains, but transformed the profile of the riverbed. In his "Account," read before the Royal Society by Joseph Banks, Rennell noted how "All the rivers that are situated within the limits of the monsoons, or shifting trade winds, are subject to overflowings at annually stated periods, like the Ganges: and these periods return during the season of the southerly wind, that being the only wind which brings vapours from the sea; and this being periodical, the falls of rain must necessarily be so too."[38] In the second half of the eighteenth century, Rennell's interest in better understanding the monsoon was driven by the needs of the Company, which was bent on creating accurate maps for commerce and security. Rennell had been named Captain and surveyor general of Bengal on January 8, 1767, a decade after Robert Clive's decisive victory at the Battle of Plassey in 1757. That battle, which occurred during the Seven Years

PICTURE

A wonderful sight.

Page 67.

Jules Verne, "A wonderful sight." **The Demon of Cawnpore**. New York: Charles Scribner's and Sons, 1881. (Princeton University).

The electric light streaming through the banyan forest.

Page 185.

Jules Verne, "The electric light streaming through the banyan forest." **The Demon of Cawnpore**. New York: Charles Scribner's and Sons, 1881. (Princeton University).

An awful clap of thunder.

Page 196.

Jules Verne, "An awful clap of thunder." **The Demon of Cawnpore**. New York: Charles Scribner's and Sons, 1881. (Princeton University).

Jules Verne, "The horizon darkened towards evening." **The Demon of Cawnpore**. New York: Charles Scribner's and Sons, 1881. (Princeton University).

The horizon darkened towards evening.

Page 177.

War (1756–1763), took place at a time when "the global rivalry between the English and the French spilled over into competition for the Carnatic."[39] As the geographer Matthew Edney has noted, this produced a paradoxical conquest by the English of Bengal and the heart of the Mughal Empire in the north, and not the south.[40]

Rennell's position was to bring coherence to an otherwise idiosyncratic set of survey practices and records of India. "So much depends upon accurate surveys, both in military operations and in coming at a true knowledge of the value of your possessions, that we have employed everybody on this service who could be spared and were capable of it," wrote Robert Clive to the Court of Directors of the Company. "But as the work must ever be imperfect while it is in separate and unconnected plans," he continued, "we have appointed Captain Rennel [sic], a young man of distinguished merit in this branch, Surveyor General, and directed him to form one general chart from those already made, and such as are now on hand as they can be collected in."[41] This map was much needed, because in 1765, the same year Clive commissioned Rennell to create a single map of Bengal, he had negotiated for the Company to assume the role of *diwan*, or chief financial and administrative officer, over the entire region. Clive also wished to produce a map of Bengal for his friend Robert Orme, a historian who wished to include a map of Bengal with his second volume on the history of the Company's military victories in India.[42]

Rennell, along with his assistants, made surveys using only a compass and chains along with Hadley's Quadrant to determine latitude. This practice of walking the land and taking measurements, known as a traverse survey or route survey, was the most popular method of topographic

James Rennell, "A Map of the Provinces of Delhi, Agrah, Oude, and Ellahabad, comprehending the Countries lying between Delhi, and the Bengal-Provinces." London: Robert Sayer, 1886. (Argosy Gallery, New York).

survey in the eighteenth century.[43] Even though he was limited in the instruments made available to him—the chronometer was not used to determine longitude until 1761— Rennell earned the title "Herodotus of India" due to the accuracy of his surveys. [44] While Rennell spent a large portion of his time in a base office, compiling maps, his work also involved extensive fieldwork for surveying, which could some times be quite dangerous. While surveying near the frontier of Bhutan in 1776, Rennell was seriously wounded by a group of Sanyasi Fakirs and almost died from the loss of blood. The attack permanently weakened his constitution.[45] By 1776, he had retired from the Company and returned to London, where along with becoming a member of the Royal Society in 1781 and an associate of the Institute of France in 1800, he would continue his work on cartography and oceanography for years to come.

Rennell's work put the Ganges on the map in more ways than one.[46] In the same year that he left India and returned to London, at the behest of the Court of Directors of the Company, Rennell published 'A Map of the Provinces of Delhi, Agrah, Oude, and Ellahabad, comprehending the Countries lying between Delhi, and the Bengal-Provinces.'[47] Compiled in the late-1760s from field surveys, Rennell's map was the first approximately correct European map of the interior of the Ganges River Basin. Prior to this map, the most accurate map was French cartographer Jean Baptiste B. d'Anville's *Carte de l'Inde* from 1752. When d'Anville published his map, the southern peninsula of the Indian subcontinent was well traversed and mapped while the north remained less so. This is clear in d'Anville's map, where much of northern India is shown full of significant patches of white with a mythical Ganges River running across the whitest section of the map. On the other hand, Rennell's map of the northern stretches of the Ganges region ends at Hardware [Haridwar] and includes topographic, political, and historical layers. In the 1786 reproduction of his 1777 map, line widths and weights along with typographic sizes all contribute to a specific topographic hierarchy. For instance, the Ganges River, like the Jamunah (Jamuna) river, is outlined by a thick black line with its bed composed of finer, parallel lines; while the majority of the tributaries to each of these rivers are but thin squiggles of single or parallel lines. Amidst the regional names in bold, major roads run along the rivers and crisscross the plains. Towns and hamlets, notated as circles, dot the map and are differentiated from the iconography of significant forts, like those at Agrah (Agra) and Delhi. In 'A Map of Bengal, Bahar, Oude & Allahabad with Part of Agra and Delhi Exhibiting the Course of the Ganges from Hurdwar to the Sea' from 1786 by Major Rennell extends from Hurdwar (Haridwar) to the Ganges delta, showing roads and outposts and political divisions of the Ganges basin.

Rennell's impact on science in colonial India extended beyond his cartographic rendering of the subcontinent. Geologists cited his writings on how the Ganges delta changes and moves for another eight decades. In *A Geological Manual* (1833), when English geologist Henry Thomas De La Bache detailed the "Delivery of Detritus into the Sea," he cited Major Rennell's observations on the modes by which the Ganges delta changes over time as "always valuable" because they show the "mode in which they have been carried."[48] Sir Charles Lyell not only discusses Rennell's work on the Ganges in *Principles of Geology* (1865), but also cites his observations on ocean currents or "oceanic rivers" that came to be called "Rennell's Current." [49]

What Rennell achieved visually in terms of space, Jules Verne achieved textually in terms of time. Both display a layered understanding of the Ganges River Basin. If we combine their spatial and temporal layers with those of the *Mahabharata* and pilgrim maps, we can see a layered, stratified nature of experiencing the Ganges: a thick visual and textual cross section that provides a rich overall picture of the Ganges River.

In the decades after the publication of Rennell's maps and Fraser's illustrated book, a new actor comes on the scene: the engineer. Much like Jules Verne's novel from 1880, the engineer of the nineteenth century does more than walk the land and mark it. As we shall see, the engineer reshapes land and redirects rivers, creating an entirely new, artificial *topos*. The map proves his most important tool to do so.

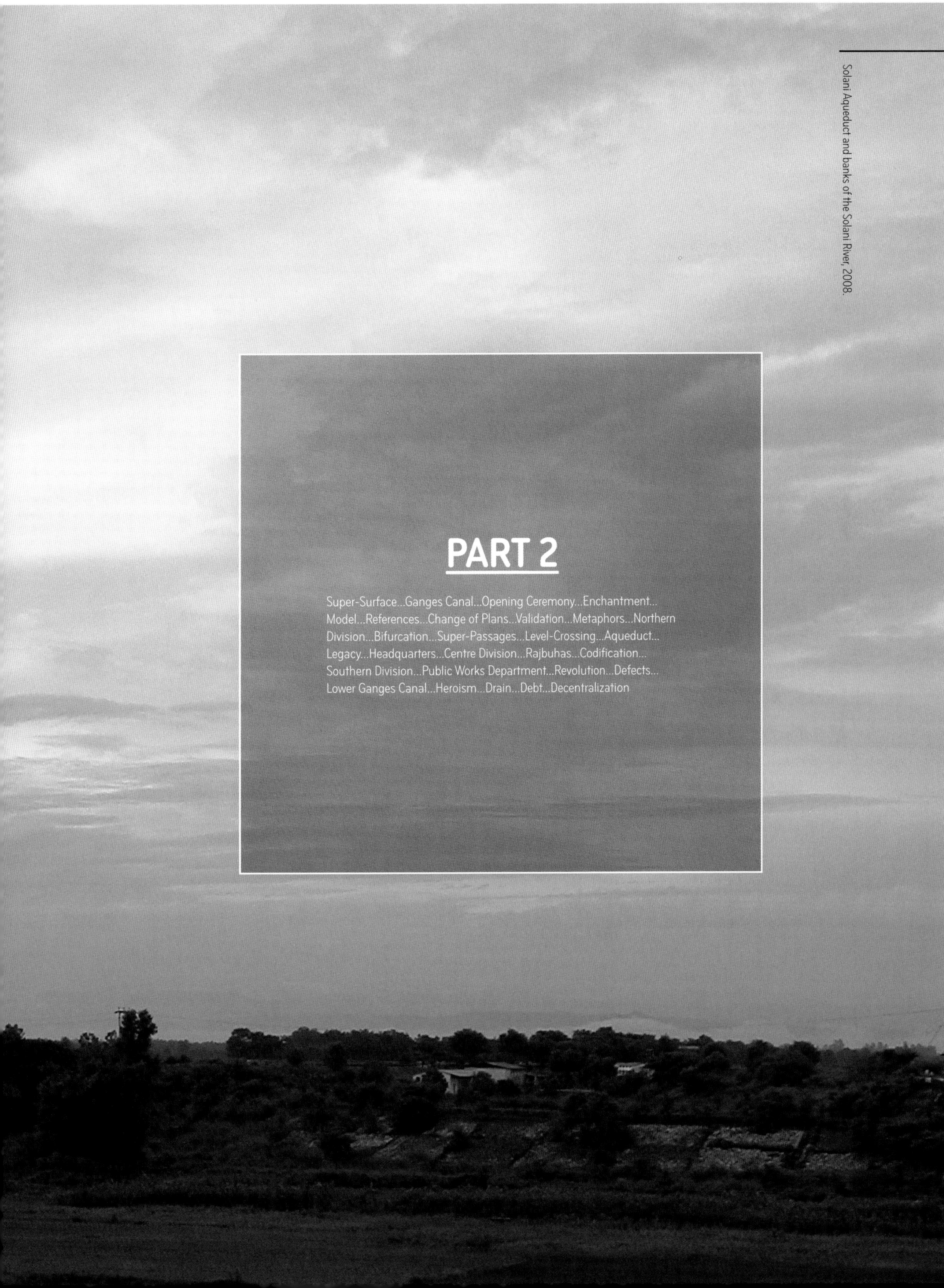

Solani Aqueduct and banks of the Solani River, 2008.

PART 2

Super-Surface...Ganges Canal...Opening Ceremony...Enchantment... Model...References...Change of Plans...Validation...Metaphors...Northern Division...Bifurcation...Super-Passages...Level-Crossing...Aqueduct... Legacy...Headquarters...Centre Division...Rajbuhas...Codification... Southern Division...Public Works Department...Revolution...Defects... Lower Ganges Canal...Heroism...Drain...Debt...Decentralization

GANGES CANAL

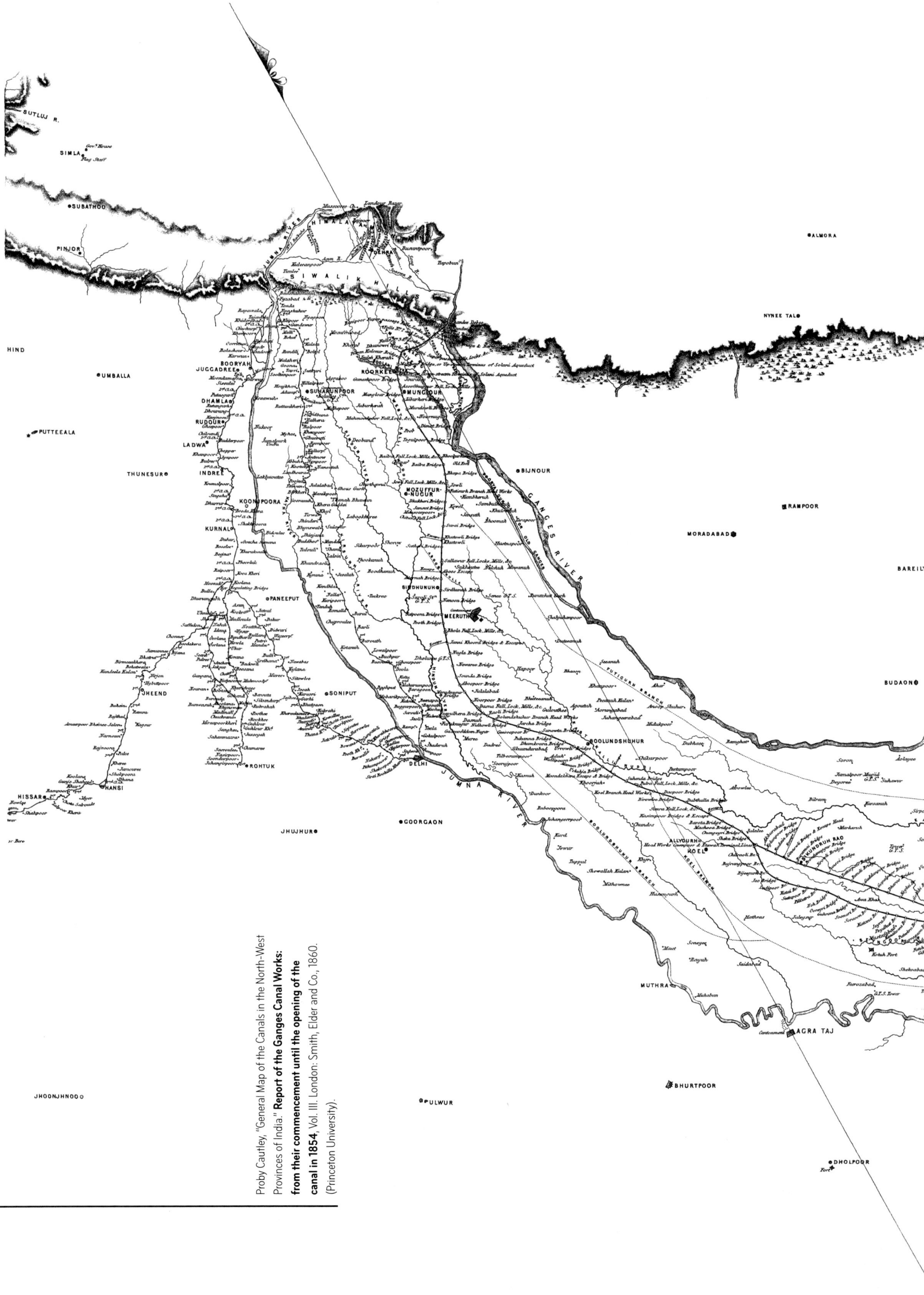

Proby Cautley, "General Map of the Canals in the North-West Provinces of India." **Report of the Ganges Canal Works: from their commencement until the opening of the canal in 1854**, Vol. III. London: Smith, Elder and Co., 1860. (Princeton University).

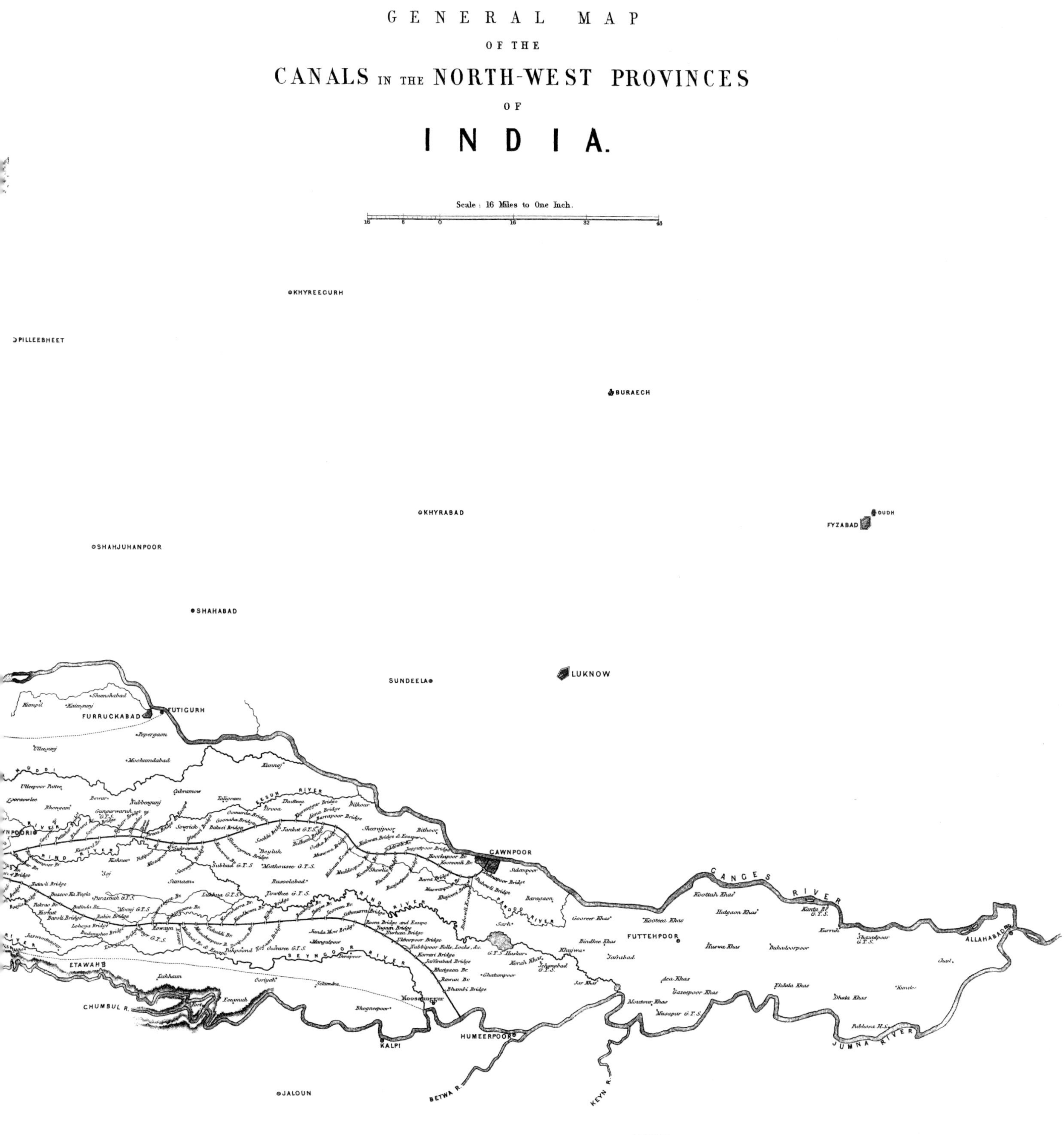
GENERAL MAP
OF THE
CANALS IN THE NORTH-WEST PROVINCES
OF
INDIA.
Scale: 16 Miles to One Inch.
KHYREEGURH
PILLEEBHEET
BURAECH
KHYRABAD
SHAHJUHANPOOR
FYZABAD
OUDH
SHAHABAD
SUNDEELA
LUKNOW
FURRUCKABAD
FUTIGURH
CAWNPOOR
GANGES RIVER
FUTTEHPOOR
ALLAHABAD
ETAWAH
CHUMBUL R.
KALPI
HUMEERPOOR
JUMNA RIVER
JALOUN
BETWA R.
KEYN R.
BANDAH

Solani Aqueduct crossing the Solani River, 2008.

The direction of the drainage on the southern slopes of the Himalayas throughout the provinces of Upper India lying to the north-west, the character of the rivers which constitute this drainage, and the topographical features of the country through which it passes, point out by unmistakeable [sic] signs a region designed by nature as a great field of artificial irrigation. These rivers after their debouche from the mountains enter upon a champaign country with a rapidly decreasing slope, and from the parallel direction which they assume in their passage to the Indus on one side, and to the Ganges on the other, divide the country into sections of a larger or smaller width; all of which, both in soil and declivity of surface, possess in themselves every requisite for irrigation. The steep slopes that exist at the immediate debouche from the mountains enable the engineer, by a proper adaptation of levels in his artificial channels, to obtain a command of water, which places the whole country under his control...[1]

Sir Proby Cautley, 1860[2]

SATURDAY APRIL 8, 1854

At a quarter past six-in-the morning, Proby stood up beneath a great awning in the shadow of the Solani Aqueduct in Roorkee, North-West Provinces. Just completed, it was the longest aqueduct in the world. The sounds of ceremonial artillery fire were heard as Lieutenant-Governor John Russell Colvin arrived to join several hundred distinguished guests. At long last, the English soldier-turned-hydraulic engineer-extraordinaire, Lieutenant-Colonel Proby Cautley, was to see this "region designed by nature as a great field of artificial irrigation" come under his control.

But all eyes were not trained on the engineer. Nor were they on Reverend Hamilton, who, there to consecrate the event, read aloud from the book of *Isaiah*, "I will open rivers in high places, and fountains in the midst of the valleys: I will make the wilderness a pool of water, and the dry land springs of water." The crowd, unable to hear or even see Reverend Hamilton, was there to witness the hydraulic architecture.

Following the Reverend's reading, the official party of "English ladies and gentlemen, and a few distinguished native guests" ascended a staircase to the top of the aqueduct where "The infantry were on the tops of the aqueduct parapets, stretching down their length in two red lines." There they witnessed Lieutenant-Governor Colvin and Captain A. G. Goodwyn, of the Bengal Engineers, pull levers to release the pent up waters of the Ganges into the channels of the aqueduct over the Solani River. "The band instantly struck up, 'God save the Queen'; the artillery commenced its royal salute... and half a million voices shouted welcome to Gunga-jee..."[2]

It was a proud day for Cautley. Known as a modest man with a penchant for paleontology—in 1837 he and geologist Hugh Falconer won the prestigious Wollaston Medal of the Geological Society for their work on fossils in the Siwalik Hills—he had, alongside his design of the Ganges Canal, been elected a fellow of the Royal Society in 1846. Working in India proved a foundation for honors in metropolitan London, but by 1854 his time in India was coming to a close. And what a relief it must have been, for this day was nearly fifteen years in the making. Cautley's work on the Ganges began in December of 1839, when, armed with a theodolite and level, he was commissioned by Governor General Lord Auckland to trek along the sacred Ganges River to the Siwalik Hills in order to make a survey of the *khadir* (floodplain) in and around the sacred city of Haridwar. The goal of the survey was clear: to see how to channel the Ganges' waters into an artificial canal that would distribute water to millions of acres of farmland. Cautley measured the flow of the Ganges at 8,000 cubic feet per second (226.5 cubic meters), an ample amount of water for a canal. He proposed calibrating the new canal to hold 6,750 cubic feet per second (191.1 cubic meters/second)—"or very nearly the supply in the river at Hurdwar during the driest months"—a volume three times that of any contemporaneous artificial canal system.[3] Haridwar was high enough above the tableland to ensure an adequate flow, but getting it to Roorkee was a challenge. Steep slopes and tempestuous mountain torrents in the first eighteen miles (29 km) of the canal presented formidable challenges. But the government was highly motivated to artificially irrigate the drier lands further south.

Spurred on by "the calamitous events of the year 1837-38," in which a great famine had occurred in the central provinces, Lord Auckland funded a study on the possibility of irrigating the most devastated areas with a canal drawn from the Ganges. "The exertions of the European community in alleviating the sufferings of a starving population were carried on with true Anglo-Saxon energy," wrote Cautley of the Company's relief efforts. He continued, "and the sacrifice of a million sterling in revenue, either in remission or in distributing food, was a guarantee of the interest that the government felt in this frightful calamity."[4] Captain Colvin had suggested a canal drawn from the waters of the Ganges as early as 1836, but the government was reluctant to entertain a project of such unprecedented cost and enormity. Only after the loss of thousands of lives and considerable treasure did the prospect of a Ganges Canal take hold. Canals provided insurance against famine, or so it was lauded in the *Delhi Gazette* of April 8, 1854: "immunity from the pains and losses that famine brings with it."[5]

Cautley submitted reports in June and August of 1840, estimating the cost of the Ganges Canal to be 2,591,158 rupees. On September 1, 1841, the Court of Directors in London replied: "'There are few measures connected with our revenue administration in India more calculated to contribute to the general improvement of the country, the amelioration of the condition of the people, and to raise the character of the Government, than those of the nature now under our consideration.'"[6] The canal was an act of reclamation of land and water for agriculture as much as it was rehabilitation for a population impoverished, it seemed to the British, by a lack of artificial irrigation. Engineering, if properly applied, could quell the destructive nature of the southwest monsoon. What was more, according to men like Cautley, nature had designed the region for artificial irrigation. This was not a plan to alter reality so much as reveal it. The engineer would be the great revealer of this latent truth.

Artificial irrigation, of course, was not new to the region, but it had remained at the scale of individual wells and water tanks, prevalent in most villages in the Ganges River Basin. In contrast to the Ganges Canal system, where a centralized authority like an irrigation department controlled water across the whole region, decentralized pockets of authority (such as prominent local families) had controlled wells and water tanks individually. In order to make improvements to these decentralized, private types of irrigation, cultivators could access the *takvi*, an institution pre-dating Company rule that provided government loans to assist cultivators to finance improvements. In the case of the Ganges Canal, the Company used this same system to finance its distribution channels, rather than providing takvi to individual farmers to make local improvements. This was an attempt to transition from local decision-making to a centralized system, controlled by the Company, which would regulate water on a grand scale.

Given the fiscal and social stakes of such a project, decisions were made deliberately. The Company formed a committee, recommendations were made, and on April 16, 1842, work commenced on the Ganges Canal.[7] After producing countless drawings, traveling extensively to study the great works of hydrology in Italy and Egypt, and employing over 35,000 men and women (and children) to dig over 898 miles (1,445 km) of trenches, Cautley's plans were finally realized. After first coming to Bengal at the age of 17 in 1819 to serve as an artilleryman in the service of the Company and rising to the rank of Superintendent of Canals, "his immortal work," as Governor-General Dalhousie called it, was made a reality: the great Ganges Canal.

Such an accomplishment warranted a theatrical spectacle like few organized by the Company. Hosted by Colvin, guests included English civilian and military officers from Calcutta and Indore, along with Jayajirao Scindia, Maharaja of Gwalior. Unable to make the ceremony, Lord Dalhousie received word of the event by means of the newly installed electric telegraph. Dalhousie responded in kind: "I have rejoiced over your message. All honour to Colonel Cautley.'"[8]

By most accounts, the town of Roorkee was teeming with people during the celebration. Estimates of the size of the crowd ranged from 50,000 to 500,000. Following the pomp and circumstance of the salvos and cannon fodder in the morning, the afternoon was devoted to games to entertain the "mass on mass of gayly clad natives," with hundreds if not thousands queuing up to see the Ganges "leave her ancient and hallowed channel for one formed for her by the hands of strangers."[9]

To mark this monumental shift in reality—an entire river rerouted without catastrophe—one hundred or so guests assembled for a dinner, presided over by Colvin, to honor both Cautley and members of his team. By nightfall, fireworks radiated over the multitudes gathered. The fifteen arches of the aqueduct, each with a span of fifty feet (15.2 m), "were illuminated with innumerable lights, and stood out in lines of fire against the dark sky."[10]

Although the waters of the Ganges entered the canal at Haridwar, it was the once remote town of Roorkee that took on great symbolic significance. Roorkee had become an unalloyed beast of hydraulic architecture, a technological outpost. Here, adjacent to the Solani Aqueduct and the Ganges Canal headquarters, the first college of engineering in the British Empire—the Roorkee College of Engineering was established to train both European and Indian engineers in the construction and maintenance of the great Ganges Canal using the massive constructions of the Canal nearby as an open-air classroom.

The region provided students with more than just the great aqueduct. Before arriving at the Solani River, the Ganges Canal passed through a series of "super-passages," a term Cautley gave to the great flumes he designed to either carry the canal above, below, or through three major river torrents (*raos*) of the Siwalik Hills. While the Solani River is either a dry, sandy bed or a small, sluggish stream throughout most of the year, during the southwest monsoon it not only floods, but also hurtles boulders, trees, sediment, and debris at great speed. Building the super-passages was no small feat, and the aqueduct to carry the waters of the Ganges over the Solani River required the most labor—in both its design and construction—and the most cost—at nearly one-quarter of the total. Cautley employed 6,000 men daily, required an estimated 90,000 bricks and one million cubic feet (28,317 cubic meters) of lime.[11] It was said, "that its piers and arches have an almost Cyclopean character."[12] The Ganges Canal, both in its magnificence and immensity, reached mythological proportions.

But the ceremony, coupled with the monumentality of the canal itself, was thought not enough to impress the thousands in attendance. Pamphlets, printed in Hindi and Urdu, "intended for the information of the native spectators," detailed not only the most important parts and why "such an immense expenditure of money and labour" was undertaken to create the Ganges Canal, but also the "result which it is expected will flow from it to the advantage not more of the state than of the people."[13] The original, written in English by Captain Richard Baird Smith (Cautley's successor), was reprinted as "The Ganges Canal" in *A Supplement to The Delhi Gazette*. On May 14, 1854, days before he was to depart for London from Calcutta, Cautley mailed ten copies of Smith's pamphlet to a Boston Brahmin named Charles Eliot Norton. Norton had met Cautley and Smith in December of 1850, while staying in residence at the new College of Civil Engineering in Roorkee, and inspected the progress of the Ganges Canal.

In October 1855, Norton published "The Opening of the Ganges Canal" in the *North American Review*, a follow up to a previous essay he published in the same journal in 1853, "Irrigation in India." Based on his travels in India in 1850–51, his 1853 piece implored readers to see canals as instruments of progress coupled with agricultural returns, or, as he wrote it: "great moral agents...promoters of peace and civilization not less than of fertility and plenty."[14] For Norton, projects like the Ganges Canal were "at once with the kindest works of Nature herself—for they partake of her enduring beneficence, her free and equal generosity." It was only natural to build a canal.

But what is perhaps most intriguing is not only that Norton equates the Ganges Canal with "works of Nature herself," but also that it had cost half as much as the "most exquisite building in the world": Shah Jahan's monument to his deceased wife, the Taj Mahal. And whereas the Taj Mahal was a monument of great wealth and design, the Ganges Canal served an immediate function beyond monumentality, spreading the plentiful waters of the Ganges "over the land through innumerable channels" and proving "of incalculable benefit to the people."[15] Norton designated it a "diffused welfare" for "the whole huge mass of population in these densely crowded provinces," again emphasizing its spatial and moral dimensions.

Ever mindful of criticism leveled at the Company in India, Norton often prefaced his praise of British public works with a soft criticism of colonial conquest. Reiterating this point in his 1855 essay, Norton noted, "The victories of peace, harder and nobler though they be, are not, however, so loudly applauded as those of war."[16] The canal was a victory over the cyclical caprice of the rains. It was a victory against the threat of famine. It declared a new era in India, one of internal development. It was one—monumental—piece of a vast program of "modernization" that included railroad lines linking the interior with the coast and telegraph lines that would connect cities within seconds. And while he was not present at the opening of the Ganges Canal—in fact he had last seen it in December of 1850—Norton declared "The night in which false religion, tyranny, and war have enveloped India, is giving place to the day of Christianity, good government, and peace… the glow of the morning is in the East, and the first streaks of light are reflected brightly in the flowing waters of the great Ganges Canal."[17]

ENCHANT-MENT

By the middle of the nineteenth century, engineering was enchanted with powers that went beyond mechanical efficiency. Public works embodied this sense of enchantment, particularly in their capacity to simultaneously re-form relationships between state, space, and labor. It was during the 1850s when the Company's public works went from an ad hoc system of preserving existing roads and lines of communication to "a separate branch of internal administration": first in 1850 with the creation of a public works commission in every presidency of India, and ultimately as a separate department in 1854. Models of governance and machinery resonated with each other, and yet, to what ends they would lead remained untested.

In 1853, Karl Marx, writing on "The British Rule in India," declared that the "working of English steam and English free trade" would do more to alter Oriental despotism than the soldier or tax gatherer ever did. Although he would have mocked Norton's wholesome notion of public works as "peaceful conquest," he saw English interference, which amounted to "blowing up the economic basis" of India, as a means to produce "the only social revolution ever heard of in Asia."[18] Observing these developments from the outside, for both Norton and Marx, English engineering and English economics stood poised to radically rehabilitate India. For Norton, these developments had intrinsic value; for Marx, they were a necessary step in a longer historical evolution that would lead to the overthrow of capitalism.

Overseeing the radical restructuring of public works during this time was Governor-General Lord Dalhousie. Dalhousie was a fanatic when it came to public works. He was President of the Board of Trade when "railroad mania" hit England in 1845–46. Rather than allow private development of the railway system in Great Britain, Dalhousie proposed greater state control over this "system of national communication" to then-Prime Minister Peel. Though Peel did not support Dalhousie's proposal, he was still known as the "virtual author of the English railway system."[19] By the end of 1846, over 600 proposals for the expansion of railways in England required his review. He attacked the huge pile with an energy which amazed and wearied out his subordinates and coadjutors," setting the stage for what would become the "foundation of painful future disease," working past the point of exhaustion.

Shortly thereafter, in 1847 he was made Governor-General of India—at 35, he was the youngest man to ever hold the position. Arriving at Government House in Calcutta (Kolkata) on February 7, 1848, he oversaw "far-reaching schemes of railways, roads, canals, and public works, he inaugurated the great revolution which has converted the agricultural India of antiquity into the manufacturing and mercantile India of our own day. Expansion of territory, unification of territory and the drawing forth of material resources."[20] In other words, Dalhousie was a revolutionary: a development revolutionary. He not only "added to the British territory an empire as large as [Robert] Clive's" as his biographer Dr. George Smith noted, but he also revolutionized how territory was administered by way of public works.[21] The Ganges Canal figured prominently within his revolutionary ambitions.

But what of those who were present at the opening ceremony of the Ganges Canal? Did they also feel the dawn of a new age? Henry George Keene tried to record his experience in a poem, "The Opening of the Ganges Canal," published in *Blackwood's Edinburgh Magazine* in 1854. A district officer in the service of the Company in Muzzafarnagar and guest at the opening of the canal, Keene urged natives to not only lift their "hoar heads with laughter to the skies," but to imagine in years to come an age old man, eyeing "the green and affluent ground":

I stood among them on the shining morn,
I saw the ruler of the land
Let loose the waters with an easy hand;
The river, vainly, idolised of yore,
Now first her servants blessed;
The White-topped mountains never bore
Us benefit before,
Till taught by those wise strangers of the West.[22]

If in Keene's poem the "natives" have, belatedly, won their "heritage from time," the editors of the London-based *The Civil Engineer and Architect's Journal* articulate a slightly different claim of conquest. In 1854, three months after the Ganges Canal opened, a report in the journal noted that, "The construction of the Ganges Canal is, in a word, the most profitable acquisition that has been made by the British empire [sic] since the conquest of Bengal, and affords a convincing proof that India is not behind the rest of the world…It was at one time said, and perhaps truly then, that if the English were expelled from India, they would leave no traces of their sojourn behind them, but a few gaols and empty bottles. Now, however, the joke may pass harmless. The Ganges Canal is the standing answer to the sarcasm."[23]

When Lieutenant-Governor Colvin was toasting Cautley in Roorkee on April 8, he too pronounced "that here at least we have an answer, which no detractor can gainsay, to the old reproach, that the British have left no permanent mark upon the soil of India to attest the power, the wealth, and the munificence of their nation."[24] For many, the ingenuity of English engineering had turned a sarcastic joke about the East India Company into laughter at the monsoon and the gods. With the main trunk, branches, and terminal lines, the Ganges Canal was 898 miles (1,445 km) long. At the time of its opening, it was five times longer than all of the canals of Lombardy, Italy and Egypt combined.[25] Extended, modified, and subdivided into the Upper Ganges Canal (UGC) and Lower Ganges Canal (LGC), which opened in 1878, the Ganges Canal system today irrigates one million hectares (10,000 square kilometers) of land and stretches over 12,240 km (7606 miles), making it the longest canal system in the world.[26]

Before we turn to the physical and technical details of this project (and there are many), let us consider one more idea central to these accounts—that of Norton's "peaceful conquest." Like much of the writing on public works in India at this time, which focused on the internal improvement of the country, the statement emphasized how the Ganges Canal charted a very different course, not only for the river and the millions of farmers partaking in its perennial waters, but also a new course of liberal governance: "a culture of peace."

Built on "peaceful conquest," the Ganges Canal was to become a model of this new culture of peace. By re-engineering the flow of the river, the Company was able to multiply its territorial holdings without having to fire a shot or invade another state. Expanding the utility of the Ganges River by means of artificial irrigation did three things at once. First, it increased the carrying capacity of existing agricultural plots. Second, it transformed uncultivated land into arable land. Third, it forced the Company to take a more active role in governing populations that were previously outside its direct control, thus taking on greater financial and social responsibilities.

In choosing to intervene through artificial hydrology, the culture of peace and peaceful conquest inaugurated a new engineering motive: hydraulic pastoralism. This pastoralism included landscape, but was not limited to it. "Pastoral" refers to a mode of living—a lifestyle and an agrarian

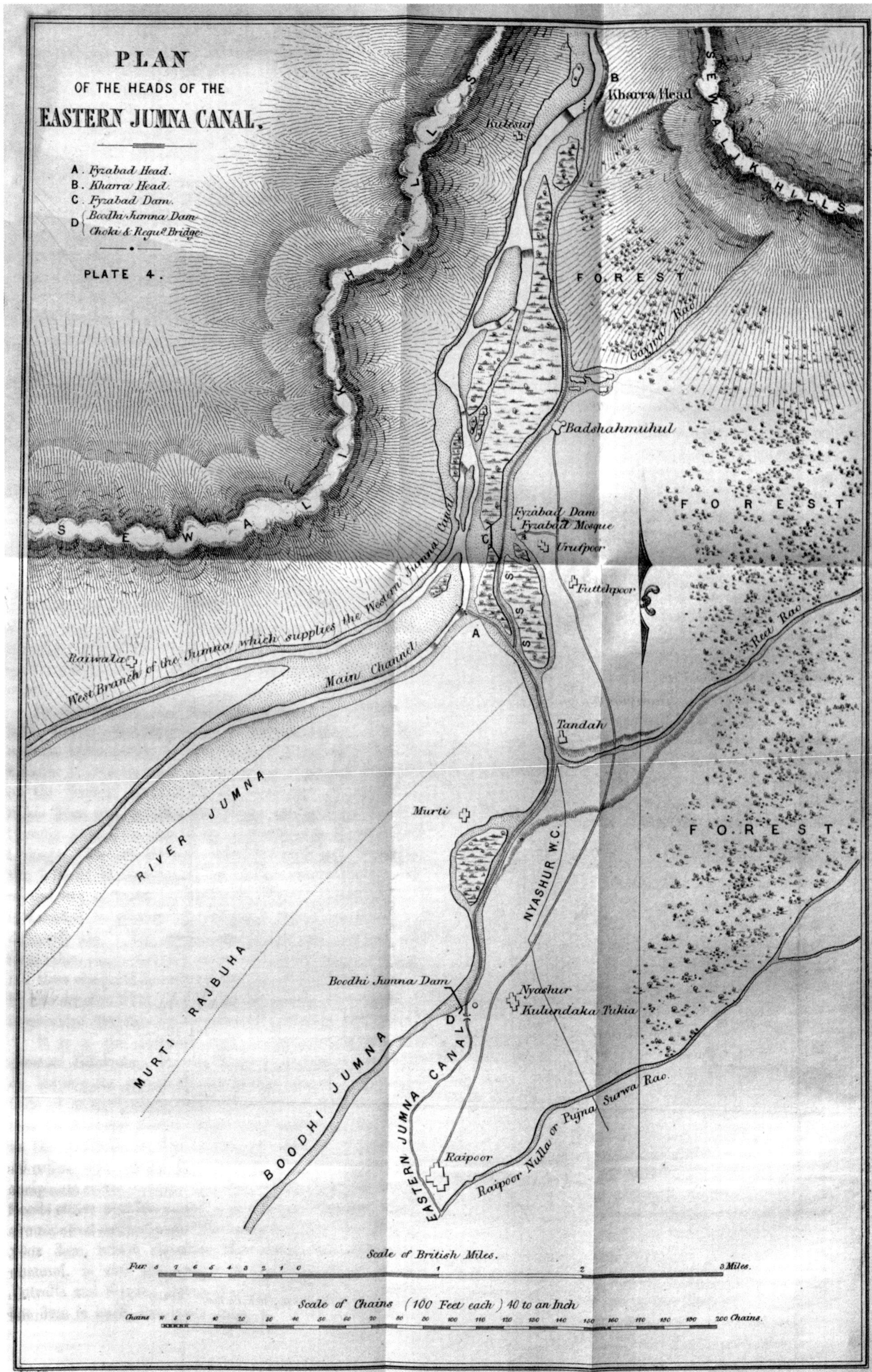
PLAN
OF THE HEADS OF THE
EASTERN JUMNA CANAL.
A. Fyzabad Head.
B. Kharra Head.
C. Fyzabad Dam.
D { Boodha Jumna Dam
Choki & Regur Bridge.
PLATE 4.
SEWALIK HILLS
Kharra Head
Kulesur
FOREST
Gayna Rao
Badshahmuhul
Fyzabad Dam
Fyzabad Mosque
Urufpoor
Futtehpoor
Raiwala
West Branch of the Jumna which supplies the Western Jumna Canal
Main Channel
Reet Rao
Tandah
RIVER JUMNA
Murti
NYASHUR W.C.
MURTI RAJBUHA
Boodhi Jumna Dam
Nyashur
Kulundaka Tukia
BOODHI JUMNA
EASTERN JUMNA CANAL
Raipoor
Raipoor Nulla or Pujna Surwa Rao.
Scale of British Miles.
Scale of Chains (100 Feet each) 40 to an Inch

Proby Cautley, "Plan of the Heads of the Eastern Jumna Canal." **Ganges Canal: A Disquisition On the Heads of the Ganges and Jumna Canals, North-Western Provinces, In Reply to Strictures by Major-General Sir Arthur Cotton**. London: Private Circulation, 1864. (Harvard University).

culture built around access to expansive artificial irrigation. The imposed infrastructure provided the perfect alibi in order to incorporate a population into a colonial government and international market. Irrigation did not alter the population's self-image—the communities would remain rural and agriculturally based, even as they were folded into the most engineered stretch of land in the world. And if engineering could supply freedom from the monsoon, it promised economic liberation, as well. And yet, paradoxically, this "liberation" was supplied from above—from the actions of an unyielding, expanding central authority.

Cautley recognized this paradox: simultaneously maximizing regional equity and mechanical governance. He and his team of thousands had been entrusted with inscribing new channels of wealth across Upper India. They cut across existing private properties and intersected long-established roadways and paths. Aware of the spatial changes these interventions produced, they built new bridges and roadways for the benefit of greater communication (both for cultivators and for the Company). Still, the attempts to see the Ganges Canal through the eyes of local farmers and *zamindars* (land-lords) had little success. What use was it anyway? Irrigation engineering, imposed by the British though it was, was understood as technical progress based on the mechanics of gravitational flow. Increasing crop yields would directly improve the economy, and indirectly improve society. Technical innovations vindicated the social and political changes produced by colonial rule. Thus, most people at the time focused only on the mechanics and ingenuity of the Ganges Canal, as if the canal itself had intelligence.

Understanding how this canal—an "irrigating machine" as engineer J. Crofton called it in 1865—came into being requires knowing something about what inspired the English in the nineteenth century to wage such epic contests between nature and machine.[27] Understanding the history of the canal system—initiated by the Company, extended under the Raj, and still under construction to this day in independent India—means understanding how and why irrigation was part of the conquest of nature and people, resources and space. It also means tracing how machines—particularly canals and wells, along with dams and barrages—were not always conceived in opposition to nature, but often as resembling and even imitating natural processes. Cautley, Norton, and others believed it was so. Machines displaced nature—that is, this process transfigured an extant nature into a nature that would exist solely to empower people—in other words, a human nature. The engineer was at the center of this dramatic change.

The site that would be transformed by this unprecedented reformation of the Ganges River was the Ganga-Jamuna Doab in northern India. This Doab, or land between two rivers, is "a large irregular horn-shaped tongue of land..."[28] situated between the two axial rivers of northern India—the grey-brown water of the Ganges (to the north) and the azure-colored water of the Jamuna (to the south). Both rivers descend from the Himalayas, but where the Ganges' tributaries also descend from the Himalayas and spread alluvium across the plains, the Jamuna's tributaries come from the Vindhya Range, which sheds far less alluvium.

Just south of the Himalayas, the Doab extends from the sub-Himalayan Siwalik range of northern India, gently slopes and narrows over 500 kilometers of fertile plain, until finally the two watercourses form a confluence known as the *Triveni* or *Sangam* at the historic city of Allahabad (Prayag).[29] When the Ganges Canal opened in 1854, the Doab was said to be "inhabited by nearly 6,000,000 of souls; it is thickly covered with villages occupied by industrious agricultural communities dependent on the soil for their daily bread. Many towns and cities are scattered over its surface and filled with active merchants, busy tradesmen, and industrious mechanics, of no mean skill in their crafts."[30] As Captain Richard Baird Smith put it in 1854, the Ganges Canal would not only improve "the general condition of the surface of the country" by removing the swamps and *jheels* (ponds) so as to "purify the air they now taint" and "to subject every town to a process of cleansing and purification through the agency of the canal water."[31]

"Index Plan Showing Hydel Channels & Irrigation Schemes on River Yamuna in U.P., H.P., & Haryana State," at Government Guest House adjacent to Hathni Kund Barrage, 2010.

Echoing the words of Reverend Hamilton at the opening ceremony of the Ganges Canal, Richard Baird Smith, quoted later that year in *The Illustrated London News*, remarked that, "...great tracts of country, once a desert, and without inhabitants, are now filled with people; the crops that were liable to perish every year from the want of rain, are now secure and every man on the banks of the Canals can sleep soundly without dreaming of the loss of all his year's toil because showers do not fall; and he knows that his family is sure of bread, even though the rains should fail altogether."[32] The canal was to manage and insure risk for both the villagers and the government; it would serve as a hedge against the uncertainties of agricultural yield and the unpredictability of the southwest monsoon. Moreover, Smith pronounced what had been a latent desire: that the Ganges Canal is at once a conquest of space—"great tracts of country, once a desert"—and a conquest of people—not just of bodies and their basic needs, but of the inhabitants' desires and dreams.

MODEL

Essential to this conquest of space and people, dreams and resources, was the Company's rehabilitation and extension of defunct canal systems in the Ganga-Jamuna Doab. Between the sixteenth and eighteenth centuries, a series of Mughal Emperors had commissioned these waterways throughout the region, an area that had been the heart of the Mughal Empire. By the 1850s, the Company either directly controlled most of the Ganga-Jamuna Doab or indirectly controlled the principalities and satrapies of the former empire. In the decades between the Battle of Delhi in 1803 (Battle of Patparganj)—when British troops defeated the Maratha Confederacy at Patparganj, a town on the eastern bank of the Jamuna River—and the opening of the Ganges Canal in 1854, the Ganga-Jamuna Doab was the site of major excavations and experiments in large-scale irrigation. Before constructing the Ganges Canal, the majority of British canal work in and around the Doab consisted of rehabilitations and extensions of anterior canal systems linked to the Jamuna (Jumna) River, and not the Ganges.

Modernizing the works of the Mughal Empire along the Jamuna River irrigated more arable land, which in turn led to increases in revenue through water and land rents collected by the Company. Though the Mughal Emperors "had in their uncertain, spasmodic way, addressed themselves from time to time to the consideration of this great prophylactic measure" of canal irrigation, according to Sir John W. Kaye, such projects did not compare to those "which the British Indian Government is now carrying out on a scale far beyond the speculations of the most large-minded of the Mogul princes."[33] Superseding the works of their imperial forbearers, the Western Jamuna Canal (WJC) and Eastern Jamuna Canal (EJC) not only showcased British technical dominance, they also exhibited British humanity and civilization. Even an engineer like Richard Baird Smith puckishly proclaimed that: "The moral triumphs of England in Hindustan are a thousand times more valuable than all of the public works, past or present, between the Ganges and the Indus."[34]

Rehabilitating and extending the Western and Eastern Jamuna canals also served as a preliminary exercise for Cautley and his fellow artillerymen and engineers. The Jamuna canals, like the Ganges Canal, start at the base of the Siwalik Hills. The surface of this country, particularly the knot of torrents fueled by the southwest monsoon, presented many of the same challenges faced by Cautley in the first eighteen miles (29 km) of the Ganges Canal: namely, inundation, both at the head of the canals and from *raos* (torrents) activated by the monsoon. It was "with the example of the Jumna Canals before us," proclaimed Cautley, "that the idea of taking a line from the Ganges originated."[35]

Built, abandoned, rebuilt, and abandoned again, the alternating states of the Jamuna Canals came to represent the extremes of canal irrigation in northern India. These waterways also represented the two competing uses of surface water in the region: urban and agrarian. When the Company captured Delhi in 1803, both sets of canals were in disuse. It was not until 1817, when Captain George R. Blane of the Bengal Engineers restored the oldest of these imperial precursors: the Western Jumna Canal (WJC) in east Punjab. Originally completed in the second half of the fourteenth century, the canal had fallen into disuse by the middle of the eighteenth century, if not before. Following Blane's renovation, water again flowed into Delhi by 1820. Restoration of another branch began in 1823 and it, too, was flowing with water by 1825.[36] Three miles (5 km) south and on the opposite bank of the Jamuna, the Company also worked to repair the Doab Canal, later called the Eastern Jumna Canal (EJC). Certain accounts date its initial construction to the reign of Mohammad Shah (1719-1748). Captain Robert Smith of the Bengal Engineers began this restoration in 1823 and by the 1830s, both the eastern and western canal systems had been repaired and made operational again.[37] The WJC supplied water to residents in Delhi and the EJC served the farmers of the Ganga-Jamuna Doab. Company engineers sought to not only rehabilitate and learn from the works of their imperial forbearers, but also to control and reshape the newly conquered lands and people. As Cautley's biographer Joyce Brown noted, with engineering "a new kind of military duty had come into being" in the 1820s. A steadily growing team of cadets trained in England who were

Panorama of Hathni Kund Barrage on the Jamuna (Yamuna) River, 2010.

"greatly instrumental in the success of the British arms...as well as settling and improving the British territories after their acquisition" reshaped urban and agrarian communities in Upper India in the first half of the nineteenth century.[38]

The restorations and extensions of the Mughal canals were the first products of a nineteenth century affair between the Company and hydraulic architecture, one which tied both local farmers and the Company to the proliferating infrastructure of water delivery. The Jamuna River had been the logical choice to test this scheme: the channels of the Mughal Empire were etched in the ground, empirical evidence of a culture attempting to counteract the enervations of the southwest monsoon. Between 1817 and 1837, as portions of the Ganga-Jamuna Doab and the lands to its west changed from rain-supplied to canal-supplied agriculture, this infrastructure had a multiplying effect—the need to draw yet more water from an expanding canal became urgent.

In 1825, Lord Amherst appointed Proby Cautley to work with Captain Robert Smith on the Eastern Jumna Canal.[39] It is not known why Cautley was assigned to work with Smith; by all accounts, he and Smith knew next to nothing about hydraulic engineering. Cautley was a fairly undistinguished graduate of Addiscombe Military Seminary—the Company's military college near Croydon, England, which had been established in the early-nineteenth century to train the Company's armed forces. Addiscombe cadets could be no younger than fourteen and no older than eighteen, with good penmanship, along with sufficient knowledge of arithmetic, and English and Latin grammar; once enrolled, they studied mathematics, fortification, French, Hindustani, and drawing.[40] Cautley's final examination did not qualify him for the rank of engineer, and so he was commissioned as a Second Lieutenant artilleryman in April 1819. He had, however, received a prize for drawing while at Addiscombe, which may be the reason he was assigned to join Captain Smith. Smith was known for his skills as a draughtsman and enthusiasm for watercolors, as well as his work restoring Mughal monuments.

Though both men lacked specific knowledge of hydraulic engineering, their shared ability to survey land was an important prerequisite for undertaking such a project in British India. Digging canals and rerouting rivers demanded the expertise of someone trained in technical drawing and the science of the traverse survey. Robert Smith was well acquainted with these techniques. While a Field Engineer in the Bengal Division in 1810, his superiors noted his exemplary drafting skills. The Surveyor-General not only considered him "well qualified to conduct and survey" and "the best draughtsman" he was acquainted with, but "His masterly rapid pencil particularly qualifies him for survey work of the Frontier, as he will be able to delineate the passes and surrounding country with the greatest correctness."[41] Shortly thereafter, from 1812-14, he surveyed and mapped the forests of west Bihar and the Central Provinces, parts of Punjab, and the North-West Provinces.[42] By 1814 he was posted as "Superintending Engineer to Prince of Wales Island, later known as Penang."[43]

Smith was one of many Company engineers around this time making surveys, painting watercolors, and being called upon to "act as engineers, architects, builders, artists and archaeologists."[44] He was, in the words of art historian Mildred Archer, an "artist engineer." This gradual accumulation of knowledge was, in part, driven by the paucity of experience with canal irrigation in Great Britain.[45] Understanding irrigation in this new territory required spending time in the field, drawing maps and perspectives; through these practices, engineers would acquire an "instinct," as one canal commissioner in Punjab called it. Having an "instinct" meant forgetting one was an engineer and being "thoroughly acquainted with the practical geography of his circle" of responsibility.[46]

Smith was also fortunate, in the case of the Ganga-Jamuna Doab, that there was a precedent in the region to study: the abandoned Mughal canal. Tracing its line proved a useful guide, drawn into the landscape. For the most part, Smith's plans for the canal followed that original course, only moving the head of the canal slightly downstream. By the time Cautley arrived in 1825, Smith had already surveyed the canal's course and begun excavating portions of it. Cautley's tasks for the next five years or so included clearing out parts of the channel, and overseeing masonry works and planning foundations for fifty some bridges. A true "artist engineer," he worked simultaneously as surveyor, draughtsman, engineer, and builder.[47]

By the time the EJC opened on January 3, 1830, Smith had retired and Cautley was the unofficial Superintendent of the project. Seventeen days after the opening, he wrote to Captain John M. Colvin, Superintendent of Canals of the Delhi Territory, to report a problem: rapids had formed in three locations, threatening to destroy the adjoining bridges by exposing their masonry foundations.[48]

Cautley re-surveyed the canal and found that the rapids had formed due to an excessive slope. At a length of 134 miles (216 km), the change in grade was just over 421 feet (128 m), with almost half the descent occurring in the first 29 miles (47 km) or so, and 45 feet (14 m) occurring in the last 11 miles (18 km). To counteract this problem, Cautley planned and constructed a series of artificial masonry waterfalls, which had to be carefully calibrated.[49] If the falls slowed the velocity of the water too much, aquatic plants would grow in the canal; if the velocity remained too great, erosion would continue. These interventions, coupled with tens of miles of embankments, abated erosion over the course of the canal.

Beholden to the characteristics of Indian topography as much as to the traditions of Indian building, Cautley was unable to dig and build caissons and piles to support the falls and bridges. Where "the badness of the soil and the height of spring water rendered excavation impracticable," he employed *Chah-kuns* (local well-sinkers), a method of sinking masonry-lined irrigation wells in order to build the foundations of bridges, artificial falls, and other structures. This process, which he described as "undersinking wells," involved laying out a grid of partially built brick piers that would sink into the ground by virtue of their own weight. Once at a stable depth, the tops of the piers became bases to build the piers even taller, and again they would sink. This process was repeated again and again until the piers reached a stable depth. He described it as a, "homage to this admirable native conception" used for ghats and "Mussulman buildings at Agra," as well "as the standing resource in the system of hydraulic architecture of Upper India."[50] Cautley soon appropriated this same method of sinking wells for foundations for the great works of hydraulic architecture along the Ganges Canal. His system became the basis for most road and railway bridges throughout northern India during the nineteenth century.

In short, the Jamuna Canals become training grounds: spaces to develop building techniques and establish forms of labor relations.

REFERENCES

The renovations and extensions of the WJC and EJC, when taken together, were the first manifestation of an obsession that would turn the entire Ganga-Jamuna Doab into an agro-industrial estate. As pieces of machinery, they were costly; every season they required improvements and adjustments to their flow and embankments. But they generated income for the government and a steady flow of water to farmers. The canal easily supplanted the rain and local tanks and wells.

Euphoric from the fiscal successes of the WJC, in 1836 Colonel John Colvin envisioned a canal drawn from the Ganges that would irrigate the eastern portion of the Ganga-Jamuna Doab. The Military Board, which oversaw all public works projects at the time, considered the plan a useless expense and chose not to act. After devoting most of his life to the Company, Colvin retired and returned to Britain, leaving Cautley to assume his duties as Superintendent of Canals in the Delhi Territory. It would be several years until Cautley was asked to seriously investigate the possibilities of a Ganges-fed canal. First, he focused his energies on the management of the ever-expanding irrigation works of Upper India, hunting for fossils in his spare time.

Around this time, there were other epicenters of irrigation activity: in the Madras Presidency, Arthur Cotton (another Addiscombe graduate and eventual adversary of Cautley) was hard at work restoring the great 'Grand Anicut' (weir) across the Kaveri (Cauvery) River, believed to be built by the Chola Kings in the second century C.E. It proved one of the most remunerative canals systems in British India. By the late 1840s, after Lord Dalhousie annexed the Punjab, Company engineers found examples of working inundation canals (a type of canal that provides water only when a river floods).[51] The Madras and inundation canals, however, did not command the level of notoriety of the WJC and EJC. One reason may be that these canals were renovations and extensions of projects originally completed during the zenith of the Mughal Empire, rather than novel productions of the Company and British government.

"With so many examples before us, no great exertion of intellect was required to arrive at the conclusion that the system might be generalized," noted Cautley in his opening paragraph in *The Ganges Canal* (1860).[52] Generalized though it may have been to Cautley six years after the opening ceremony of the Ganges Canal, "taking a line from the Ganges" was anything but straight or direct. From April 16, 1842, when digging began, the canal underwent numerous revisions and stops.[53]

CHANGE OF PLANS

As Cautley planned the "peaceful conquest" of the Ganga-Jamuna Doab, Lord Auckland's successor, Lord Ellenborough, arrived in Calcutta in February of 1842. A former President of the Board of Control—a cabinet post in London, responsible for overseeing the Company—he came to India "to emulate the magnificent philanthropy of the Mahomedan Emperors."[54] Upon assuming his new post in Calcutta, Lord Ellenborough suspended "the great undertaking of the Ganges Canal, which his predecessor had sanctioned."[55] By 1843, he had ordered a "total modification of the design" in both scope and scale. Cautley, no doubt bewildered, recorded the dispatch he received from the Supreme Government: "it should be in the first instance a canal of navigation, and all the water not required for that purpose may be distributed for the purposes of irrigation."[56] In other words, the main trunk of the canal, originally designed to end at Cawnpore (Kanpur), half-way down the Doab, was now to extend to the tip of the tongue of the Doab at Allahabad, the point furthest up the Ganges River to which steamers ascended from Calcutta. This strategy was a peculiar one in what was to become "a railway epoch." As Richard Baird Smith put it already in 1852, a canal for navigation "can be considered as only a second-rate means of transport to its commerce."[57]

"The projectors [Cautley and Colvin] stood aghast at this decision," remarked John W. Kaye, in his 1853 history of the Company, since "irrigation was the great object of the contemplated works."[58] Still, they followed Lord Ellenborough's orders—in late-1843 to early-1844 Cautley surveyed the canal line all the way to Allahabad. However, in an implied protest against the modifications, he submitted three design proposals to the government. Each scheme included branches of irrigation in conjunction with navigation so as "not to lose sight of the purposes for which this canal was originally projected, and to provide means for irrigation as extended as the security for navigation to Allahabad would admit of."[59] Scheme number one terminated at the Jamuna River at Allahabad with irrigation lines provided from Roorkee to Kanpur (estimated at 10,236,643 rupees); scheme number two included a series of locks and reservoirs and irrigation in and around Kanpur (estimated at a cost of 9,703,558 rupees); scheme number three extended the navigable line to the Jamuna but not as far as Allahabad, with a branch of irrigation extending from Kanpur (estimated at a cost of 9,339,746 rupees).[60]

Before a decision could be made, Lord Ellenborough's administration ended. "He returned to England to speak of the humiliation he had experienced on witnessing the great works of native rulers, and contrasting them with" those in England.[61] Sir Henry Hardinge assumed the role of Governor-General of India in July of 1844, and Cautley soon met with him in Calcutta, where he learned of new objections to the Ganges Canal—namely, that the canal would extract too much water from the Ganges River at Haridwar and impair navigation up the river. Additionally, the probability of irrigation producing insalubrity (malaria) in the highlands of the Doab was a growing concern.[62]

A committee began investigating the matter in November 1845—an effort interrupted by the Sutlej campaign of 1845-46 in which the Company went to war with the Sikh kingdom in the Punjab—eventually submitting their final report on March 3, 1847. The committee, headed by Major Baker, deemed the extraction of water from the Ganges negligible in terms of its impact on navigation up the Ganges. Likewise, the committee did not find ample proof that stagnant water, created by leakage from the canal, would create an epidemic of malaria. With this approval and the support of the Governor-General, on May 1, 1847, the Supreme Government approved the construction of the Ganges Canal.[63] Five years after Ellenborough's reversal, irrigation was again the primary object of the canal.

VALIDATION

Two years earlier, in February 1845, Cautley had left India for England on a three-year furlough, his health failing after twenty-six years of service to the Company. In his absence, he appointed his assistant and fellow fossil enthusiast Major Baker in charge of the canal. In London, Cautley met with those "whose views on the navigation and malaria questions led them to doubt the policy of continuing the canal works."[64] As he weighed these questions, Baker and the members of the committee prepared their pro-Ganges Canal report for Governor-General Hardinge. Word reached Cautley of the May 7, 1847 approval. He must have been relieved and excited, but also, one imagines, cautious about returning to India with his diminishing health.

Proby Cautley, "Diagrams of the flow of surface and sub-surface water" **Report of the Ganges Canal Works: from their commencement until the opening of the canal in 1854**, Vol. I, London: Smith, Elder and Co., 1860. (Princeton University).

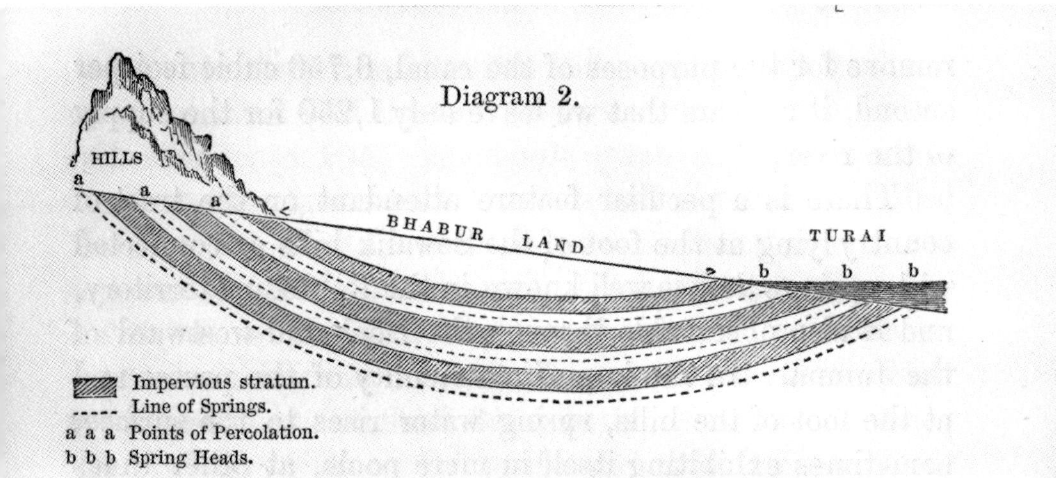

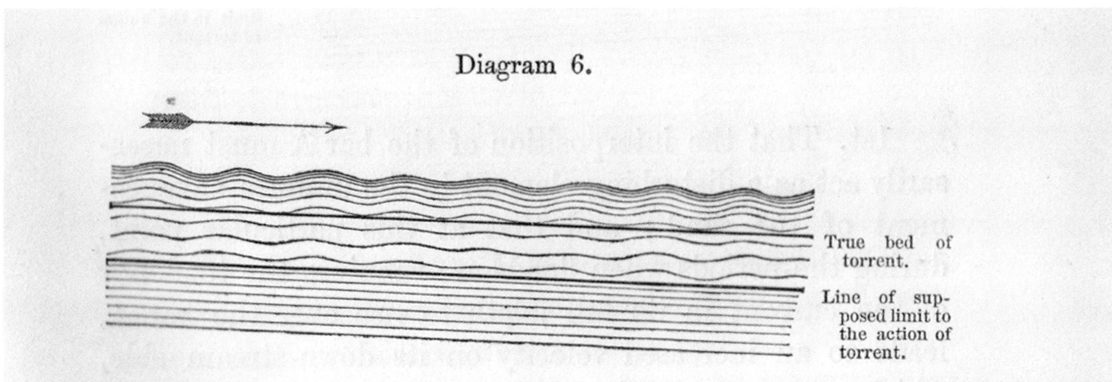

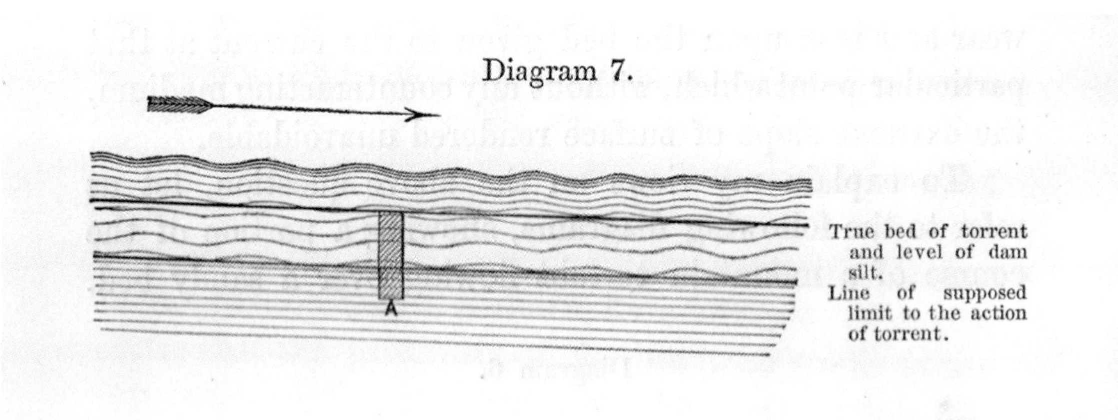

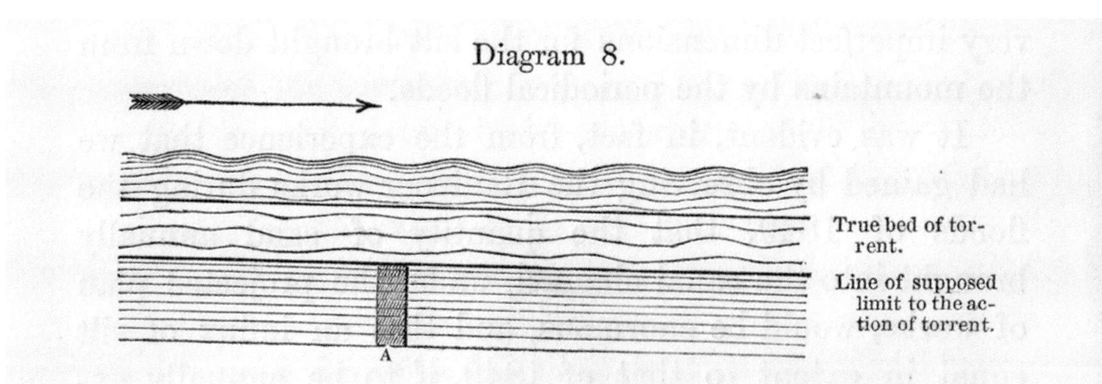

In August 1847, Cautley set out to resume his position in India and release Major Baker from his post. He left England many months in advance, in order to make a trip to Lombardy and Piedmont, Italy, as well as a planned stop in Egypt. In Italy, he studied the Italian method of embanking rivers and methods of directing watercourses. Over six weeks, Cautley inspected the canals and head works near Milan, the locks at Pavia Canal, and in Lombardy he witnessed the *fontanilli* "method of collecting water for irrigation by tapping springs and by these means providing irrigation for considerable tracts of country."[65] He observed systems put in place to rid canals of silt in Bergamo and techniques to construct *bundhs* (embankments) near Mantua and Cremona, along the river Po. He found resemblances between the conditions in the North-West Provinces in India and the canals from the Dorea Balten to their heads near Ivrea: "connected with mountain rivers subjected to violent floods, and therefore liable to evils similar to those we have to contend."[66] He noted that conditions at the Stura River resembled those found in the mountain torrents at the foot of the Siwalik Hills. He also purchased descriptive plans from the Board of Works at Turin for the Ganges Canal Library in Roorkee.

But Cautley confessed disappointment in the works he encountered in Italy. In Lombardy, the mountain torrents "are received in the bosom of great lakes, where they exhaust themselves over very extended surfaces before the effects of the influx of the canals can be felt."[67] The Siwalik Hills provided no such easy exit for the proposed canal's torrential flows. And where the Lombardese engineer had access to the Bavino Quarries, which held some of the finest granite in the world, no such sturdy material existed in northern India.[68] In Piedmont, Cautley felt he searched in "vain for examples of canals crossed by mountain torrents after their departure from the source of supply"; only along the Agogna River did he find a "*a salto di gatto*" (a cat's leap) or a masonry channel passing a river—not unlike his later designs for super-passages in the first eighteen miles (29 km) of the Ganges Canal. Otherwise, he wrote, "I saw nothing approaching to the great evil which renders our canal engineering in India a matter of such serious difficulty."[69]

Cautley also left with a sense of surprise and satisfaction because, though he was "separated from all communication with the Italian engineers," he "had by the mere process of simple reason, arrived so frequently at precisely the same results, and in so many cases had adopted the same expedients." Previously unaware of Italian irrigation, he felt a sense of validation.[70] Standing at a weir along the Irvea River, he could not help but imagine that he was in India standing atop one of his own works.[71]

On December 14, 1847, Cautley reached Bombay. He replaced Captain Baker as head of the Ganges Canal Works on January 11, 1848.[72] For the next six years he worked tirelessly to bring the Ganges Canal to completion. Years later, in 1860, Cautley compared the hierarchy of the Ganges Canal system to the hierarchy of distribution of water to cities and towns in Europe in the nineteenth century. In Europe, the system begins with a reservoir where water is collected. Mains "or great pipes" deliver water to the outskirts of the town. From there, distribution pipes follow street patterns and service pipes connect to homes. In the case of the Ganges Canal, "the trunk and terminal lines, with the great branches, play the part of the reservoir; the rajbuhas [small branch canals] hold to these the relation of 'distributary pipes,' and the village watercourses represent the 'service pipes' in the disposal of the supply in detail."[73]

To the crowded and unruly tracts of land in the Doab, a spatial and economic order was imposed by the Ganges Canal—a series of mechanisms, governed by the central goal of irrigation. But the canal was at once creation and obliteration: having conquered the caprices of unreliable rainfall, the river was transformed into plumbing. The Ganges, once looked upon with trepidation and despair, was rendered a technical amenity to service millions of souls (including those of the Company).

For the first eighteen miles (29 km), from Haridwar to Roorkee, known as the Northern Division, the architecture of the Ganges Canal increases in size and complexity from the head works to the Solani Aqueduct. The next 160 miles (257 km), from Roorkee to the village of Nanoon, known as the Centre Division, consists of numerous bifurcations and engineered falls to deal with changes in slope. The last leg, the Southern Division, continues from Nanoon to Cawnpore (Kanpur) where it terminates at the Ganges River. To understand the promise and potential of the Ganges Canal, to really see how it was built and evolved over time, it is best to examine the portions of the canal separately, in order to see their integration into a larger irrigating machine. The engineer oversaw the maintenance and repairs of the canal, managing the flow and distribution of water, while at the same time managing European and Indian subordinates, as well as the farmers supported by the project. Here, one begins to see that the Ganges Canal is not a pure mechanism—many overlapping populations animate this machine.

View from Bhaderabad Power House, above Bhaderabad (Bahadoorabad) falls, facing south above Ganges Canal, 2008.

METAPHORS

Structurally, the Ganges Canal system may have resembled the hierarchy of water distribution in European cities, but in terms of labor, it could not be more different. As early as 1876, British engineer J. S. Beresford theorized both the mechanical and social aspects of the canal in his "Memorandum on the Irrigation Duty of Water, And the Principles on Which Its Increase Depends." Dry formulas peppered what was otherwise a set of bullet points organized into sections. In his section on the efficiency of a canal, he proposed studying the Ganges Canal "as a great machine composed of many parts...much in the same way as a that of a steam engine." Accordingly, for him:

> *This irrigating machine is made up of four important parts which are quite separate, and, as things stand at present, at least two of them depend on different interests. They are as follows:*
>
> **1.** Main canal
> **2.** Distributaries
> **3.** Village water-courses
> **4.** Cultivators who apply the water to the fields.[74]

This breakdown of the "machine" into four parts follows Cautley's urban plumbing metaphor of the same system. Beresford's fourth point, however, is crucial: it adds human labor into the functioning of this instrument. On the one hand, this should come as no surprise. For if the concern of hydraulic engineering is efficiency of water distribution, and cultivators distribute water directly onto their fields, then the efficient distribution of water by a cultivator concerns the engineer. On the other hand, managing the cultivator to make sure he and his family use water judiciously, requires the engineer to manage labor in addition to water. The engineer becomes a social manager as well as a technical one.

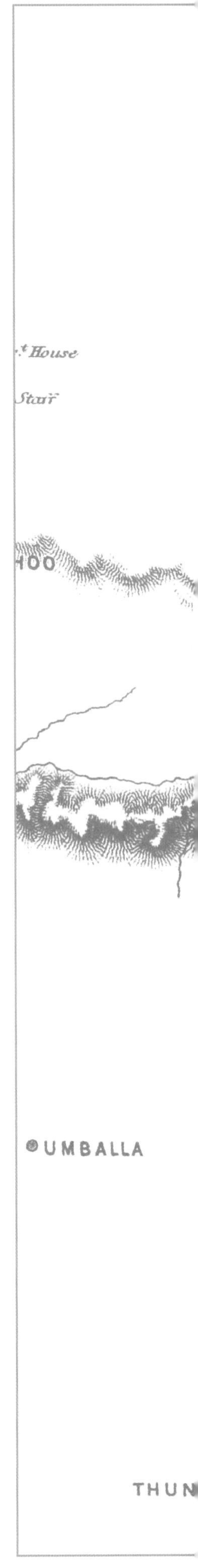
UMBALLA

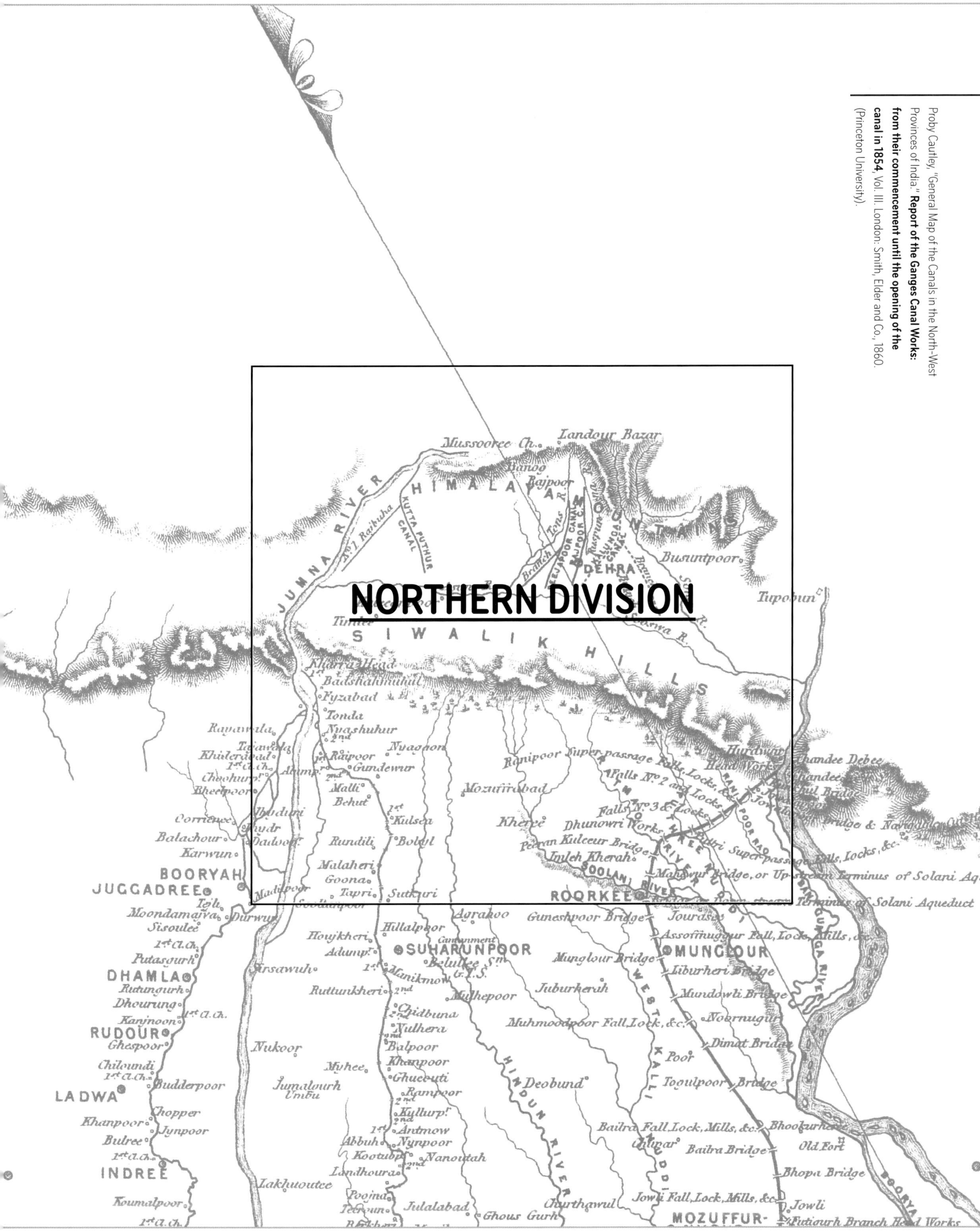

Proby Cautley, "General Map of the Canals in the North-West Provinces of India." **Report of the Ganges Canal Works: from their commencement until the opening of the canal in 1854**, Vol. III. London: Smith, Elder and Co., 1860. (Princeton University).

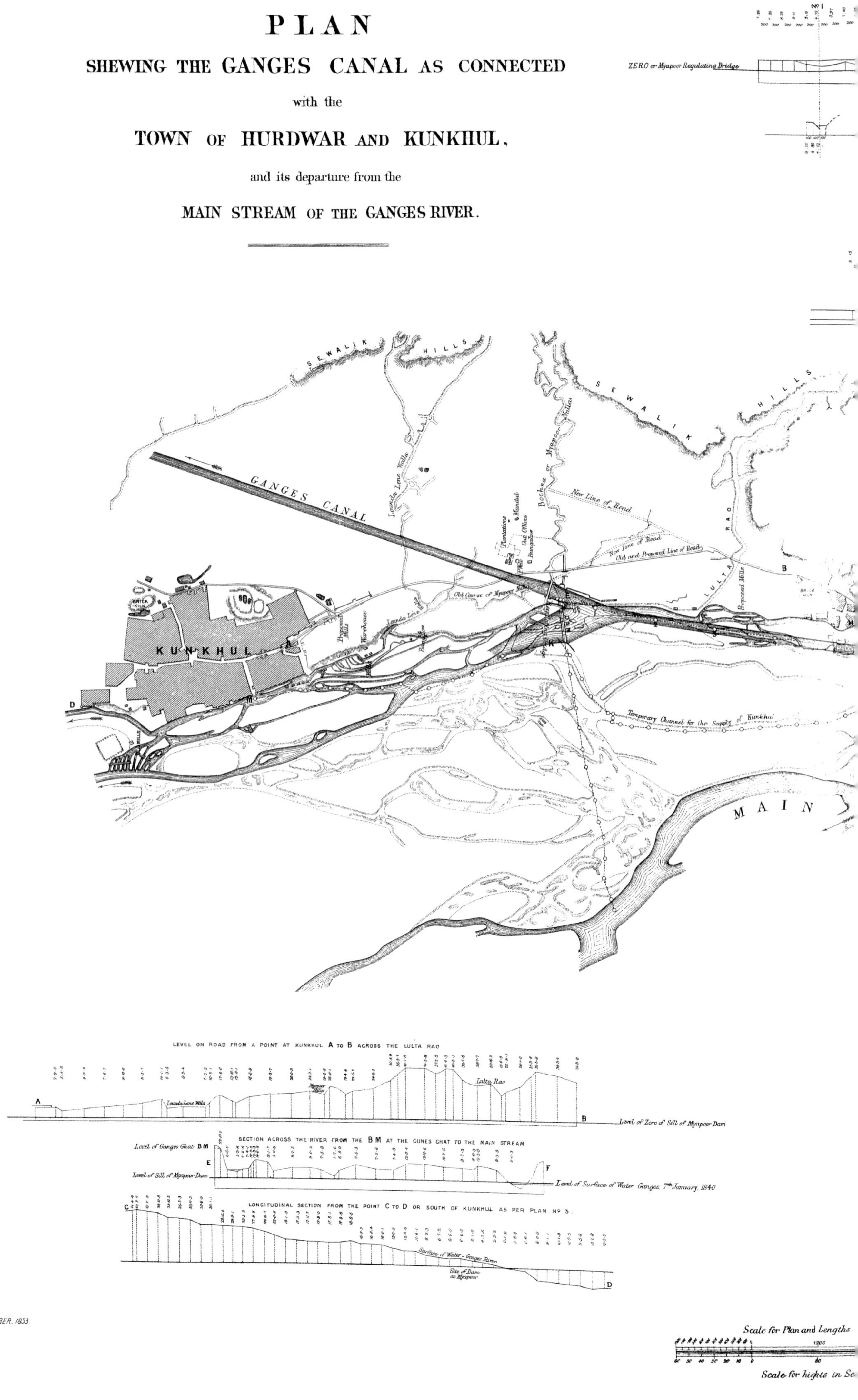

Proby Cautley, "Plate No. 13. Plan Shewing the Ganges Canal as Connected with the Town of Hurdwar and Kunkhul and its departure from the Main Stream of the Ganges River." **Report of the Ganges Canal Works: from their commencement until the opening of the canal in 1854**, Atlas. London: Smith, Elder and Co., 1860. (Princeton University).

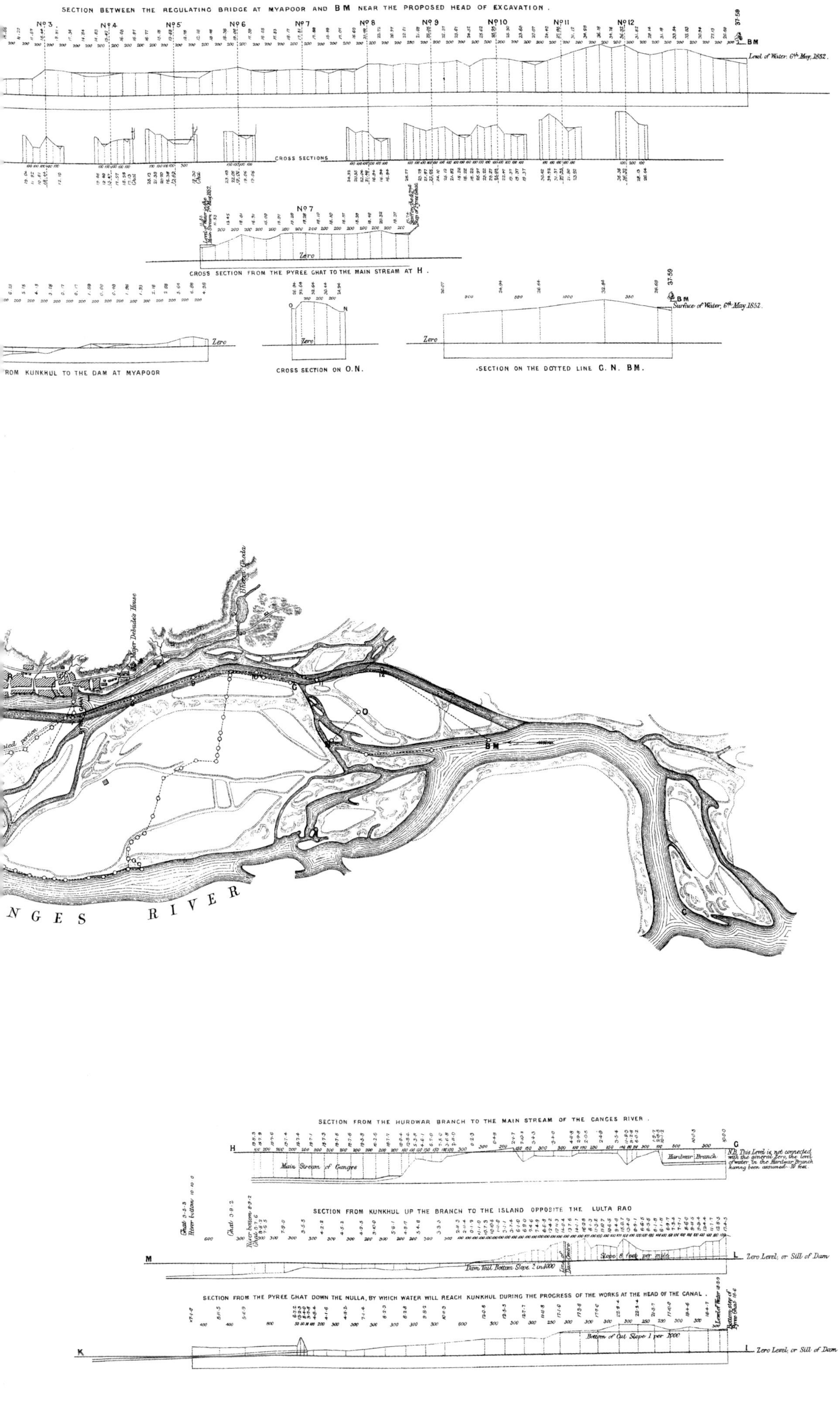

London, Smith, Elder & Cº 65, Cornhill

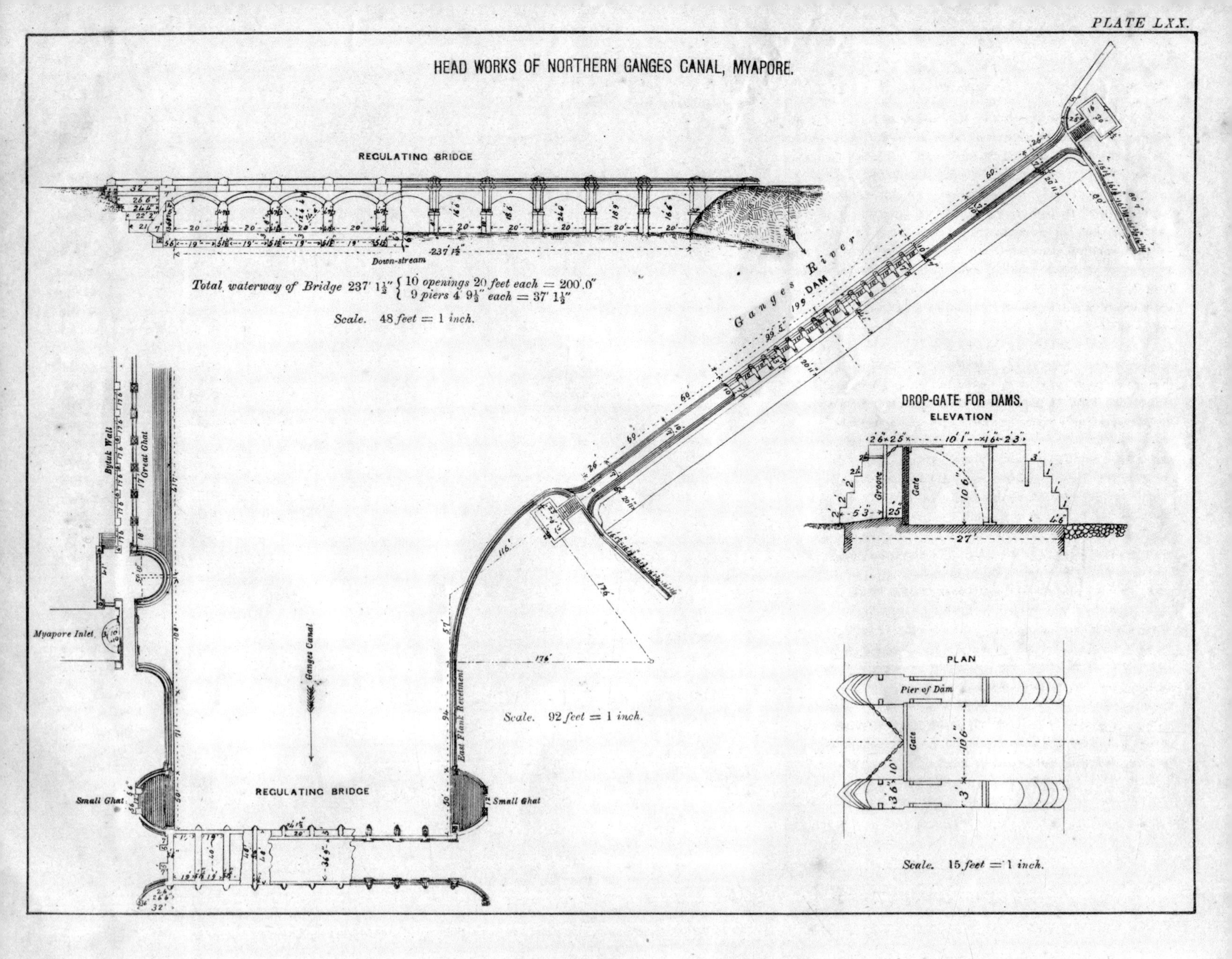

FIG. 17.—Myapur dam, Ganges canal.

Top: "Plate LXX: Head Works of Northern Ganges Canal, Myapore." **The Roorkee Treatise on Civil Engineering in India**, Volume 2, compiled by Julius George Medley, third edition enlarged and improved and edited by A. M. Lang. Roorkee: Thomason College Press, 1877. (AA).
Bottom: Herbert M. Wilson, "Fig. 17. Myapur dam, Ganges canal." **Irrigation in India**. Washington, D.C.: Government Printing Office, 1892. (AA).

Cautley's great Ganges Canal offers many proofs of his obsession. At irregular intervals he must make compromises: how should the canal cross paths with a river (flow above, below, or mix with it)? Where and how should runoff during the monsoon rains intersect with the canal? How does the canal pass through a village or under a roadway? To maintain a constant flow and volume, he engineered a range of intersections. Each intersection represents a crossing of three kinds of space: one infrastructural and mechanical in character, one topographic and fluvial in character, and one urban and traditional in character. At every intersection, there developed a distinct setting that was shaped by these layers of infrastructure, topography, and agriculture.

Managing the technical and social dimensions of each of intersection presented challenges to Cautley's design. Nowhere was this more visually apparent than in the first eighteen miles (29 km).

Once the Ganges River breaches the Siwalik Hills and enters a large gorge, it throws out a branch to the right, two-and-a-quarter miles (3.6 km) above Haridwar. "From this branch the inhabitants of Hurdwar and Kunkhul had been in the habit of bringing a watercourse for the purpose of supplying the towns," noted Cautley, "the priests being interested in maintaining water at their ghats or bathing places; the community in general depending upon this supply for water both for drinking and for culinary purposes, for turning corn-mills, and to a moderate extent for the irrigation of the gardens."[75] He goes on to say that "The head works of the Ganges Canal, therefore, were, *in limine*, determined by the priests and the people' so far as supply was concerned, it was only a matter of degree." This was indeed an intersection, not just between canal and town, but also between canal, town, tradition, and topography.

A closer look at the plan (Plate No. 13) shows the construction of these intersections. Nodes dotting the landscape and riverbeds, points from where Cautley surveyed both land and stream, form a constellation across the drawing. These nodes designate the peaks and valleys of longitudinal and cross sections he cut to determine the high water marks of the stream. In other words, everything "point[s] out by unmistakeable [sic] signs a region designed by nature as a great field of artificial irrigation."

Cautley selected Gunes Ghat, just upstream from the Bochna Nulla (torrent), as the site for the head works of the Ganges Canal. A head work or regulator is "the structure in which are fixed the sluice gates to control the water-supply of the canal."[76] Located a mile (1.6 km) or so below Haridwar, a peepal tree (ficus religiosa) and a lingam (ceremonial marker) at its base marked the spot. Otherwise, "there was neither building, nor temple, nor flight of steps, to show that the locality in question was remarkable for its sanctity," wrote Cautley.[77]

The head works at Myapoor (Mayapur) were modified in sporadic phases from 1854 to 1992. Its metamorphosis, along with its infrastructural offspring of earthen embankments and permanent embankments, presents a prototypical distinction that was repeated over and over again throughout the Ganges' development. The earthen embankment (*kachcha* bundh) and the masonry/concrete embankment (*pucca* bhund) represent the two extremes of artificially controlling the flow of the Ganges River. Because the flow of the Ganges varies considerably over the course of a year, the head works had to be designed to manage dramatic changes in flow.

Earthen embankments, made of compacted earth or stones cribbed in wooden cages, are inherently temporal and melt into the river with the monsoon. They must be replaced every year and served as the traditional mode of managing the flow of the Ganges in and above Haridwar.

Permanent embankments, made of masonry and later concrete, immutably alter the flow of the river. Nineteenth and twentieth century engineers preferred such hard, stable structures. The development of the Ganges Basin is a dialectic between these two forms of edges; one hard and fixed—transforming the river into hydrographic plumbing—and one soft and temporal—accounting for the amorphous edges and changing flows. Each kind of edge fundamentally changes the choreography of the river and how different groups, from pilgrims to engineers, perceive its sanctity and flow.

Cautley's decision to place the head works south of Haridwar was as much a social decision as it was a topographic one. The Mayapur head works did not obstruct the flow of water at the ghats in Haridwar. Even though late-eighteenth and early-nineteenth century British descriptions of Haridwar described it as "a small place" or "inconsiderable, 'having only one street,'" observers recognized the significance of the city and the flow of the Ganges to Hindus. This was particularly apparent during one of the great festivals (melas), when, "Every twelfth year, the planet Jupiter being then in Aquarius, a feast of peculiar sanctity occurs, known as a Kumbh-Mela, which is attended by an enormous concourse of people."[78] Pilgrims from across India undertake a long journey to Haridwar

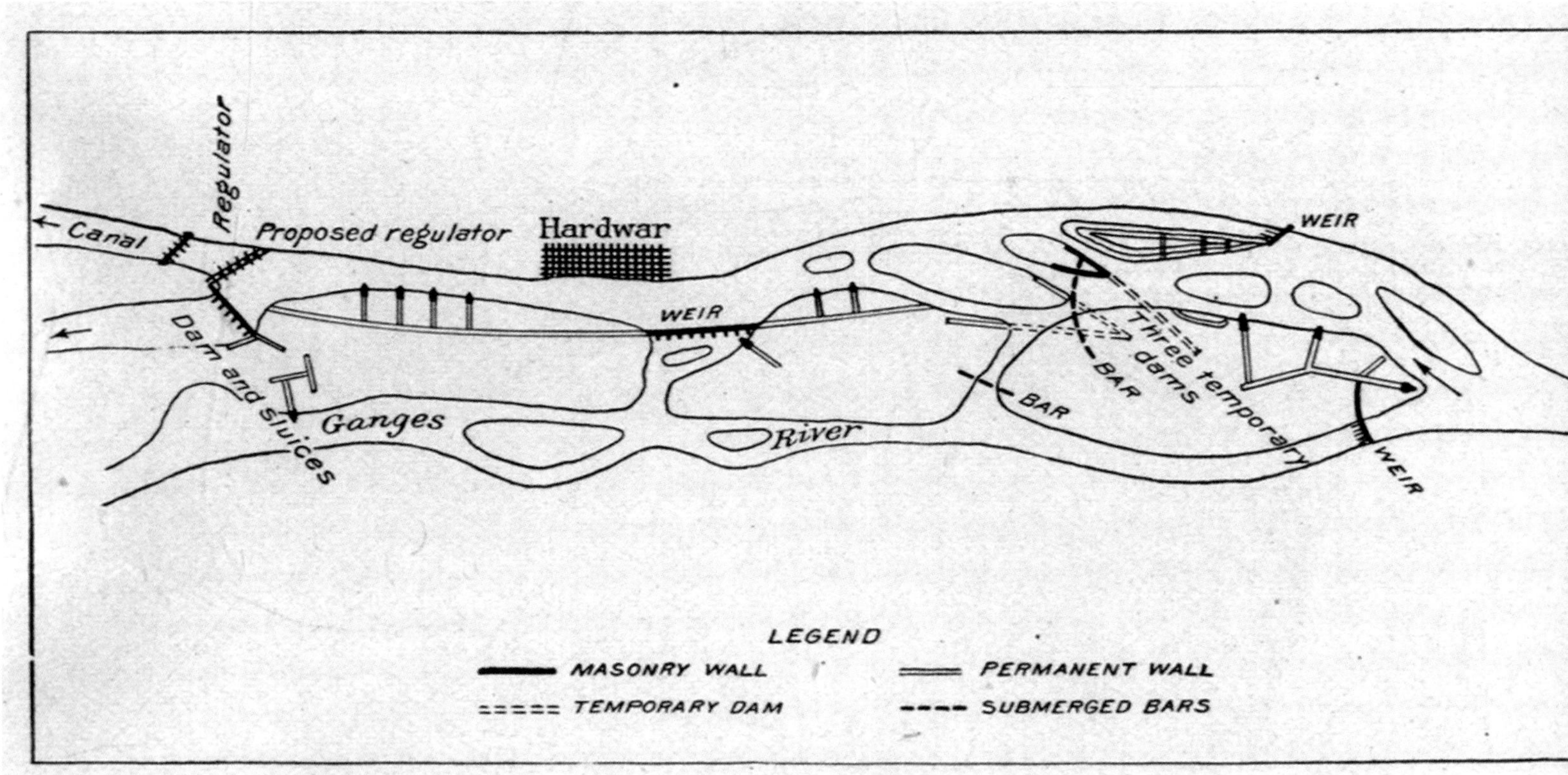

Fig. 18.—Head-works and weir training works, Ganges canal.

Herbert M. Wilson, "Fig. 18. Head-works and weir training works, Ganges canal." **Irrigation in India**. Washington, D.C.: Government Printing Office, 1892. (AA).
Wilson's plan drawing defines the sites and materials that will come to produce tensions between the irrigation department and the All-India Hindu Sabha.

to bathe in the Ganges during this auspicious time. Managing hundreds of thousands of people during the Kumbh-Mela became a concern of the Company from their acquisition of the region in 1817. After 437 people died from a stampede at the main bathing ghat in 1819, the Company widened the 60 steps to 100 feet (30 m). Cautley was wise to not only place the head works south of the sacred city, but also renovated Har-ki-pairi ghat in Haridwar and constructed a platform in the middle of the stream. As a consequence, there were few, if any, objections to the construction of the head works, which consisted of a brick "bridge with ten bays or openings across the mouth of the canal, with shutters, each bay being 20 feet [6.1 m] in width" all inside the mouth of the canal.[79] The advantages of two sets of sluicegates—one set in the dam and another in the regulating bridge—"either the whole amount of water in the river branch can be intercepted and sent down the canal, or the whole, or any part of it can be shut off from the canal, and allowed to escape through the dam, to rejoin the main Ganges river below."[80]

And yet, from its opening, the Mayapur head works posed numerous headaches to engineers and pilgrims alike. The "pocket" between the dam and the regulating bridge produced considerable shoaling at the head and forced an annual closure of the canal in order to clear boulders and debris after the southwest monsoon. Shutting the canal from September to October meant blocking the flow of water from the main branch of the Ganges River above the head works, a time during which farmers lost access to water to irrigate the kharif crop or sow the rabi crop. Pilgrims also lost the "uninterrupted" flow of the Ganges to the sacred ghats of Haridwar and Kunkhul. Temporary earthen embankments were constructed as a stopgap measure, diverting waters to the Mayapur head works for pilgrims at the ghats and to supply cultivators further downstream. To capture the small shingle from the flow of the Ganges and thwart percolation, engineers wrapped the boulders with "a mat of grass or fine branches woven together..."[81] These measures helped reduce the amount of shoaling at the Mayapur head works, but only temporarily. Every year after the southwest monsoon, the embankments washed away and had to be replaced. This was considered an annual inconvenience for engineers, and so in 1864 a permanent embankment was established at Bhimgoda, 300 feet (91.4 m) long, to redirect the water of the Ganges. By 1876, it had grown to 880 feet (268 m) in length. Over the next several decades, engineers both modified the Mayapur head works and extended the permanent embankments above the sacred ghats of Haridwar.[82]

The engineers desire to make many of the embankments permanent would have social consequences at many sacred sites along the Ganges. This came to a head in the early-twentieth century. In 1909, the Irrigation Department of the United Provinces proposed building a permanent

"Postcards." **Album: Visit Hardwar-Rishikesh in 16 Picture Post Card.** Publisher Unknown. Circa 1950s. (AA). Only "Canal Bridge (Nahr Shatabdi Pul)" designates the channel and the water an object of infrastructure.

1: Panorama from bridge above Ganges Canal, southeast of Har-ki-pairi ghat in Haridwar, 2008.

NORTHERN DIVISION

2. Panorama from bridge above Ganges Canal, south of island opposite Har-ki-pairi ghat in Haridwar, 2008.

NORTHERN DIVISION

GANGES CANAL AT HARIDWAR

Between the new head works at Bhimgoda and the original head works at Mayapur, pilgrims bathe surrounded by colossal infrastructure and monumental topography. Below, a diagram of the descent of the Ganges Canal, Ganges River, and their intersections, 2010.

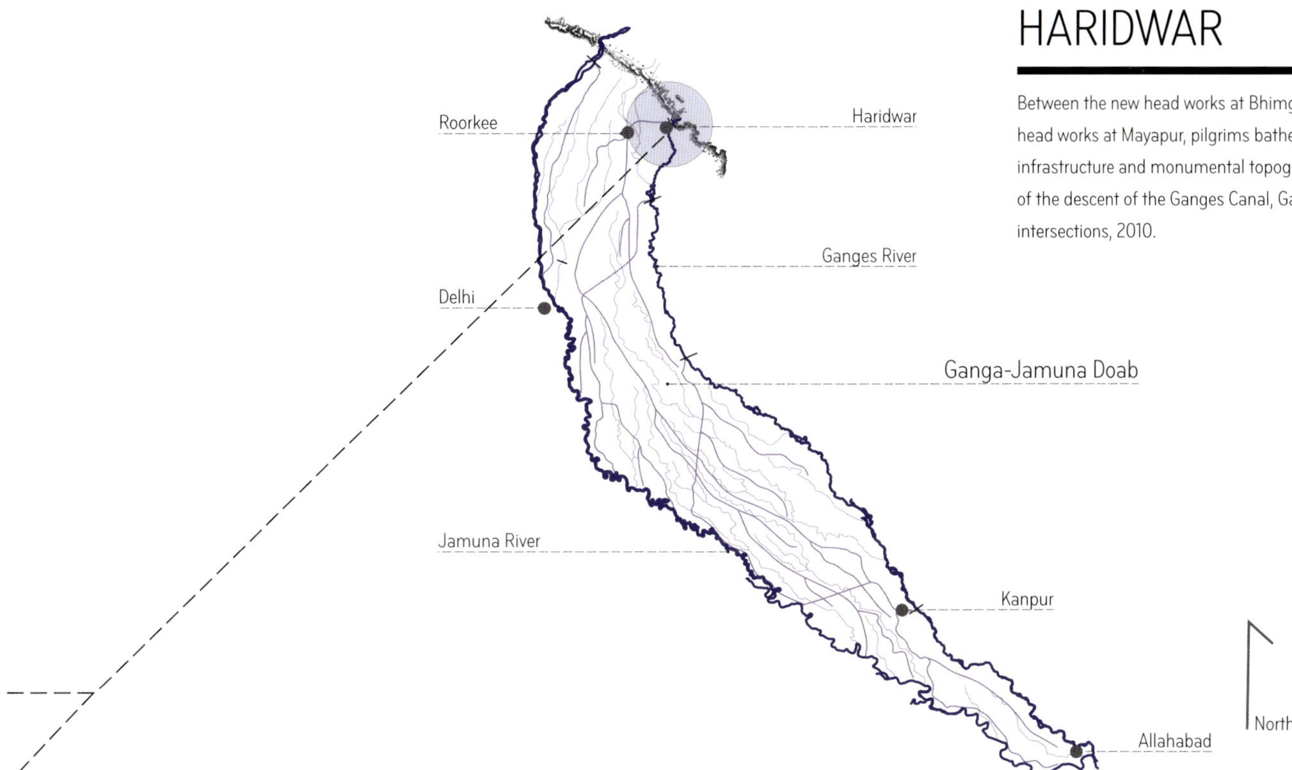

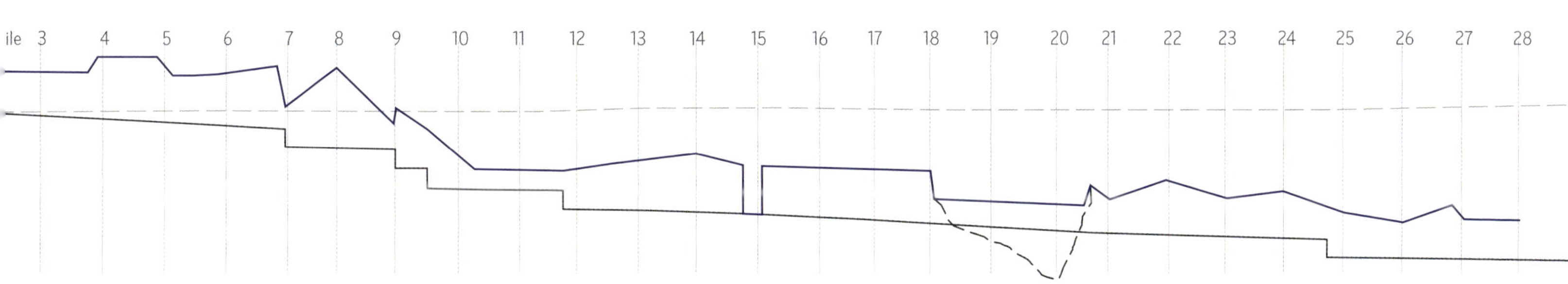

embankment to replace the temporary bunds built at Bhimgoda—the site just above Har-ki-Pairi ghat—moving the head works of the Ganges Canal from Mayapur to Bhimgoda. Engineers in the Irrigation Department, oblivious to why pilgrims or Hindu leaders might object to permanently altering the flow of the river above Haridwar, proceeded without caution in their systematic conversion of the waters of the sacred Ganges into a technological amenity.

BIFURCATION

While the water of a river legally belongs to the state, it belongs culturally and religiously to the people who venerate it. "There is the Ganges—the Bhagiruthi-Gunga, and there is the Ganges-Canal—the Cautley-Gunga of the natives," wrote Bholanauth Chunder in *The Travels of a Hindoo to Various Parts of Bengal and Upper India* (1869).[83] By the late-nineteenth and early-twentieth century, Indians, especially Hindus, viewed irrigation with a mixture of admiration and suspicion.

These evolving views coincided with the rise of Indian nationalism in the early-twentieth century, and public works that ignored long standing cultural traditions started to become the sites to challenge British authority. At first, only the All-India Hindu Sabha, a national organization offering a "Hindu alternative" to the Indian National Congress and a counter-point to the All India Muslim League, protested the new construction at Bhimgoda. At that early stage, local leaders had little reason to distrust the Irrigation Department. After all, when the canal was constructed in the mid-nineteenth century, the Company had renovated Har-ki-pairi ghat and improved access to the sacred waters with a newly constructed platform in the middle of the stream. The Bhimgoda pool was lined with masonry and a tunnel supplied it with river water.[84] These were all seen as engineered improvements to the ghats at Haridwar, not as permanent works altering the natural flow of the Ganges River.

But improving access to the waters of the Ganges in the nineteenth century paled in scope when compared with altering the flow of the Ganges in the twentieth century. Particularly at and above Haridwar, the tension between the ritual uses of the Ganges and the engineers' interventions would prove persistent. By 1916, the Irrigation Department changed the plans of the permanent works considerably. Whereas in 1914 the permanent works were designed both to contrast visibly with the head works and to avoid compromising the uninterrupted flow of Ganges' waters, the updated proposal two years later visually resembled the head works and engaged with the Ganges through a weir, new feeder channel and dam opposite Har-ki-pairi ghat (See Figure 18—Head-works and weir training works, Ganges canal). This ensemble of new works, according to the Irrigation Department, protected the ghats and head works at Mayapur from flooding

4: Walking panorama along east side of island facing west, 2008.

as well as providing year round water for farmers. However, it was to achieve all of these benefits by obstructing the flow of the Ganges—a chief objection to Hindu pilgrims and leaders.[85] Religious mores intersected with the larger aspirations of the desires of British engineers.

As construction of the Regulator at Bhimgoda continued and protests mounted, in December of 1916 an agreement was struck between the government and representatives of the All India Hindu Sabha and other prominent Indian national figures. The government made five concessions:

1. The main supply channel would remain as is with no sluice;
2. An opening in the weir will never be closed to ensure water for bathers;
3. The new supply channel would not be made of masonry or be called a canal;
4. An earthen embankment, rather than a concrete structure, at the supply channel;
5. A permanent opening in the weir and in the Mayapur Regulator.[86]

An earthen embankment, unlike a concrete or masonry embankment, was thought less of an interruption to the natural flow of the Ganges River. The new weir at Bhimgoda was completed in 1920.

The 1916 agreement remains in place to this day. It guides the practices of the Northern Division of the Ganges Canal in Roorkee and can be found in Appendix I of the Standing Manual for River Ganga and Ganges Canal.[87] Even still, additions and changes to the Ganges Canal did not end in 1920. In 1980, the government began construction on the Eastern Ganga Canal Project (A Kharif Irrigation System) on the left bank of the Ganges at Bhimgoda, an irrigation canal designed to utilize surplus water during the southwest monsoon to irrigate 105,000 hectares.[88] While the Eastern Ganga Canal Project remains unfinished due to incomplete funding, in 1991-92, a new barrage (head works) replaced the old weir at Bhimgoda.[89]

By virtue of the Standing Manual and the continuing changes to this site, Cautley's claim remains prescient: "the engineer, by a proper adaptation of levels in his artificial channels, to obtain a command of water," does indeed place "the whole country under his control..." Negotiating the intersection of all of these interests remains a chief concern of the engineer working in the Ganges Basin even today.

3: Walking panorama along east side of island facing east, 2008.

SUPER-PASSAGES

"Engineering in Asia may usually need less money," mused Alfred Deakin in the 1890s, "but always implies a greater tax upon the health, patience, and brains, of those responsible for it."[90] At a loss to "recapitulate" the "gigantic system" of the Ganges Canal, Deakin explained that, "Its colossal character can be best conceived when a picture is presented to the mind of an artificial river sometimes carried over what in the rains are rivers, sometimes having those rivers carried over it, and at other times, taking them into its course." Cutting deep into the topography for the first twelve miles (19 km), sometimes at a depth of 30 feet (9 m), Cautley slit inlets into his "artificial river" to allow water to pass into it, while significant falls (similar to those he developed for the Eastern Jumna Canal) helped negotiate steep descents in elevation. Allowing water to pass through was potentially destructive. Silt and boulders threatened to clog the canal or ruin its bed. Carrying water above or passing water below the canal was costly and required large works of masonry. In the first twenty miles (32 km), Cautley built only one level crossing at the Ratmau River and one aqueduct across the Solani River—which happened to be the longest aqueduct in the world at the time. A look at the designs of these four types of structure merit inspection, if for no other reason than because, as Deakin put it, they must withstand the "Titanic onset" of water hurled down the mountains, which "threatens even the most massive masonry."[91]

A super-passage—the term Cautley gave to his flumes to distinguish them from an aqueduct—is when a river or torrent crosses over a canal. As P. J. Flynn, Late Executive Director of Public Works in the Punjab observed in 1892, the super-passage has the "advantage of keeping the canal completely free from any influx of flood-water from the torrent, which is always more or less heavily charged with silt."[92] Super-passages were meant to be a time-and-labor-saving mechanism. Unlike a level crossing, where the torrential waters and silt mingle with a canal and must be cleared after the floods, a super-passage allows a canal to flow without interruption.

Two of Cautley's super-passages—Ranipoor (Ranipur) and Puttri (Phatri)—are low, straddling structures that allow the Ranipur rao and the Phatri rao to cross the canal. Channeled between parapets, these "are very massive structures, as they have to withstand the whole force of rivers swollen to a great size, and rushing down beds having excessively rapid slopes."[93] They allow the torrents to leapfrog over the canal (similar, but on a much larger scale, to what Cautley inspected in Italy: "a salto di gatto," which means "cat's leap").

The two super-passages, level-crossing, and Solani Aqueduct were all considered feats of engineering in the mid-nineteenth century. Engineers from around the world traveled to see these works.

1. The Ranipoor (Ranipur) super-passage occurs at mile six (9.6 km) below Mayapur. It is 200 feet (61 m) wide and 14 feet (4.3 m) deep. The super-passage is 450 feet (137.2 m) long and is supported by eight arches with a 25-foot (7.6 m) span. It carries the torrent over the canal during the southwest monsoon.

2. The Puttri super-passage carries the Puttri (Pathri) rao over the canal at mile ten (16 km) below Mayapur. It is 300 feet (91.4 m) wide. Extensive training works along with an excavated channel to carry the torrent over the canal and discharge the waters into "the swamps of the Ganges khadir."[94]

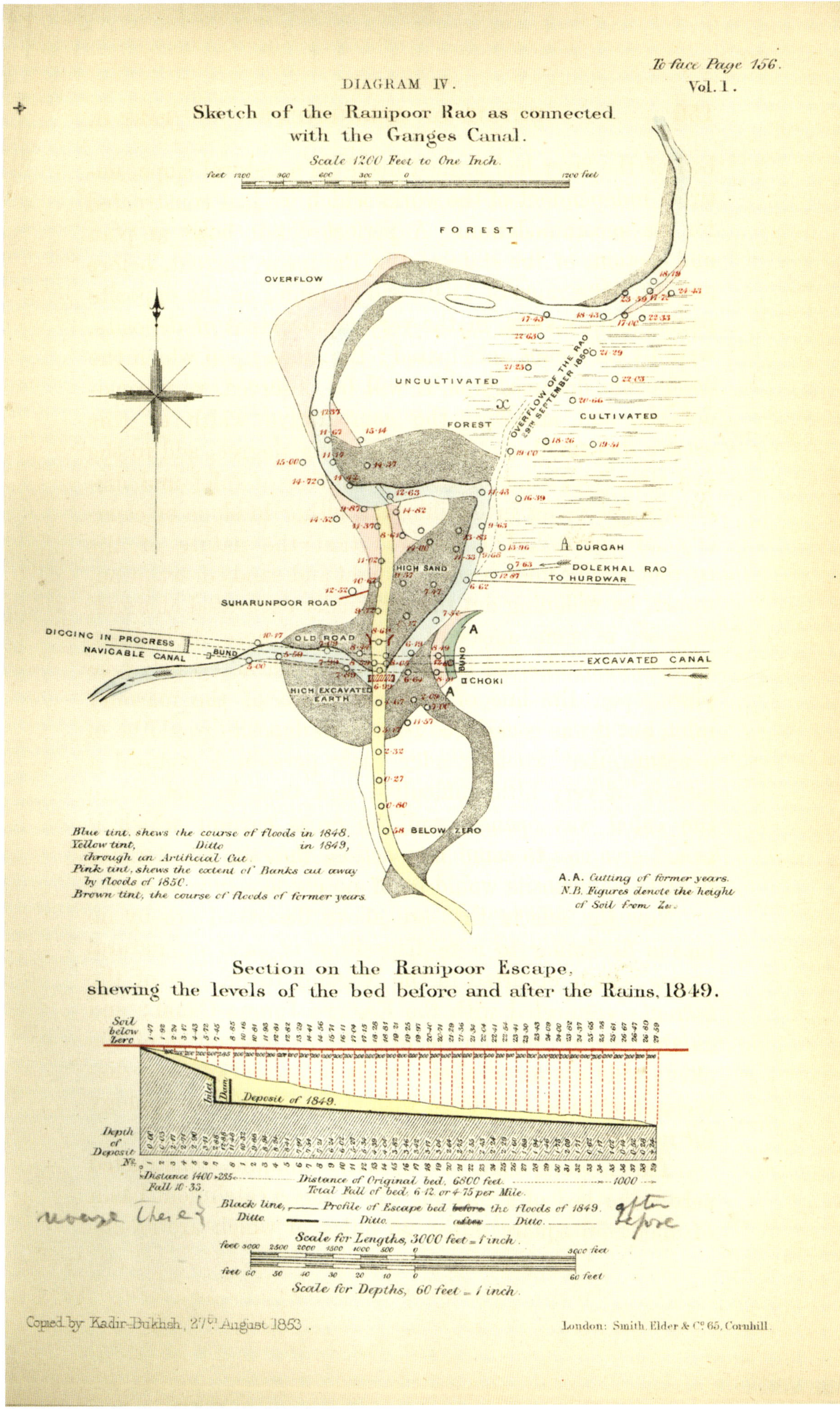

Proby Cautley, "Diagram IV. Sketch of the Ranipoor Rao As Connected with the Ganges Canal." **Report of the Ganges Canal Works: from their commencement until the opening of the canal in 1854**, Vol. I, London: Smith, Elder and Co., 1860. (Princeton University).

SUPER-PASSAGES

Ranpiur Rao (torrent) crossing above Ganges Canal, 2008.

Panorama of Ranipur Rao Super-Passage, 2008.

Ranipur Super-Passage, Ganges Canal passing beneath Ranipur Rao (torrent), 2008.

Herbert M. Wilson, "Plate XVII. Ranipur Superpassage, Ganges Canal." **Irrigation in India**. Washington, D.C.: Government Printing Office, 1892. (AA).

1: Panorama Looking North at Ranipur Rao Super-Passage, 2008.

SUPER-PASSAGES

2: Panorama Looking South at Ranipur Rao Super-Passage, 2008.

SUPER-PASSAGE AT RANIPUR RAO

Ranipur Rao (torrent) is carried over the Ganges Canal, 2010.

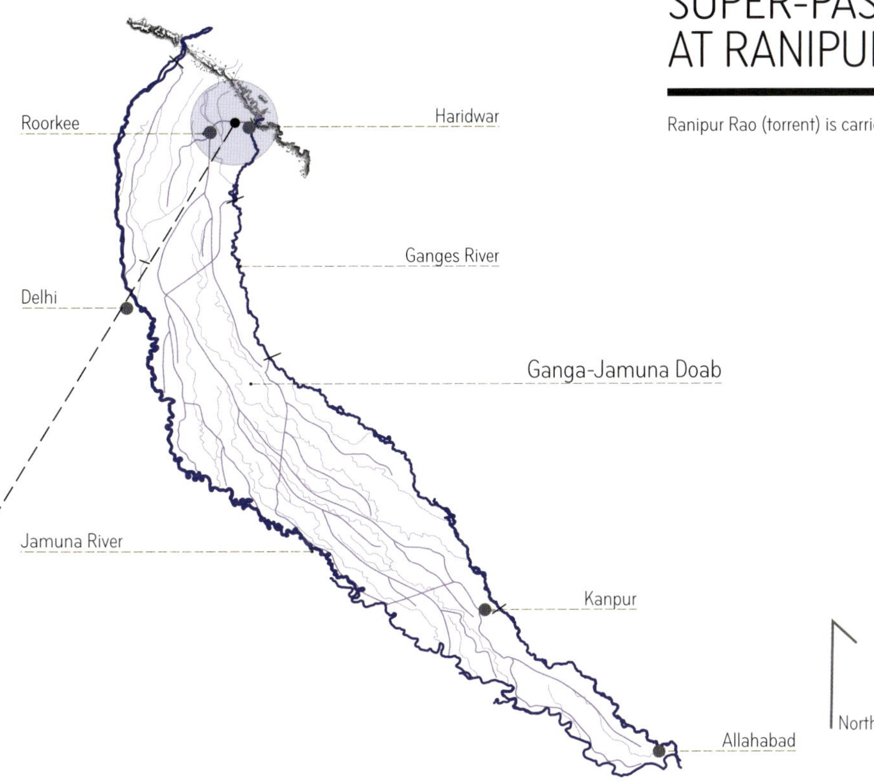

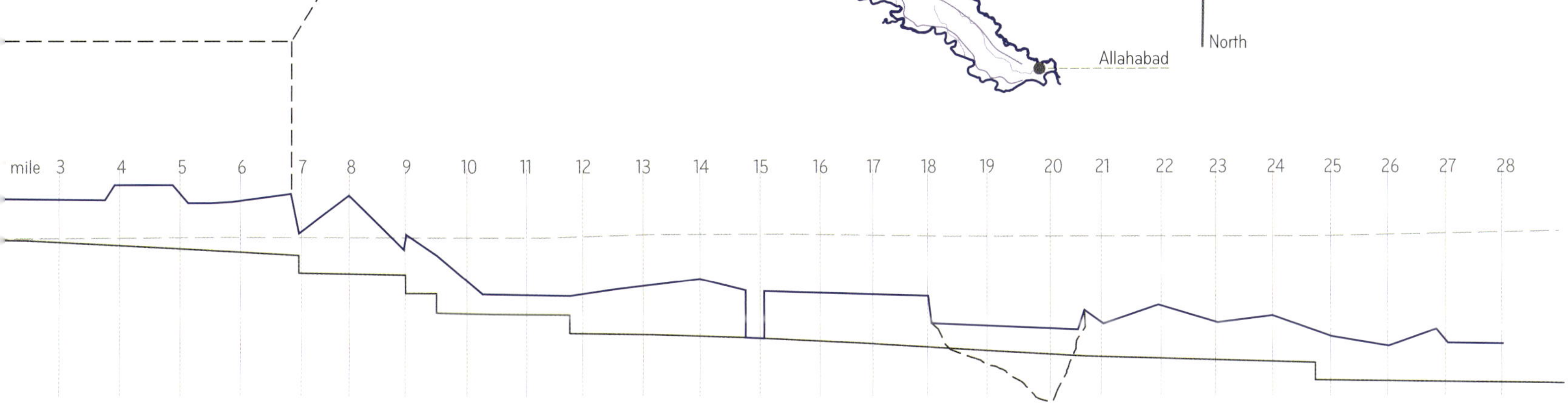

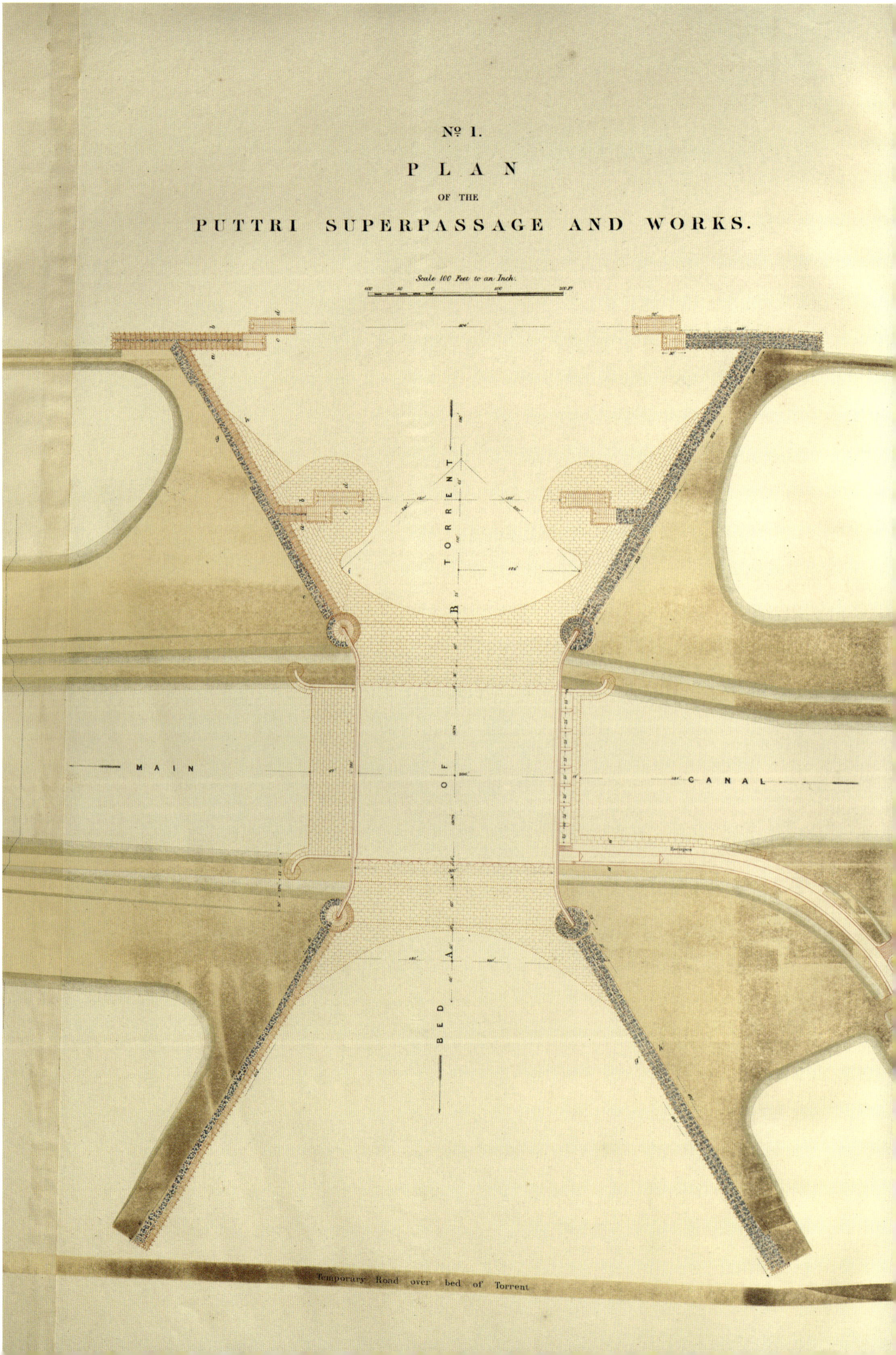

Proby Cautley, "No.1. Plan of the Puttri Superpassage and Works." **Report of the Ganges Canal Works: from their commencement until the opening of the canal in 1854**, Atlas. London: Smith, Elder and Co., 1860. (Princeton University).

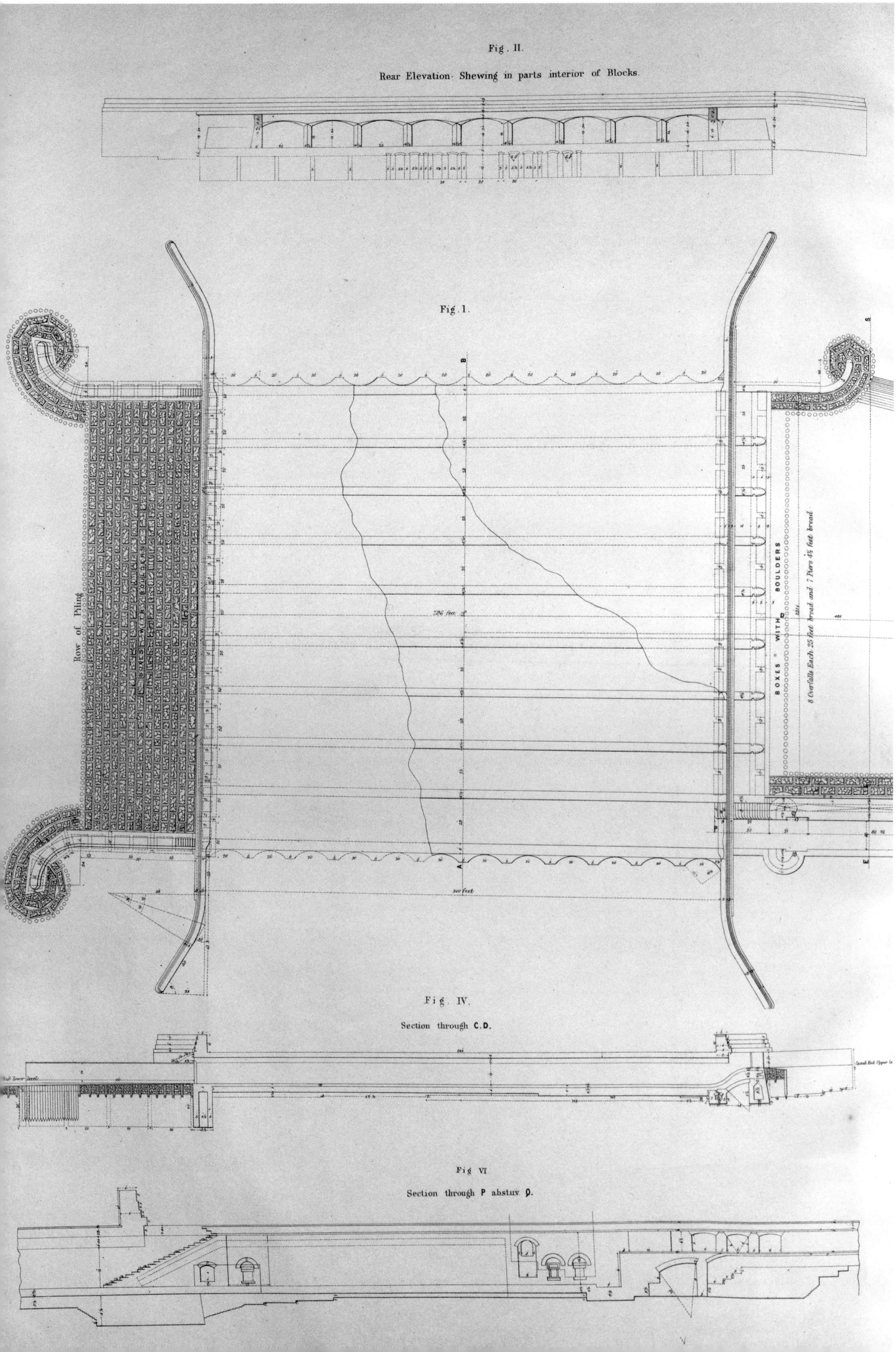

Proby Cautley, "Putti Superpassage." **Report of the Ganges Canal Works: from their commencement until the opening of the canal in 1854**, Atlas. London: Smith, Elder and Co., 1860. (Princeton University).

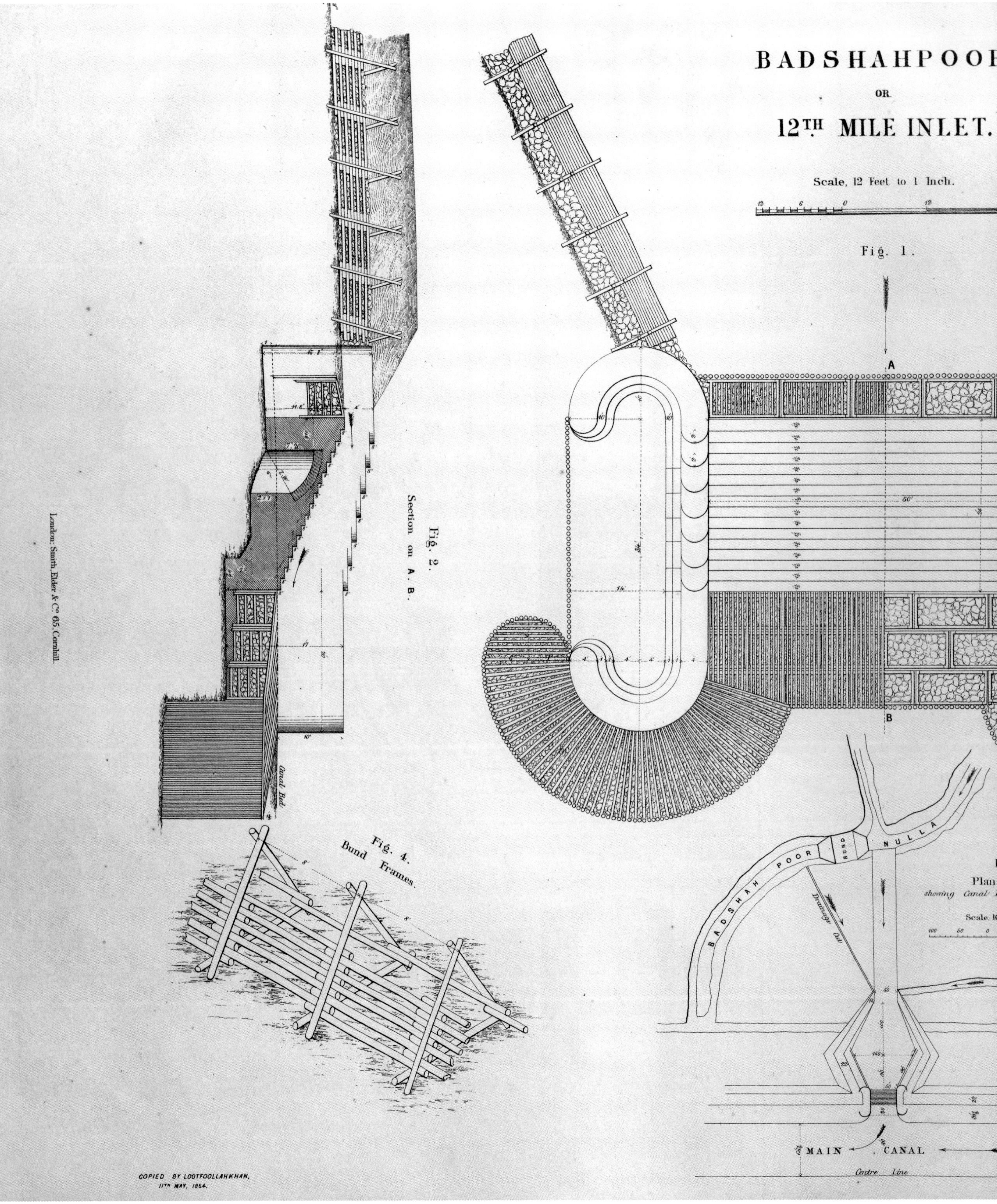
BADSHAHPOOR
OR
12TH MILE INLET.
Scale, 12 Feet to 1 Inch.
Fig. 1.
A
B
Fig. 2.
Section on A. B.
London: Smith, Elder & Co 65, Cornhill.
Canal Bed
Fig. 4.
Bund Frames.
BADSHAH POOR NULLA
BUND
Drainage Cut
Plan
showing Canal Be
MAIN CANAL
Centre Line
COPIED BY LOOTFOOLLAHKHAN,
11TH MAY, 1854.

Proby Cautley, "Bhadshahpoor or 12th Mile Inlet." **Report of the Ganges Canal Works: from their commencement until the opening of the canal in 1854**, Atlas. London: Smith, Elder and Co., 1860. (Princeton University).

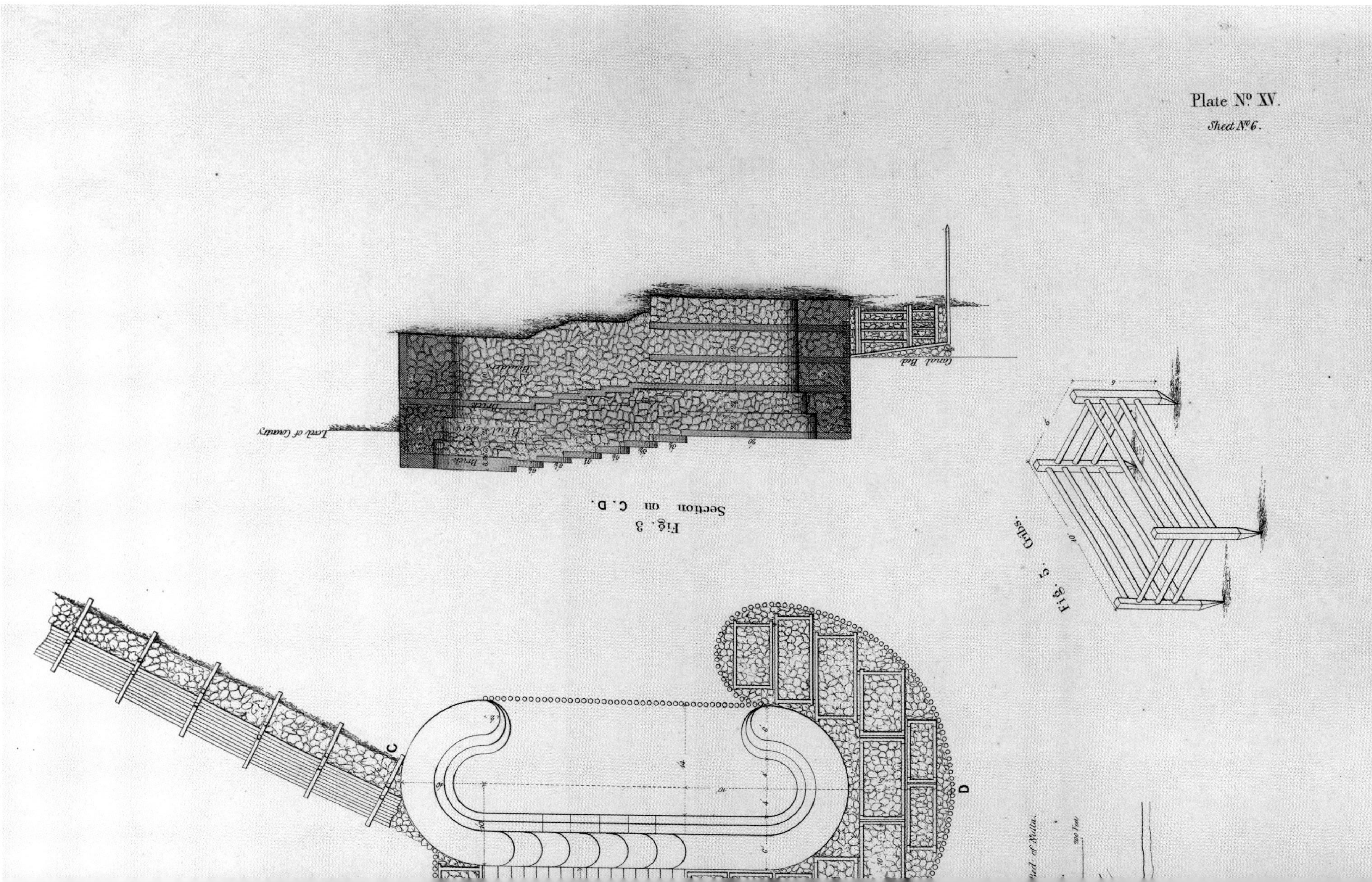

Proby Cautley, "Ganges Canal, Looking Upstream. The Peeran Kuleeur Bridge in the front, & the Rutmoo Regulating Bridge in the distance." **Report of the Ganges Canal Works: from their commencement until the opening of the canal in 1854**, Vol. I. London: Smith, Elder and Co., 1860. (Princeton University).

LEVEL-CROSSING

At mile 13 (21 km), the Ganges Canal intersects the perennial Rutmoo (Ratmau) River. Cautley did not build a super-passage to circumvent the flow of the river. Instead, he constructed a level-crossing—"an astounding spectacle" according to Deakin—choosing to co-mingle waters, but not without massive masonry structures to protect the canal and channel the river. Before reaching the Ganges Canal, the crossing is reinforced with one mile (1.6 km) of revetment walls, supported by well-foundations 20 feet (6 m) deep below the bed of the river Ratmau. A "forest of piles," along with cribwork stuffed with boulders (much like the temporary embankments built at Bhimgoda) break up the flow of the water before it enters the canal. A regulating bridge with the same number of bays as the one at Mayapur straddles the canal "so as to prevent an undue supply from rushing down the canal."

Only during floods does the Ratmau flow into the canal across the level crossing. Otherwise, it is lead through a double tunnel, nearly 500 feet (152 m) long, running below the bed of the canal. Not unlike a railway crossing in the nineteenth century, forty workmen stood guard to "close the canal and open the escape," writes Deakin.[95] For this reason, in 1877, *The Roorkee Treatise on Civil Engineering* made note of: "The patent objection to this kind of work is, that it requires a permanent establishment of men on the spot to work it, and that, if they are careless or neglectful, a sudden flood may do serious damage."[96] The editors conclude that, "On this account, level crossings are to be avoided whenever it is possible." Regulation and degradation at the Dhanauri Level Crossing ranked as one of the top four problems along the Ganges Canal in the 1980s.[97] By 1994, with the support of the World Bank, the Ratmau Aqueduct replaced the Dhanauri Level-Crossing.

LEVEL-CROSSING

"Plate LXXXI: Dhanauri Works—Ganges Canal." **The Roorkee Treatise on Civil Engineering in India**, Volume 2, compiled by Julius George Medley, third edition enlarged and improved and edited by A. M. Lang. Roorkee: Thomason College Press, 1877. (AA).

Herbert M. Wilson, "Pl. XVIII. Rutmoo Level Crossing, Ganges Canal" **Irrigation in India**. Washington, D.C.: Government Printing Office, 1892. (AA).

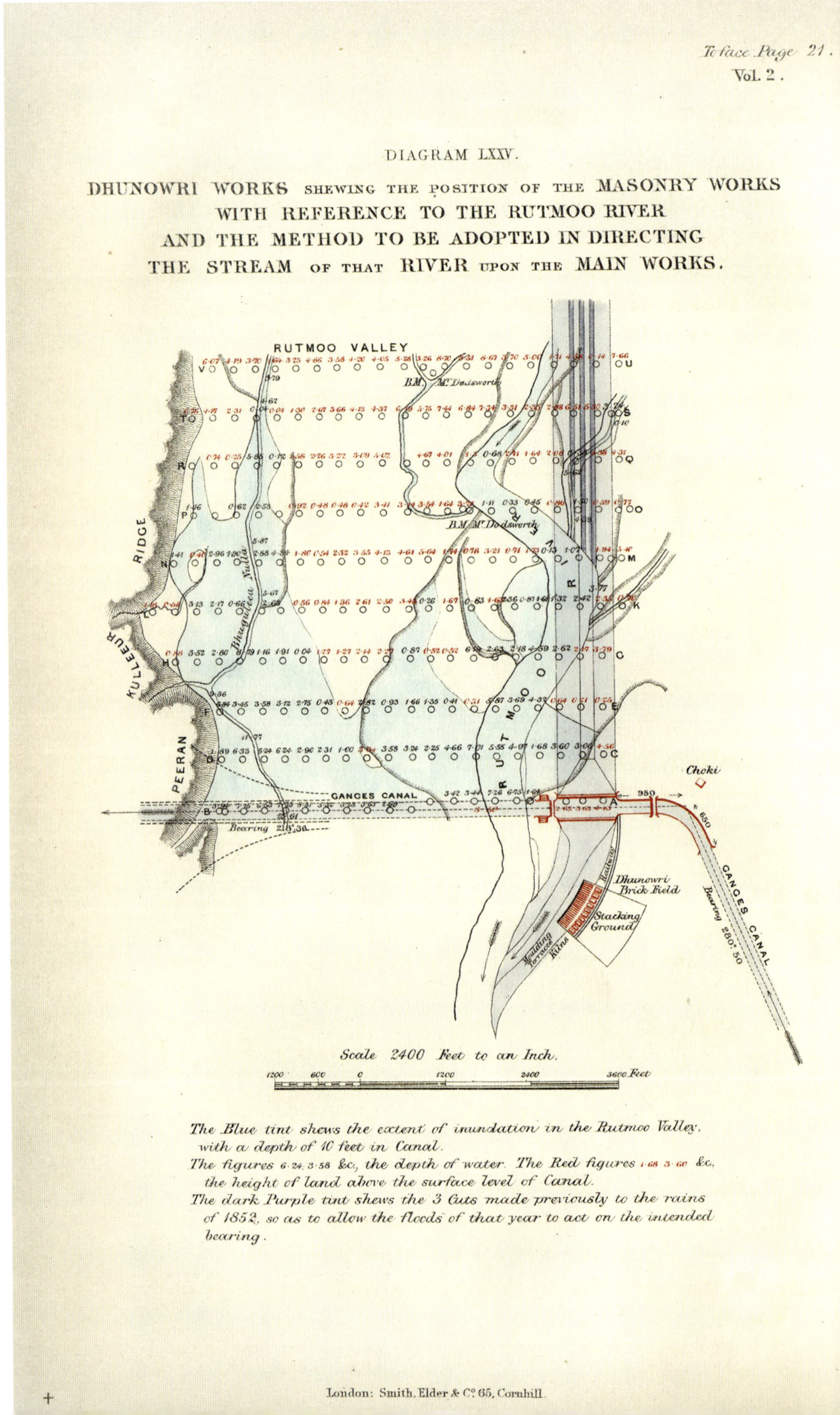

Proby Cautley, "Diagram LXXV." **Report of the Ganges Canal Works: from their commencement until the opening of the canal in 1854**, Vol. II. London: Smith, Elder and Co., 1860. (Princeton University).

Diagram 10.

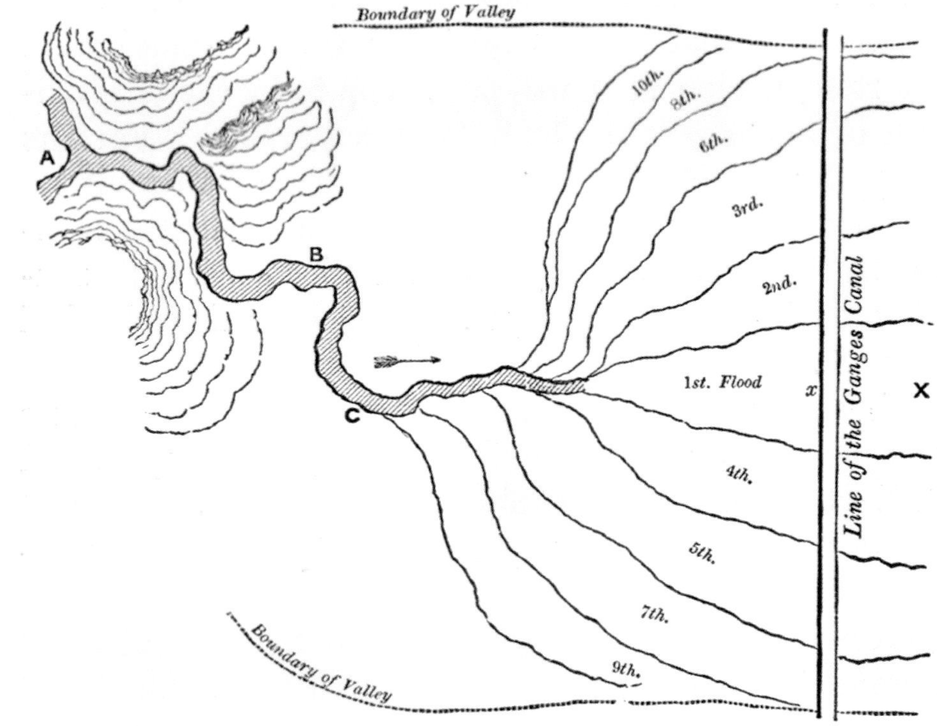

Proby Cautley, "Diagrams10-12." **Report of the Ganges Canal Works: from their commencement until the opening of the canal in 1854**, Voll. London: Smith, Elder and Co., 1860. (Princeton University).

Diagram 11.

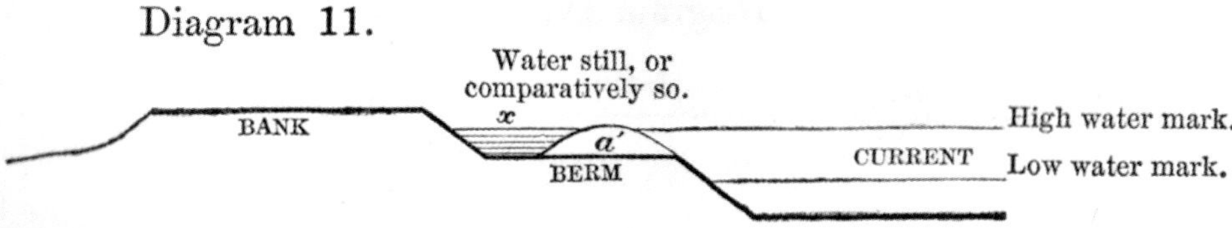

In addition to mapping how rivers expand and contract in plan and section during the monsoon, Cautley also looks to the Indus River and how it naturally forms the profile of an artificial canal. When seen in plan and section, nature provides a model for his great work of artificial irrigation.

On the Indus the most magnificent exemplification is shown of the same law, where for a mile on each side of the river's course the land is elevated, whereas the surface of the distant country is below that of the water of the river. The following section of the Indus, for which I am indebted to my friend Major W. E. Baker, of the Bengal Engineers, is literally that of an artificial canal with embankments, the latter being composed of silt deposited during inundatioms.

Diagram 12.

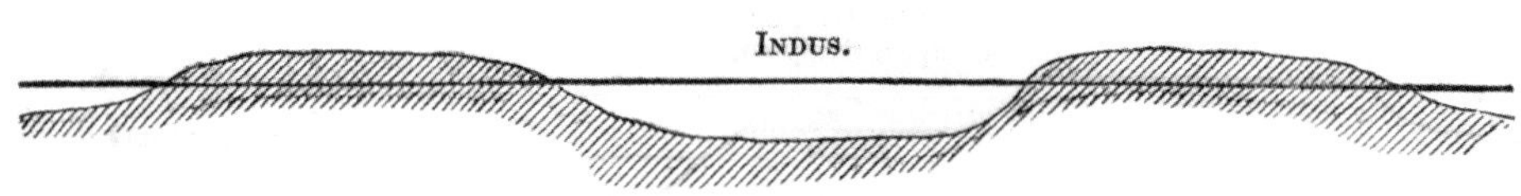

The fan-like and convexical form which deposits assume, when silt-bearing water, running on a steep declivity, escapes from its channel and spreads over the surface of the country, is too conspicuously shown at points on the Eastern Jumna Canal, where fractures have taken place in embankments much elevated above the surface of the country to have escaped the attention of men in the Canal Department; but the rationale is too clear to require further explanation, when the effects of deposit depending on the velocity of the current, and the gravity of the suspended matter, are taken into consideration.

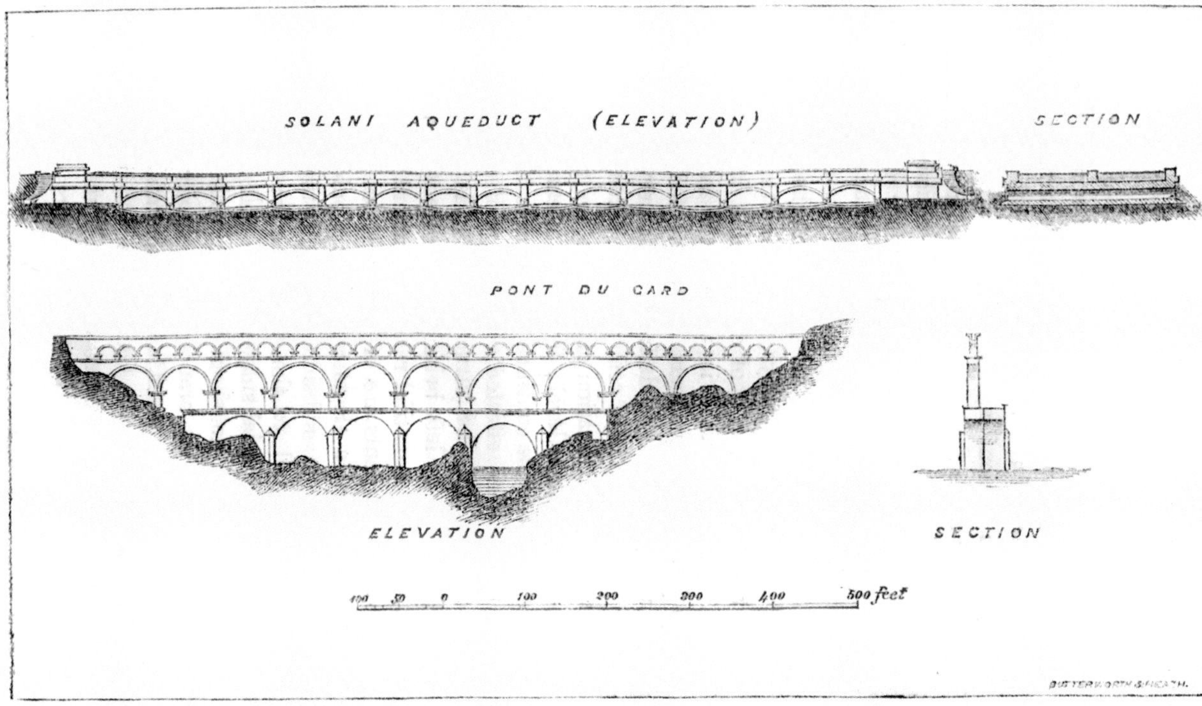

George Rowdon Burnell, "Solani Aqueduct and Pont du Gard." **The Rudiments of Hydraulic Engineering**. London: John Weale, 1859. (Harvard University).

AQUEDUCT

As noted earlier, it was with good reason that the inauguration of the Ganges Canal took place at mile 19 (30.6 km) at the Solani Aqueduct. Designed to carry 6,500 cusecs across the most dynamic of the seasonal torrents—the Solani River—the bed of the lined canal runs above the level of the surrounding land for 2.25 miles (3.6 km). At a width of 172 feet (52.4 m) and height of 24 feet (7.3 m), the Solani Aqueduct stretches 1,770 feet (539.5 m) across the Solani River. Just past the point where the aqueduct ends and the canal continues, two sitting lions, sculpted to be seen at a great distance, mark the point where the canal at long last enters its intended irrigable area.

What was tremendous about this work was not its height, but the amount of water that passes across it throughout the year. It was a source of pride for British engineers. "[I]t is upon the great Indian irrigation works that the English engineers have achieved their most recent glorious triumphs. The Solani Aqueduct, upon the Ganges Canal is in fact the most stupendous monument of that kind yet constructed," wrote S. J. Burnell in 1877.[98] In *The Rudiments of Hydraulic Engineering* (1858), engineer George Rowdon Burnell visually compared the elevation and section of the Solani Aqueduct with the Pont Du Gard Aqueduct built by the Romans. He draws them at the same scale so that "it may here suffice to direct attention to the fact, that the modern work which is comparatively speaking, so little known, has an area of water-way, not less than eighty times that of the celebrated Roman example."[99]

Comparing Imperial Rome and the administration of British India, both under the Company and later the Raj, was nothing out of the ordinary—Sir William Jones likened himself to the "Justinian of India" in the 1780s.[100] "There has never been the same misgovernment of India as there was of Roman provinces," noted Alfred Deakin in the 1890s, "but there has been, and is still, a necessary subordination of the conquered country, and a suppression of its individuality, which, though abundantly beneficial to its inhabitants on the whole, renders their relations' to Great Britain irksome, and affords a constant opportunity to those whose advantage lies in promoting misunderstanding."[101] Before the Solani Aqueduct was even completed, it was said in 1846 that the Ganges Canal will "...remain a lasting monument of British architecture and of British benevolence in India."[102] Upon arriving in Roorkee in 1857-58, following the Indian Rebellion of 1857, Julius George Medley, Captain in the Bengal Engineers, wrote how he, "admired, as every Engineer must, that monument of scientific skill, the Solani Aqueduct."[103] Even beyond India and Great Britain, the Solani Aqueduct was admired. In his essay for *Scientific American*, "What Civilization Owes the Architect and Civil Engineer," George R. Bramhall wrote of the aqueduct in 1885: "In grandeur of design, solidity, and utility, it challenges competition."[104] As late as 1966, it was said of the Solani Aqueduct: "Unequalled in its dimensions by any work executed anywhere in the world at that time, this crossing is a magnificent monument to the outstanding engineering skill of its builders."[105]

AQUEDUCT

SOLANI AQUEDUCT.

Proby Cautley, "Solani Aqueduct." **Report of the Ganges Canal Works: from their commencement until the opening of the canal in 1854**, Atlas. London: Smith, Elder and Co., 1860. (Princeton University).

"Ganges Canal Bridge—Roorkee." Moorli Dhur & Sons. Ambala Cantt. Circa late-19th or early-20th century (AA). At Roorkee the Ganges Canal enters the plains.

Proby Cautley, "Plan of Part of The Solani River Near Roorkee." **Report of the Ganges Canal Works: from their commencement until the opening of the canal in 1854**, Atlas. London: Smith, Elder and Co., 1860. (Princeton University).

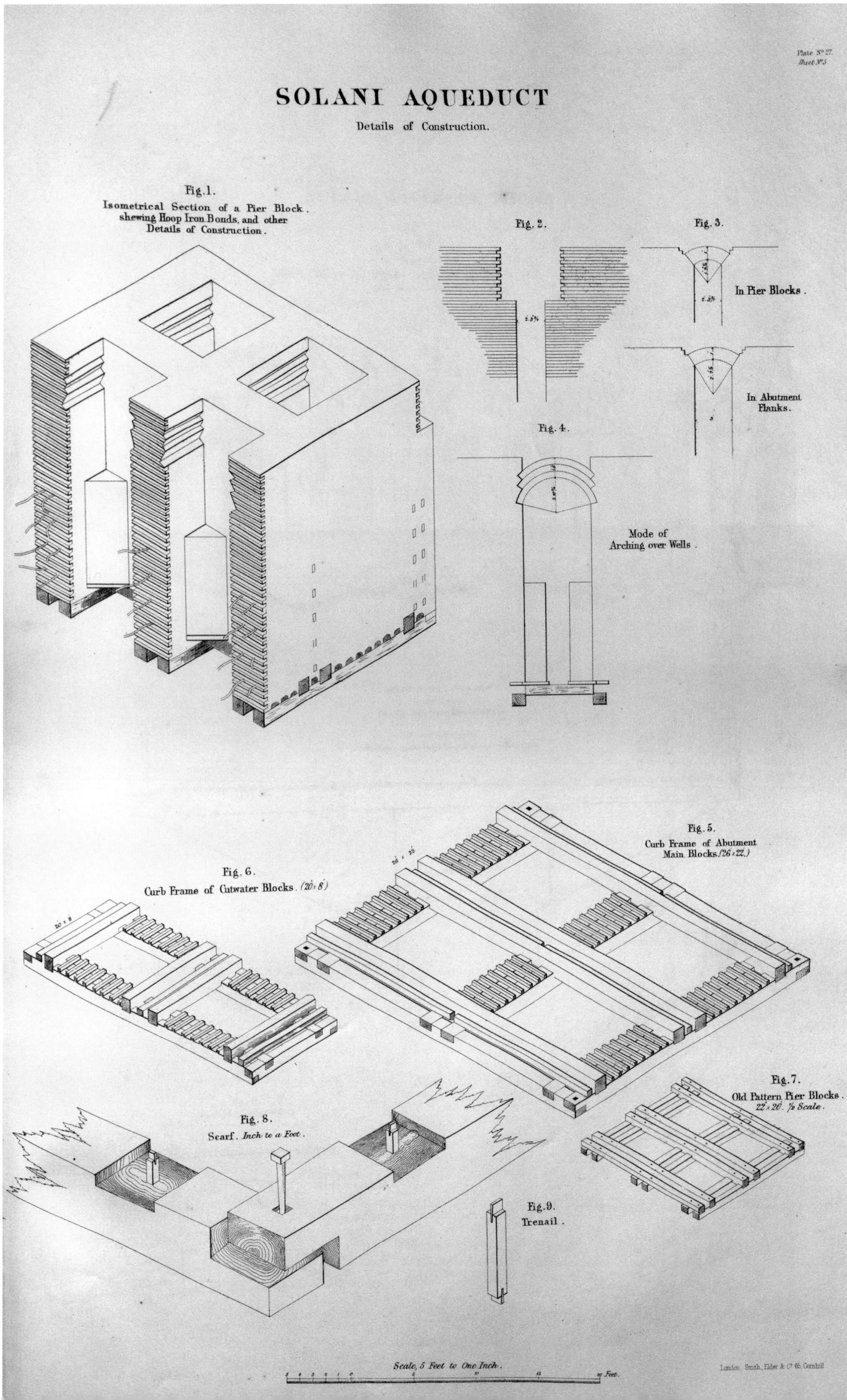

Proby Cautley, "Solani Aqueduct: Details of Construction." **Report of the Ganges Canal Works: from their commencement until the opening of the canal in 1854**, Atlas. London: Smith, Elder and Co., 1860. (Princeton University).

AQUEDUCT

1: Panorama Looking East at Original Solani Aqueduct, 2008.

2: Panorama Looking West at Solani River Valley, 2008.

SOLANI AQUEDUCT: HYDRAULIC MONUMENT

The longest aqueduct in the world at the time of its opening (1854), for over one century it was revered as one of the great monuments of British engineering in India, 2010.

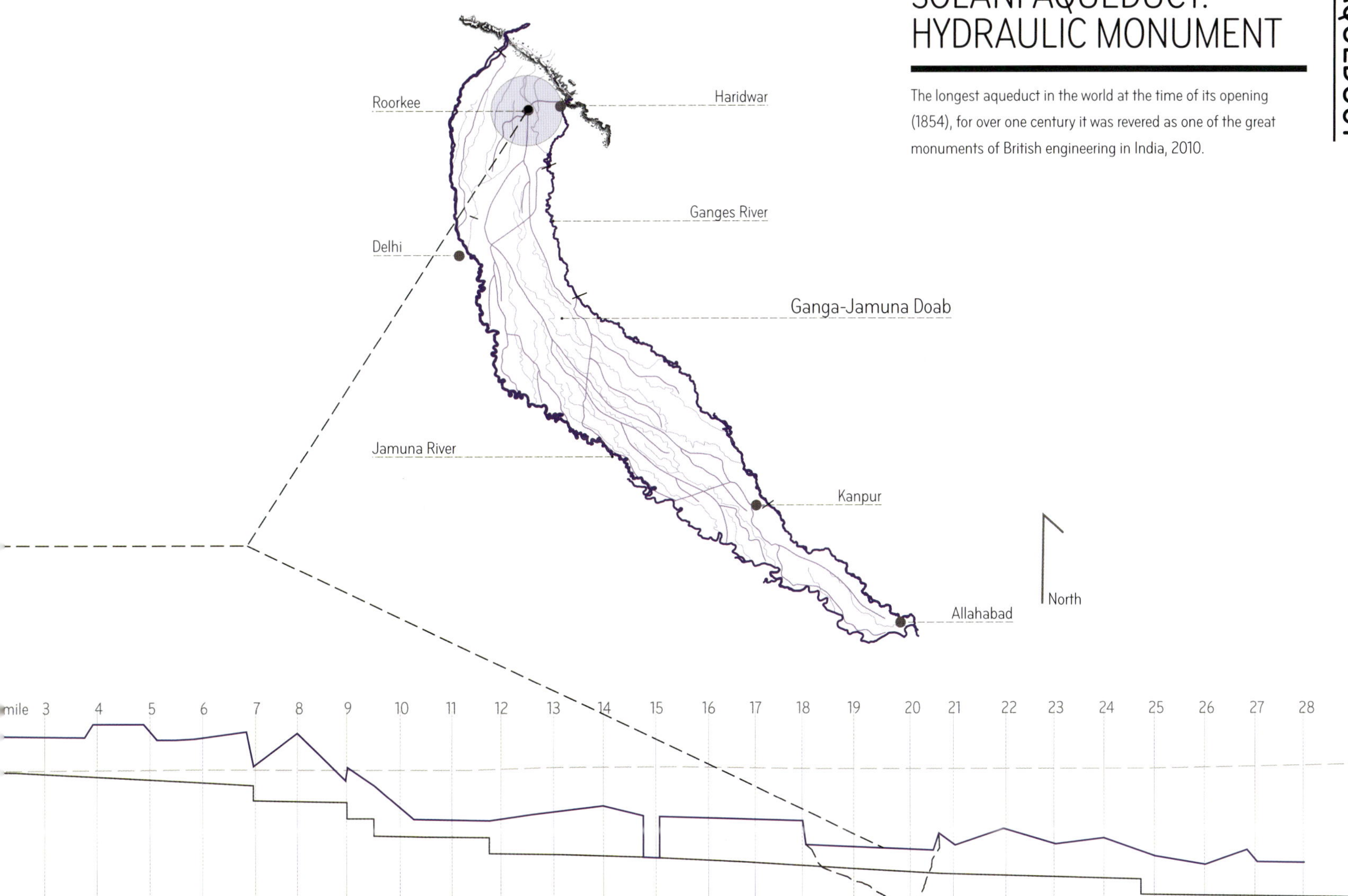

Proby Cautley, "Model of the Lion On the Termini of the Masonry Aqueduct at Roorkee." **Report of the Ganges Canal Works: from their commencement until the opening of the canal in 1854**, Vol. I. London: Smith, Elder and Co., 1860. (Princeton University).

LEGACY

In this first twenty miles [32 km] therefore are works of almost every kind upon the greatest scale," noted Alfred Deakin at the end of the nineteenth century.[106] Amazingly, all four of these works stayed in place, with few alterations, for over a century. It was not until 1984, when the World Bank loaned the government $125 million as part of the Upper Ganga Irrigation Modernization Project, that Cautley's super-passages, level-crossing, and aqueduct were either rehabilitated or replaced. In their report, the World Bank stated that the Upper Ganga Canal system had "reached the end of its effective life span by the mid 1970s." This came with a warning: "There is a very high risk of the main canal collapsing with disastrous consequences to agricultural production in the State, and hence, there is an urgent need to take measures to avert the risk."[107]

Among the five objectives of the World Bank, the most serious was to "rehabilitate and modernize the system" to enlarge the command area for irrigation.[108] To do this, they proposed and built a parallel canal to "replace the existing (dilapidated) main canal...together with modern regulating and access structures..."[109] By 1994, these objectives were completed, and a new, 23.6km-long canal was completed. While the two super-passages were rehabilitated with a new syphon at Ranipur super-passage, the Ratmau Aqueduct replaced the Dhanauri Level-Crossing and a new aqueduct runs parallel to the Solani Aqueduct. Despite delays, the World Bank deemed the project a success by 1995.[110] Its achievement was measured in how many thousands of hectares it either protected from risk or expanded in terms of irrigation. And while it continues to be measured in terms of crop yields and agricultural productivity, the spatial impact of these works, both by Cautley and those co-financed by the World Bank and the Government of India, produce economic and social changes within these areas.

New Concrete Aqueduct running parallel to the Solani Aqueduct (financed by World Bank and Government of India in the 1990s), 2008.

Proby Cautley, "Mortar Mill." **Report of the Ganges Canal Works: from their commencement until the opening of the canal in 1854**, Vol. II. London: Smith, Elder and Co., 1860. (Princeton University).

It would not be until the 1990s when concrete structures like the new aqueduct (below) would replace or run parallel to the masonry structures of the original canal works.

HEADQUARTERS

The spatial and social implications of the Ganges Canal in the Northern Division were not just limited to the disbursement of water for cultivation. They transformed villages and towns, perhaps no place more visibly than at Roorkee. "Before the commencement of the Ganges Canal works," reported the *Imperial Gazetteer of India* in 1887, "Rúrki was a mere mud-built village on the banks of the Soláni; it is now a flourishing town with metalled roadways, meeting at right angles, and lined with excellent shops."[111] The Ganges Canal brought order and improvement to the form of the town, replacing its dirt roads and irregular blocks.

The engine of this change was not the Ganges Canal itself. Roorkee is not only the site of the Solani Aqueduct; it is also home to the Headquarters of the Ganges Canal. Besides being "beyond the influence of the hill drainage" on the tableland, Roorkee is just 60 miles (96.6 km) north of Meerut, while Dehra Dun and Mussorie are due north, and it was said to be "more accessible from the Punjaub than any other place."[112] Nineteenth century descriptions and illustrations of the town are telling, perhaps none more so than the "Opening of the Ganges Canal," belatedly published in *The Illustrated London News* on August 4, 1854, (nearly four months after it opened). The accompanying engraving, "'The Ganges Canal, at Roorkee,'" does not feature the Solani Aqueduct. It is nowhere to be seen. It is blocked behind a bridge crossing the canal. Boatmen ply their skiffs across India's newest "artificial river" while townspeople descend ghats to bathe. Two buildings are pictorially sandwiched between the bridge and the Himalayas. Only their tops and smoke stacks peer out over the high embankments and compete in height with the peaks of the Himalayas.

To accompany the engraving, *The Illustrated London News* reprinted a condensed version of Captain Smith's "Opening of the Ganges Canal," originally published in *The Delhi Gazette* on April 8, 1854. Smith's extended description of Roorkee is telling: "We have now arrived at Roorkee," observes Smith, "a place...which bids fair to be one of the busiest and most interesting stations in the North-Western Provinces." He reassures readers that, "We shall not linger at it long, but there are a few things worthy of a moment's passing." Perhaps with a sketch by Lieutenant G. Price in mind (which became the basis for the engraving published months later in *The Illustrated London News*)[113], Smith writes, "The great mass of building occupying the high land on the right of the canal, and even stretching down nearly to the level of the valley of the Solani, is the Roorkee Workshop, or, as it might more properly from its magnitude and the nature of its contents be called, the factory. The high chimneys rising above the enclosure, and pouring out volumes of dense smoke, indicate the nature of the powers by which movement is given to the machinery within." He describes a steam engine constantly running, driving "saw-mills, and planing machines, and punching-machines, and lathes, and such other pieces of mechanism as are required for the varied demands of the works." In the workshop "the din of many hammers is heard," emanating from the black-smiths while the carpenters work "with equal though less noisy vigour." The building crowned with a clock-tower is the "Model-room, where some beautiful pieces of intricate and delicate work may be examined, in steam pile-engines, iron roofs, bridges, &c." Further down from the Model-room one will find "an iron-foundry, a mathematical-instrument shop," as well as store and show rooms "rapidly advancing towards completion," all of which, when combined together would "supply many most important wants both for the public service and the general community." Sextants, compasses, and other tools needed to survey and execute public works were forged at Roorkee. It is also meant to "act as a school of industry for the native workmen, in whom we have an educational material to work with, at once plastic, impressionable, industrious, and skilful [sic] in manual labour." Smith proudly notes how and why old prepossessions have "melted away beneath the influence of European science, wisely and considerately displayed" by British engineering. "Such is the machinery at work for the improvement of the working artisan," concluded Smith.[114] However, the pedagogical and material improvement of "natives" extended further.

"On the opposite side of the station, and occupying the crest of a gentle eminence, there rises another mass of buildings designed in excellent architectural taste, and imposing in its extent and general effect," writes Smith. This new building, still under construction in 1854, was the College for Civil Engineers at Roorkee.

Proby Cautley, "Plan of the Station of Roorkee." **Report of the Ganges Canal Works: from their commencement until the opening of the canal in 1854**, Atlas. London: Smith, Elder and Co., 1860. (Princeton University). With the imposition of Prices Bazaar, a cruciform market, Roorkee became, "a flourishing town with metalled roadways, meeting at right angles, and lined with excellent shops." On the opposite side of the Ganges Canal, the "machinery" of the headquarters functioned in a pastoral "nursery."

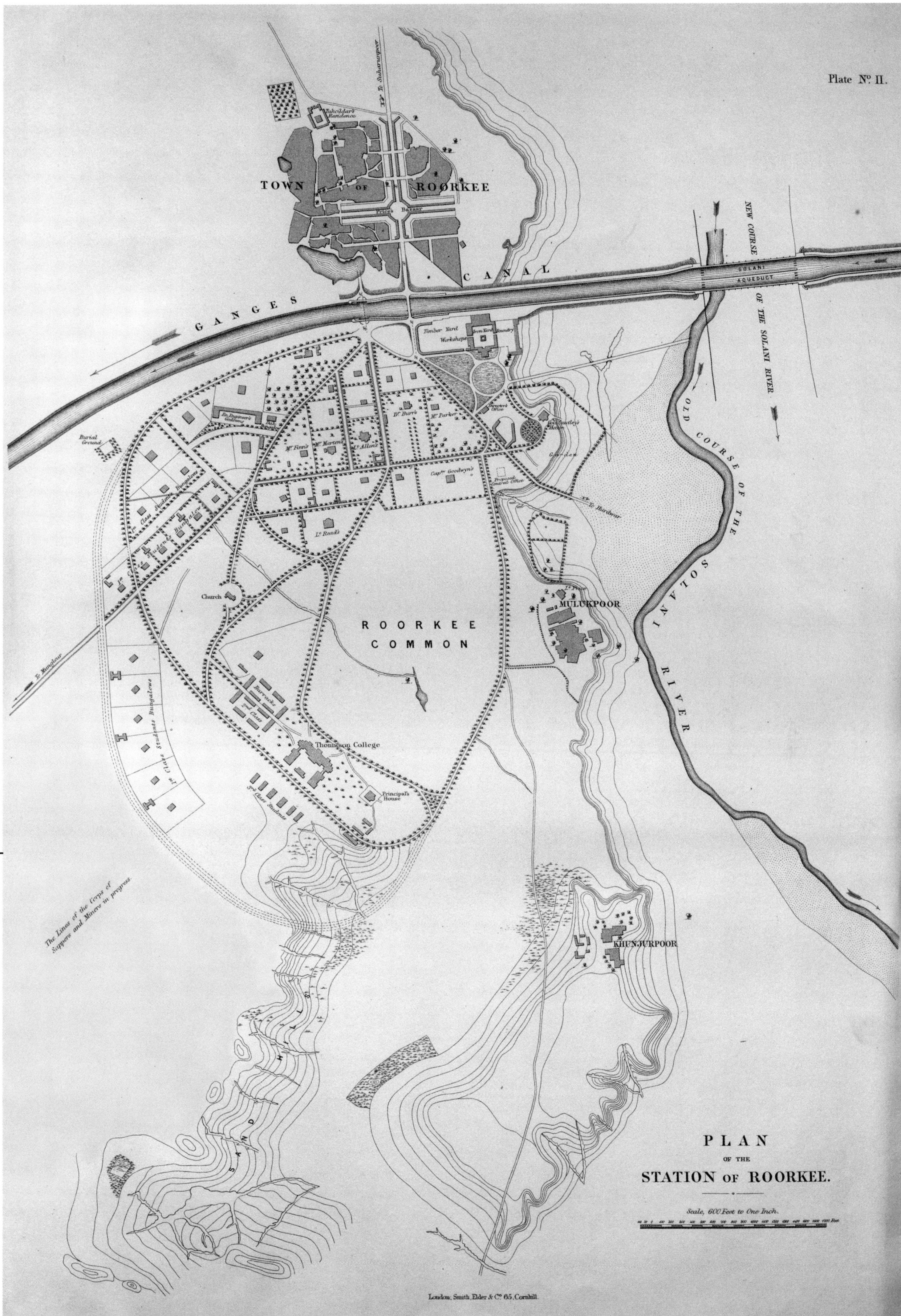

Plate No. II.
TOWN OF ROORKEE
Prison Bazaar
GANGES CANAL
SOLANI AQUEDUCT
NEW COURSE OF THE SOLANI RIVER
OLD COURSE OF THE SOLANI RIVER
Timber Yard
Workshops
Iron Yard Foundry
Burial Ground
Church
ROORKEE COMMON
MULUKPOOR
Barracks
Thomason College
Principal's House
To Hurdwar
To Munglour
To Saharunpoor
KHUNJURPOOR
SAND HILLS
The Lines of the Corps of Sappers and Miners in progress.
PLAN OF THE STATION OF ROORKEE.
Scale, 600 Feet to One Inch.
London; Smith, Elder & Co. 65, Cornhill.

"The Ganges Canal, at Roorkee." **The Illustrated London News**. August 4, 1854. (AA).
Crowned with a clock tower, the Model Room produces "beautiful pieces of intricate and delicate work may be examined, in steam pile-engines, iron roofs, bridges, &c." Volumes of dense smoke emanate from the Roorkee Workshop or "Factory," where saw-mills and lathes are driven by steam power—supplying the building materials for the canal works.

HEADQUARTERS

"Workshop—Roorkee College." Moorli Dhur & Sons, Ambala. Circa early-20th century. (AA). Years before the introduction of railways in India, the "factory" was powered by steam and operated by "native workers."

Proby Cautley, "Thomason College, For Civil Engineers. Roorkee." **Report of the Ganges Canal Works: from their commencement until the opening of the canal in 1854**, Vol. III. London: Smith, Elder and Co., 1860. (Princeton University).

On November 25, 1847, Lieutenant-Governor Thomason of the North-West Provinces authorized the establishment of the new college for instruction in civil engineering, which was to be the first engineering college in India. Aware of the need to train a class of British and Indian officers to design, maintain, and administer public works, the Company deemed it remunerative to establish a training school. The Prospectus issued by the government listed the objectives of the new school. First, "The College is designed to give Theoretical and Practical instruction in Civil Engineering to Europeans and Natives, with a view to their employment on the Public Works of the Country." Students were trained in three disciplines. The first, Sub-Assistant Civil Engineer, trained eight students with "knowledge of Geometry, Algebra, Mensuration, Plane and Spherical Trigonometry, Conic Sections and Mechanics." The second, European Non-Commissioned Officers and Soldiers, received training to serve as "Overseers in the Department of Public Works." The third, "Native youths desirous of instruction in surveying, Levelling, and Plan Drawing."[115] Once completed, the complex would include "a library, a museum, an observatory, a printing press department, copper-engraving, wood-cutting and lithography." The museum was no ordinary museum; it was the Museum of Economic Geology, where students "will find specimens of building material from all parts of the country."[116] And the Observatory "will be of great use in familiarizing the Students with the use of Astronomical and Meteorological Instruments, and in teaching them how to conduct the ordinary observations in Astronomy and Meteorology."[117] By economizing the geology of the country and training students to accurately survey, Smith predicted that "the educational machinery of the College will in course of time become as complete as that of like establishments in Europe."[118]

"Roorkee: The Workshops." **The Illustrated London News**, October 10, 1857. (AA).

Smith chose to compare Roorkee College to "establishments in Europe" as opposed to Great Britain, with good reason: there was no equal to it in Great Britain. While technical education in France dates back to the establishment of the corps of the Ponts et Chaussées in 1716 and the founding of the Ecole des Ponts et Chaussées in 1756, engineering education in Britain was conducted by means of apprenticeship until the end of the nineteenth century; in sum, formal engineering education in the nineteenth century was far more advanced in India than in Britain.[119]

William Willcocks, a graduate of Roorkee in the 1870s and designer of the Aswan Dam in Egypt at the turn of the twentieth century, noted in his autobiography that, "We were taught on the sound lines of the 'Ecole Polytechnique' of Paris and not on the ridiculous lines generally in vogue in England at that time."[120] Students studied a broad range of topics, as outlined above, and their examinations reflected an equally eclectic body of knowledge. It included questions ranging from "Mention the distinctive features of Grecian and of Gothic architecture," to "Explain the method of tracing contour lines."[121] On August 1, 1852, the Civil Engineering Press at Roorkee was established, having already begun treatises in English and Urdu "on some of the most important processes of

THE THOMASON CIVIL ENGINEERING COLLEGE AT AGRA.

"The Thomason Civil Engineering College at Agra [Roorkee]." **The Illustrated London News.** October 10, 1857. (AA).

Engineering or Architecture" in India.[122] By 1853, the Press published manuals to standardize the construction of buildings, irrigation, and earthwork.[123] In the two decades after the college was established in Roorkee, the government opened schools of engineering in Calcutta, Madras, and Poona. The Roorkee Treatises became standard manuals at these institutions and beyond, helping professionalize engineering in British India.

As historian Arun Kumar has noted, unlike engineering education in France and Germany, British Indian pedagogy was not meant to accelerate industrialization. On the contrary, it was to serve the specific needs of the colonial government, which meant the school operated as preparation for work in the Public Works Department—a branch of government primarily concerned with civil engineering as opposed to electrical, mechanical, and mining engineering.[124]

Already by 1854, less than one decade after it was founded, Roorkee College had "become so important a nursery for civil engineering," that it enjoyed a high reputation, remarked Joachim Hayward Stocqueler in *The Hand-Book of British India: A Guide to the Stranger*.[125] And yet, "the demand for engi-

EDUCATION IN THE NORTH-WESTERN PROVINCES OF INDIA.

THE Civil Engineers' College at Roorkee was instituted in 1847, by the Hon. James Thomason, Lieutenant-Governor of the North-West Provinces of India, for the purpose of educating young men, natives as well as

ANNUAL GOLD MEDAL PRIZE OF THE COLLEGE OF CIVIL ENGINEERS, ROORKEE.

Europeans, in different branches of civil engineering, surveying, architecture, &c. The original scheme embraced three classes of students: the first for young men who, from sub-assistant civil engineers, might rise to the highest grade of their profession, and whose education extended to the higher branches of mathematics; the second class for soldiers—whether sergeants, corporals, or privates,—who might ultimately become overseers and assistant-overseers in the Department of Public Works; the third class for youths who were to learn surveying, levelling, plan-drawing, &c., and to receive the education necessary to fit them for subordinate employment on district duties, where simple surveys of boundaries, roads, &c., or the simple construction of works, would be required. The above is an outline of the idea on which the College was founded; but the plan has been extended and improved since 1847.

"Annual Gold Medal Prize of the College of Civil Engineers, Roorkee." **The Illustrated London News**. February 17, 1855. (AA). The giant male figure clutches Euclid and Newton while reading and looking down on the Solani Aqueduct.

“Agricultural Exhibition at Roorkee, North-West Provinces of India.” **The Illustrated London News**. September 24, 1864. (AA).

neers, surveyors, and overseers, is greatly in excess of the means of supply.”[126] With the Public Works Department the largest employer of technically skilled men in the nineteenth century, if engineers were to lead the charge of “peaceful conquest,” then training a corps of engineers was paramount to this enterprise.

However, within three years, this new corps, their college, and their factory were under siege.

The history of the engineering school over the course of the nineteenth century can be witnessed in a series of engravings. Following the article from 1854, Roorkee reappeared in *The Illustrated London News* in 1857. By that time, it not only had a new name, The Thomason Civil Engineering College—after the late Lieutenant-Governor Thomason of the North West Provinces—but it had also “assumed a very military appearance” and become a garrison during the Indian Rebellion of 1857.

This time, Roorkee merited two engravings. The top engraving, misnamed “The Thomason College of Engineering at Agra,” showcased the “excellent architectural taste” of the new college. Made of brick and described as a “very imposing object when seen from the Solani Aqueduct,” the building boasted a roof covered in corrugated iron, “giving to the whole a very English and railway-station-like appearance.”[127] The second engraving, “Roorkee: The Workshops,” foregrounds where some “nearly 800 hands” worked before the outbreak. It is meant to present “a striking contrast to its modern and more imposing rival.”[128]

If indeed the nursery took on the appearance of a garrison, then it was short lived. Less than ten years later, the next time Roorkee featured in *The Illustrated London News*, the headline read “Agricultural Exhibition at Roorkee.” Captions did not illustrate the Ganges Canal or the Thomason School of Engineering. Instead, ‘Prize Field-Bullocks,’ ‘Wild Ass,’ and ‘General View of the Roorkee Exhibition’ appeared as captions beneath the three engravings. Perhaps it was with good reason that William Wilson Hunter would write of the Thomason Civil Engineering College in 1877: “This institution is the largest and most important engineering college in India; and during its forty years of teaching may be said to have created an indigenous engineering profession.”[129] The Ganges Canal was a central part of the development of this “indigenous engineering profession” in the service of British Public Works. By the end of the nineteenth century, engineering had fulfilled its duties—no longer foregrounded in the image, it was so omnipresent that it scarcely deserved notice.

AGRICULTURAL EXHIBITION AT ROORKEE, NORTH-WEST PROVINCES OF INDIA.

PRIZE FIELD-BULLOCKS.

WILD ASS.

AGRICULTURAL EXHIBITION AT ROORKEE.

IN the Number of this Journal for July 2 we gave some Illustrations of the cattle show held about the beginning of this year in the Belvedere Grounds, Calcutta, under the auspices of the Lieutenant-Governor of Bengal. The photographs of the prize animals, from which our Engravings were made on that occasion, were taken, at the Governor's request, by an amateur photographer, Mr. W. L. Noverre, since appointed Assistant Commissioner of Forests in Central India. We have now received a set of photographs, taken by Mr. James Freeman, of the Agricultural Exhibition which took place at Roorkee, in the North-western provinces, at the beginning of April. Our Engravings comprise a view of the entrance to the showyard from the high road outside, with representations of a pair of prize bullocks and of a wild ass, which attracted much notice in the exhibition. There were some beautiful horses from every district of Upper India, various species of oxen and cows, a two-humped camel, some elks, and even a tiger cat, which could hardly be reckoned among the domestic animals. The show of sheep and poultry was not very good. In the department of machinery there were chaff-cutting machines, seed-crushers, and cotton-gins, a large hydraulic cotton-press, and several machines for raising water, fixed beside the canal adjoining the showyard. As Roorkee is situated on the road to the great fair of Hurdwar, frequented by many thousands of the Hindoos at that season, the number of visitors to the exhibition was very great.

THE CARNARVON AND LLANBERIS RAILWAY.

THE picturesque vale of Llanberis, with the neighbouring hamlet of Dolbadarn, about nine miles from Carnarvon, is well known as the head-quarters of tourists in the Snowdon mountain district. On Thursday week a commencement was made of the proposed Carnarvon and Llanberis Railway, which will not only afford great accommodation to that class of travellers, but will be of the greatest importance in the carriage of the mineral traffic of the district, consisting of slates, ironstone, sulphur, and copper. The carriage of these products to Carnarvon, for shipment, has hitherto cost from 4s. to 9s. per ton, but when the railway is completed the cost will be limited to from 1s. to 1s. 6d. per ton. The district, although a mineral one, is not coal-producing, and all the coals consumed are obtained from Carnarvon, and carted there at a cost of little less than 7s. 6d. per ton. The railway when completed will convey coals to the principal works in the district at a cost of about 1s. per ton; and general merchandise, which costs now about 7s. 6d. per ton, will, on the completion of the line, be carried to and from the district at a cost of about 1s. 6d. per ton. The gradients of the line will be easy, and it is expected it will be opened for traffic in the July of 1866. Along a great portion of the line of its route, some dozen or fifteen years ago, there were very few houses, but within the last twelve months 106 houses have been built in the district of Llanberis alone—a strong proof of the necessity for additional accommodation in the neighbourhood. The line was projected by Mr. Rolfe, a local engineer, and successfully carried through Parliament in the last Session. The contract has been let to Messrs. Dalrymple and Findlay, and the works are now commenced. Since much interest was taken throughout Carnarvonshire in the ceremony of "cutting the first sod," three or four thousand people assembled under the shadow of Snowdon to witness the proceedings. The ground was kept clear by the 1st Carnarvon (Artillery) and 3rd Carnarvon Volunteer Corps, whose bands of music were in attendance. The children of the village schools were marched in procession to the ground. The Hon. F. Wynn, with Miss Emily Wynn and other members of the family of Lord Newborough, accompanied by Mr. Thomas Turner, the chairman, and other directors of the company, and many of the local gentry, were present. After a prayer from the Rev. Mr. Williams, the Vicar of Llanberis, and a brief explanatory speech from Mr. Turner, the lady we have named, the Hon. Emily Wynn, performed the office of first "navvy" with admirable tact and elegance. An enthusiastic cheer greeted the discharge of the first sod from the elegant barrow she wheeled to the end of the platform, and this was followed by a salvo from some 18-pounder guns which the Carnarvon Artillery had brought upon the ground. Their heavy booming shook the earth, and was responded to by the echoes along the whole mountain range till the sounds died away in the distance. The effect was inconceivably grand, the crowd cheered, and so the proceedings were brought to a close. Shortly after, a select company assembled at the Victoria Hotel, Llanberis, on the invitation of the chairman and directors, to partake of luncheon in celebration of this event.

GENERAL VIEW OF THE ROORKEE EXHIBITION.

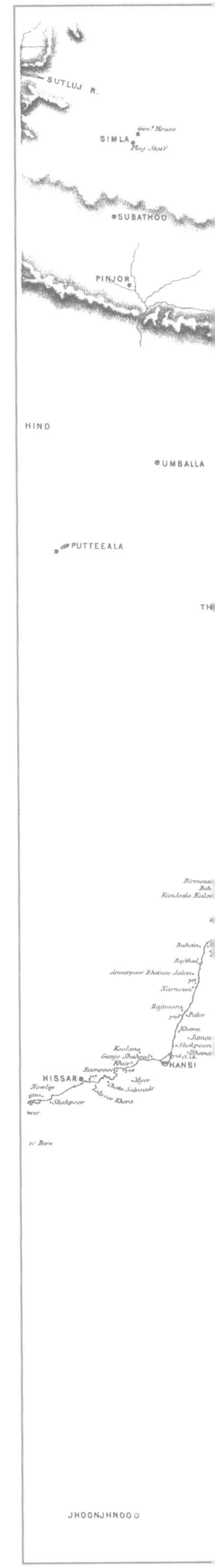
SUTLUJ R.
SIMLA
SUBATHOO
PINJOR
HIND
UMBALLA
PUTTEEALA
HISSAR
HANSI
JHOONJHNOO

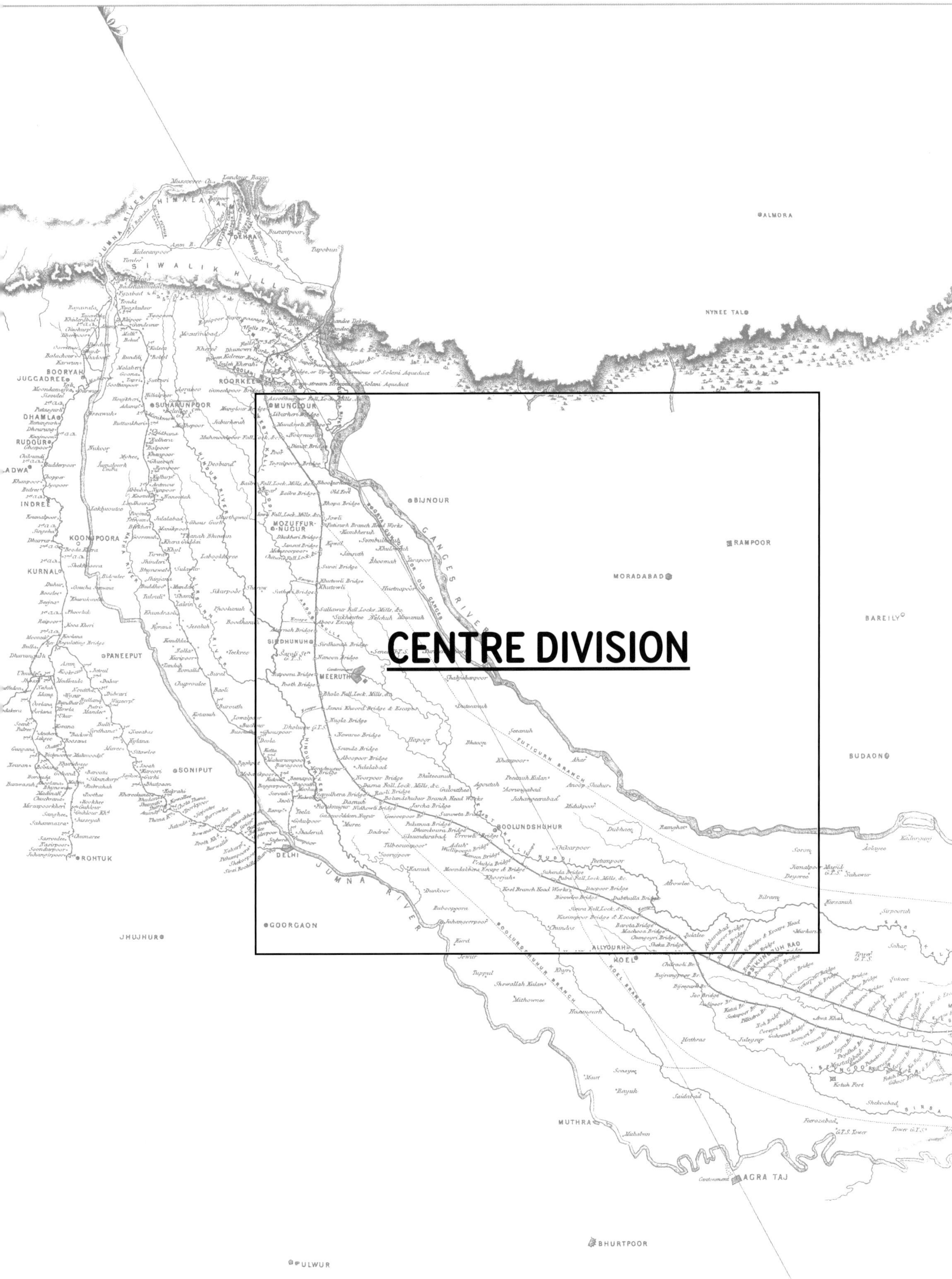
CENTRE DIVISION
ALMORA
NYNEE TAL
BIJNOUR
RAMPOOR
MORADABAD
BAREILY
BUDAON
MUNGLOUR
ROORKEE
SUHARUNPOOR
MOZUFFURNUGUR
SIRDHUNUH
MEERUTH
BOOLUNDSHUHUR
ALLYGURH
KOEL
DELHI
GOORGAON
JHUJHUR
ROHTUK
SONIPUT
PANEEPUT
KURNAL
KOONJPOORA
INDREE
ADWA
RUDOUR
DHAMLA
JUGGADREE
BOORYAH
MUTHRA
AGRA TAJ
BHURTPOOR
ULWUR
SIWALIK HILLS
GANGES RIVER
JUMNA RIVER

If the "upper, or northern, portion of this canal was excavated on an excessively steep slope in a light and sandy soil," wrote Cautley, "the lower, or southerly, portion on irregular and moderate slopes, and through good, firm soil."[130] After its debouche at Roorkee, the Ganges Canal travels another 160 miles (257.5 km) to the village of Nanoon in Aligarh District.[131] Here, there are no major monuments of hydraulic architecture. Besides two major bifurcations at the fiftieth and one-hundred-and-tenth miles, along with four shallow masonry descents to compensate for changes in topography, the canal maintained a monotonous slope of 18 inches per mile (45.7 cm/1.6 km). And yet, the Centre Division was Cautley's greatest intervention up to this point.

In order to achieve its utilitarian goal of tempering nature through the steady provision of water year round, the canal must form a new landscape. Its artificiality must be so integrated that it appears natural. In reality the Ganges Canal thinly, but systematically, re-layers the landscape. A roadway running parallel to the canal was designed "both for convenience of communication, and for the purposes of patrol."[132] Canal officers rode horseback, patrolling and inspecting the canal. Bridges accompanied the new roadway, situated every three miles (5 km). Every bridge became a crucial node along the canal, both subdividing the lands in between and also forming regular links to the other side. This system of bridges reconfigured the landscape with new roads and connections—new conduits for commerce and trade.

Cautley also proposed seeding the main trunk of the canal with mango trees every 100 feet (30 m) to provide an "ornamental quality;" though, in the days before insecticides, "there is some difficulty in the establishment of this species of tree, as it is a martyr to the whole tribe of the white ants, who attack it with remorseless severity..."[133] Nonetheless, he imagined "plantations of 'grafted mangoes'" marching from Roorkee to Nanoon village, believing that in 20 or 30 years those traveling along the Ganges Canal works will appreciate this "single line of growth."

Mindful of how difficult it would be for gardeners to maintain such a lengthy linear plantation, Cautley proposed the trees only in and around the first class chokies—the main station points of every district—which were planned every twelve miles (19 km). Second class chokies stood watch at every intermediate bridge along the canal. With the design of each class of choki to be uniform in its plan and position to the canal, Cautley created a network of ornamental monitoring stations along his new linear hydro-plantation.

Like most of the Ganges Canal works, these chokies were to be made of kiln-fired bricks, the production of which proved to be a major obstacle facing the completion of the canal. Brick production increased the cost of the canal significantly as it was used in all of the bridges as well as the hydraulic architecture. With no stone quarries to draw from for the construction of the super-passages and embankments, brick was the only option. Fortunately for Cautley, he had already set out to make millions of bricks for the work along the main trunk of the Ganges Canal, following the committee's recommendation from 1842.[134]

Cautley's *Ganges Canal Report* (1860), explained in detail what difficulties he faced with the production of bricks: "To those who are unacquainted with the part of India in which the above works are situated, and with the system which the nature of the people and of the country itself renders necessary in the Department of Public Works, it may be advisable to point out here that there are no contractors to depend upon for the supply of material. The engineer, as the architect, has to make his own bricks, and to burn his own lime; he has to purchase his supplies of timber, and, in many cases, to proceed to the forests himself to superintend its collection."[135] The engineer must become a master-builder. There was no "respectable and wealthy class of contractors which in Europe relieves the engineer and the architect from the detail of building," pined Cautley. To do most of the excavation and firing of the kilns to produce bricks, Cautley employed "tribes of a wandering class whose chief, if not entire occupation, is digging tanks and watercourses, and to these people we are indebted for the greater part of all our canal excavations."[136]

Proby Cautley, "Plan and Section of Canal, with reference to roadway, plantations, and ultimate annual clearance." **Report of the Ganges Canal Works: from their commencement until the opening of the canal in 1854**, Vol. I. London: Smith, Elder and Co., 1860. (Princeton University).

Proby Cautley, "Diagram 38." **Report of the Ganges Canal Works: from their commencement until the opening of the canal in 1854**, Vol. I. London: Smith, Elder and Co., 1860. (Princeton University).

Both diagrams depict how Cautley thought not just about how the canal and environs were spatially organized, but also how it was to look and grow over time (as well as maintained). Roadways, plantations, berms, and shops all figured into transforming the Doab.

Diagram 38.

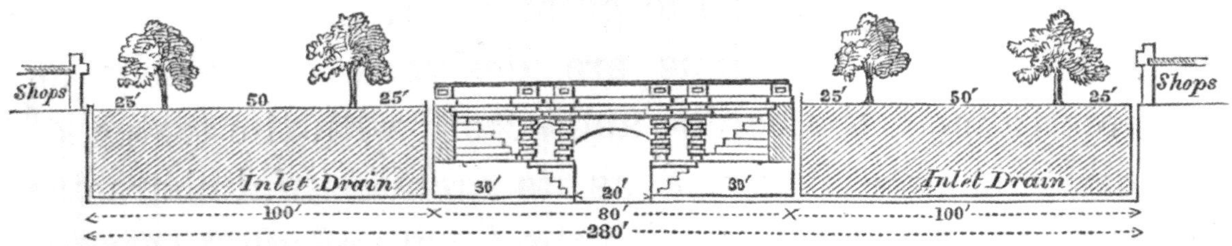

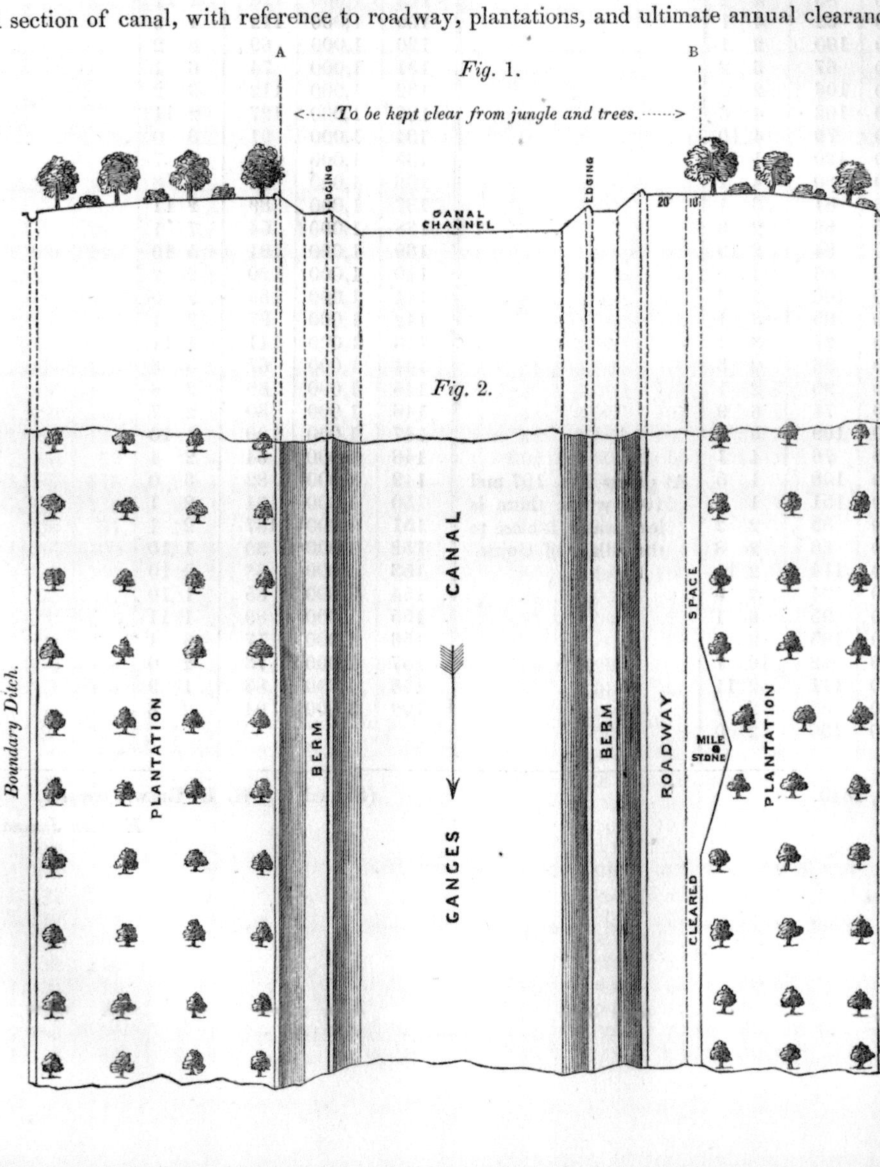

CIRCULAR MEMORANDUM *to all* EXECUTIVES *on the* GANGES CANAL.

No. 82.

Dated 1st *January*, 1852.

PLAN and section of canal, with reference to roadway, plantations, and ultimate annual clearance.

J. Crofton, "Ganges Canal. Remodeling Futtegurh Branch Head." **Report of the Ganges Canal**, Vol. III. Calcutta: Public Works Department, 1866. (Harvard University).

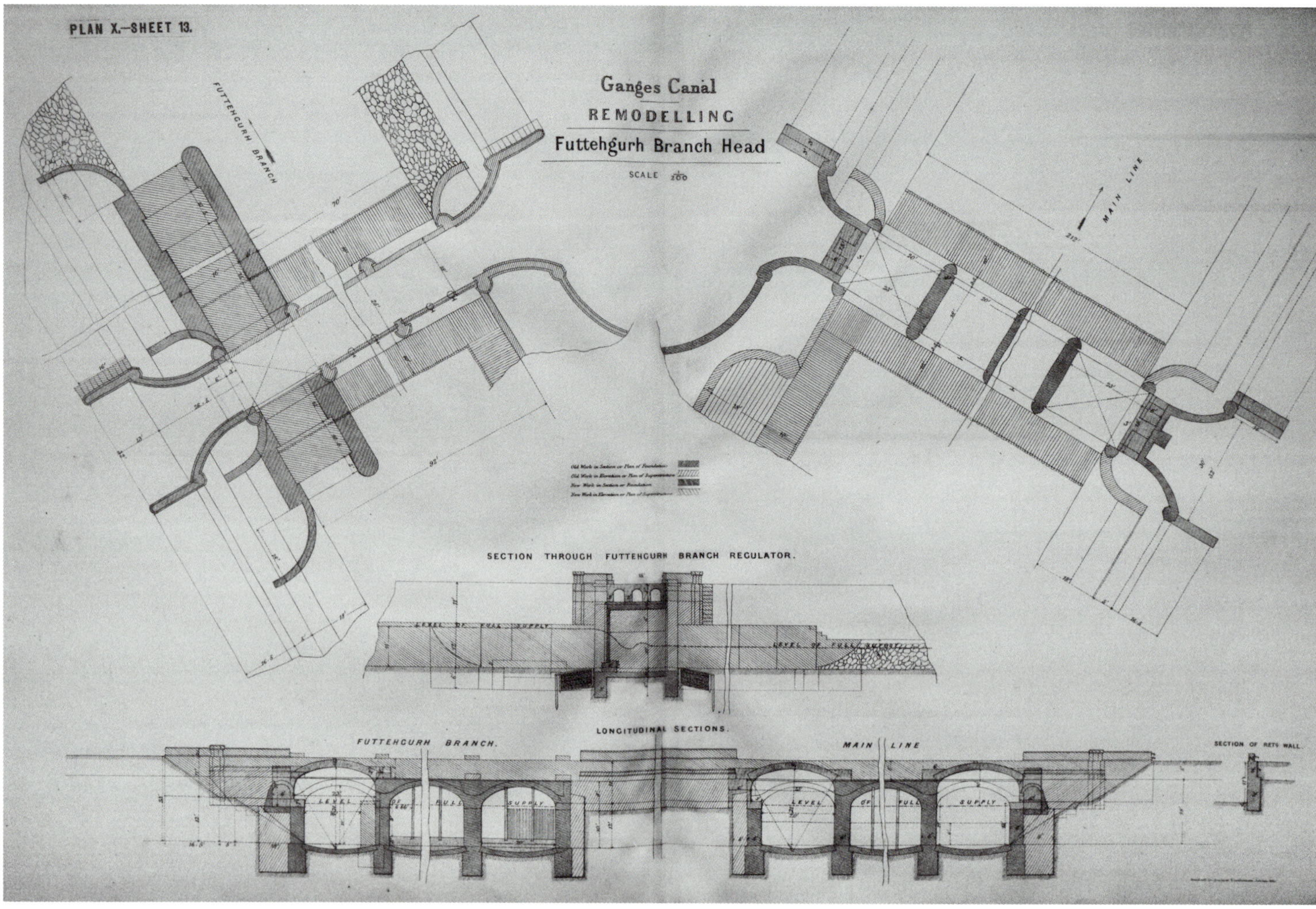

"Assafnagar Fall." Publisher Unknown. Circa early-20th Century. (AA).

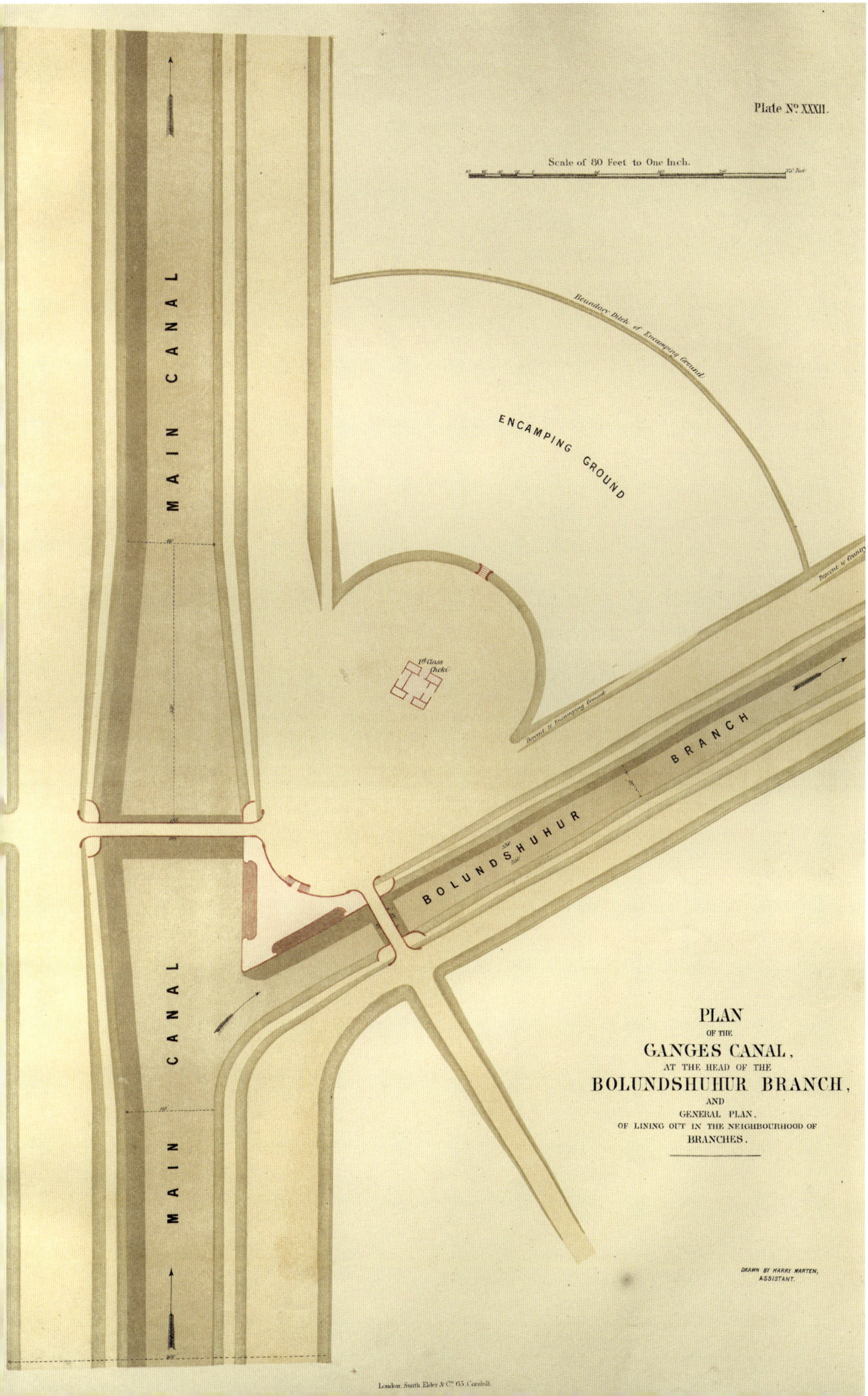

Proby Cautley, "Plan of the Ganges Canal at the Head of the Bolundshuhur Branch, and General Plan of the Lining Out in the Neighbourhood of Branches." **Report of the Ganges Canal Works: from their commencement until the opening of the canal in 1854**, Atlas. London: Smith, Elder and Co., 1860. (Princeton University).

RAJBUHAS

To make good on his system of distribution of water to millions of cultivators, Cautley needed to produce millions of bricks, not just for the trunk and terminal lines or for the Solani Aqueduct, but also for his most important innovation of all: rajbuhas. The *rajbuha*, or outlet to a village, is "the connecting link between the supply and the service channels." Unlike the main supply or trunk channel, they are private property of the irrigators, even though the government executes them, oversees them, and maintains them.[137] A rajbuha typically has "a masonry regulator at the head, from which the cultivators make their own water-courses to their fields."[138] This system of water distribution was developed years before Colonel Colvin fantasized about a canal fed by the Ganges River.

Cautley tested and developed this system of distribution along the Eastern Jumna Canal (EJC) in the 1820s and 30s before grafting it onto the Ganges Canal system. In the 1830s, he commissioned then-Superintendent of the EJC, William Elliot Morton, to codify and enumerate "the great advantages which distribution of water by Rajbuhas possesses over every other method that has been tried."[139] For Cautley, with the exception of his rajbuhas system, "the method of distributing water from the canals in these provinces has... owed as little to science as possible; it commenced in the year 1819, when the first canal was opened, and has continued up to the year 1853, much in the same fashion."[140]

But it was not without trial and error. Initially, the government advanced funds to local zamindars (landlords) to construct their own rajbuhas. This practice was stopped when Cautley standardized the measurements of the system in order to ensure proper slope and drainage. Canal officers built and maintained the engineered watercourses and remitted the cost from zamindars over an agreed-upon calendar. Or, as Cautley put it, "the government, in fact, makes the advances in 'rajbuhas' instead of in 'money.'"[141]

Before building a rajbuha in a village, the local zamindars submitted an agreement to show how the land was to be acquired and agree on the terms and assignments granted. This was done primarily because "Water-courses to distant villages were constituted in utter defiance of the levels assigned by nature to the country over which they were carried." But this legal prudence was also enacted because watercourses to villages separated from the Canal often passed through other communities, places where prior disputes and feuds could lead to the interruption of irrigation.[142] The engineer was thus both the revealer of nature's topography and an occasional conflict mediator. As Morton suggests in his memoranda, rajbuhas were to introduce new habits "for all benefitting alike."[143] Canal officers were vested with authority over the physical construction and maintenance of these channels as well as the social relations between different groups of cultivators. To achieve this goal, cultivators had to buy into the system.

In the tradition of free enterprise, in order to "To simplify accounts and to insure [sic] an uniformity of system and management," Cautley adopted "the plan of joint stock companies." This would allow the Canal officers to become "the Director and Treasurer" and the zamindars to become shareholders.[144] Through the use of this model of ownership, there emerged a braid between local trade, colonial economies, and engineering expertise.

These intertwining relationships were based on technical concerns, for the purpose of a rajbuha was "to effect irrigation at the cheapest cost, by laying out lines for irrigation on true and correct levels, and by reducing their slopes of bed to a minimum sufficient to keep up a scour, and at the same time to maintain a good head supply for irrigation."[145] It required a delicate balance, or what Cautley would more precisely call a "science," to provide a flow favorable to irrigation and detrimental to disease.

Each rajbuha was equipped with technological accessories to manipulate the flow of water and transform the surroundings, establishing new spatial boundaries and economic partnerships. On November 29, 1852—exactly one month before Morton penned his *Memoranda On Rajbuhas*—Cautley issued an order to two of his executive officers on the technical specifications of these technological accessories in the lower half of the Ganges Canal. A series of six diagrams, embed-

Proby Cautley, "Dubowli Regulating Bridge and Rajbuha Head." **Report of the Ganges Canal Works: from their commencement until the opening of the canal in 1854**, Atlas. London: Smith, Elder and Co, 1860. (Princeton University).

DUBOWLI REGULATING BRIDGE
AND
RAJBUHA HEAD.

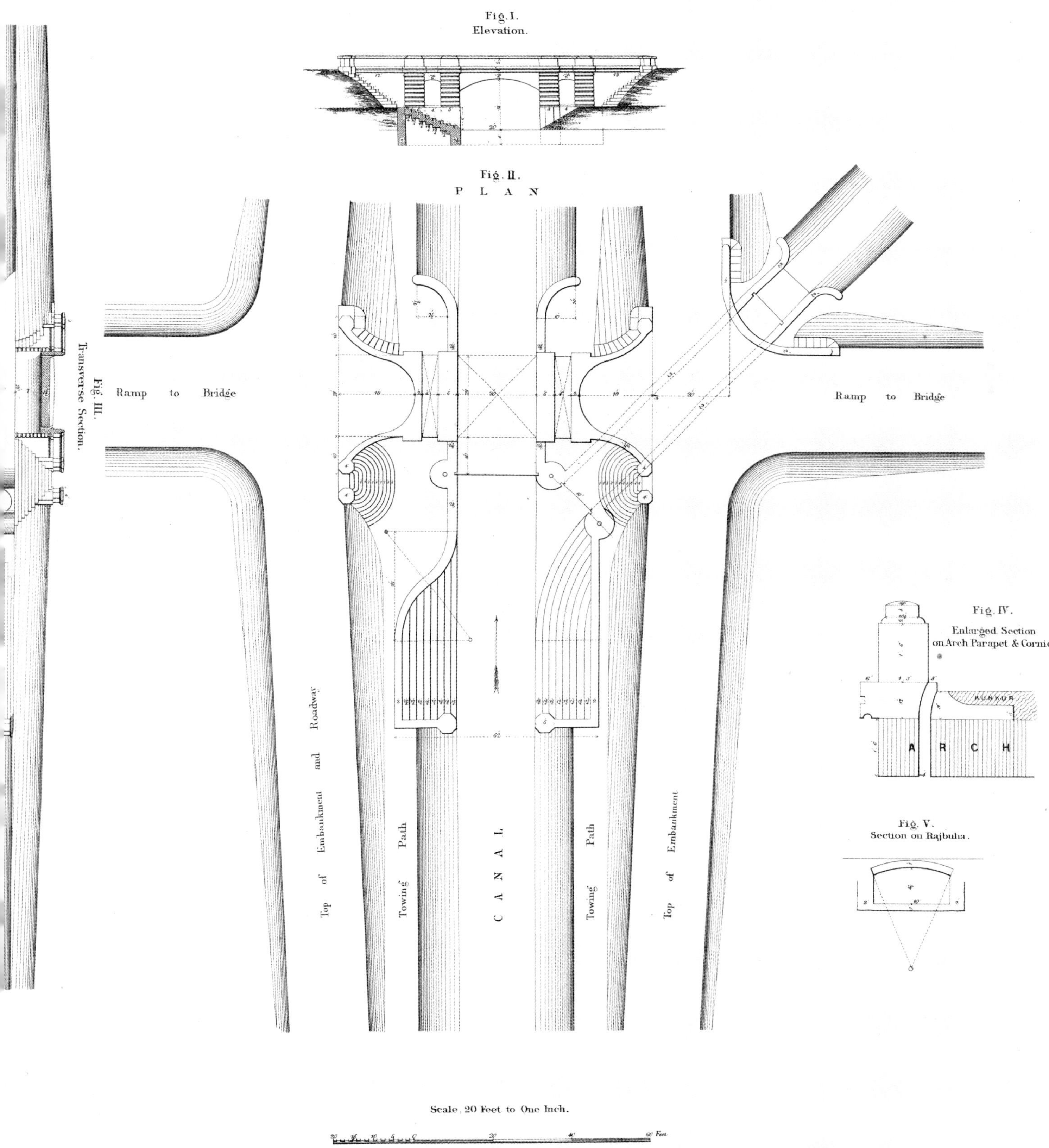

London; Smith, Elder & C° 65, Cornhill.

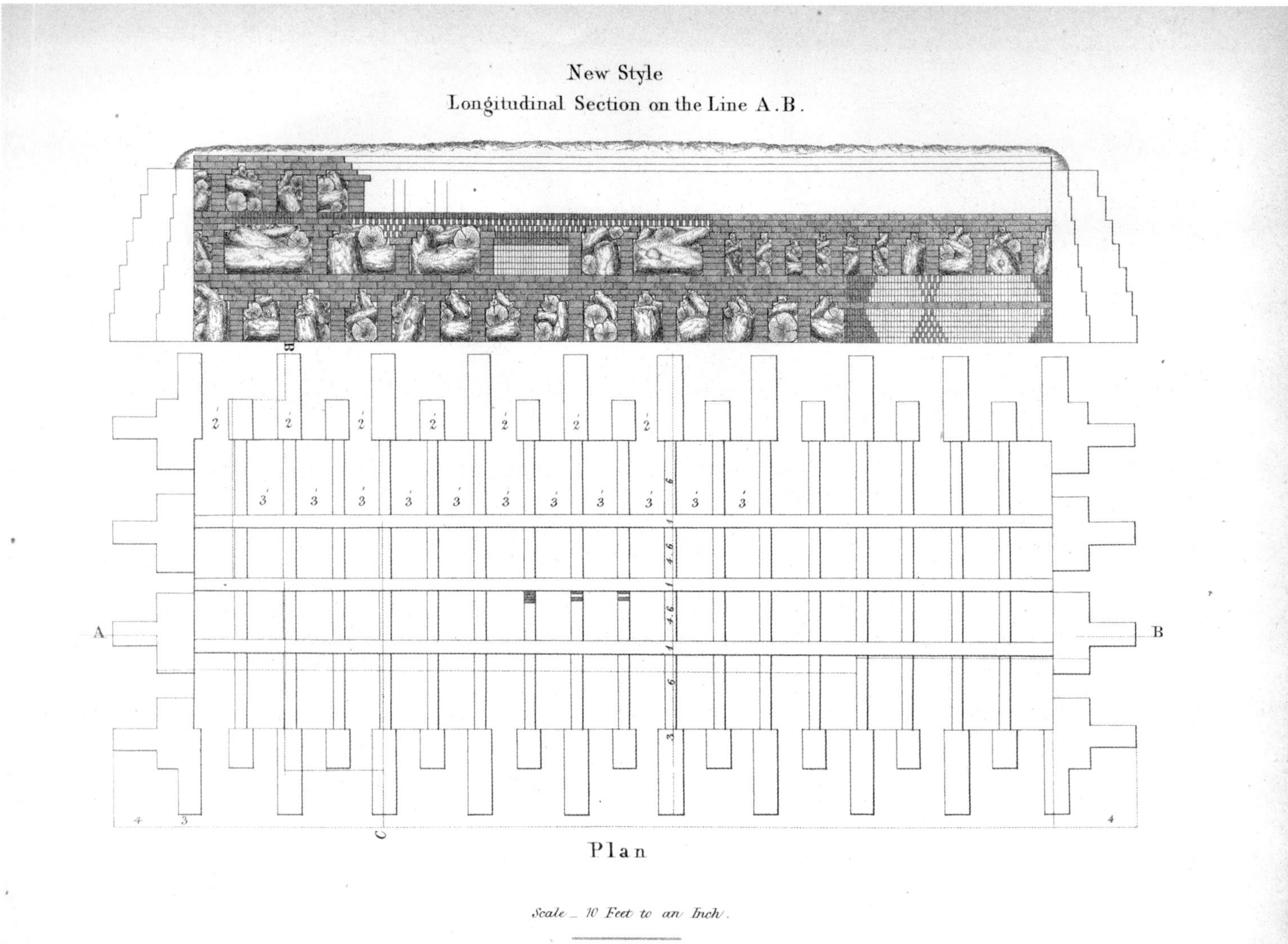

To do most of the excavation and firing of the kilns to produce millions of bricks, Cautley employed "tribes of a wandering class whose chief, if not entire occupation, is digging tanks and watercourses, and to these people we are indebted for the greater part of all our canal excavations."

Proby Cautley, "New Style: Longitudinal Section on the Line A. B. [of the Brick Kilns]" **Report of the Ganges Canal Works: from their commencement until the opening of the canal in 1854**, Vol. III. London: Smith, Elder and Co., 1860. (Princeton University).

ded in a list of twenty-two points, comprised his order. It was a concise manual on how to both extend and subdivide the Ganges Canal *ad infinitum*. Just as they were expected to multiply water rents and income for the Company, the rajbuhas multiplied territory as well. By supplying villages with water, they created new sites of revenue and control, folding villages into a larger system of irrigation and governance.

Like the other components of the canal, every rajbuha was an intersection shaped by the layering of infrastructure, topography, and agriculture. Each head and inlet of a rajbuha was "designed as a component part of a bridge," so as to "become accessible from both sides of the canal." In addition, every rajbuha head and inlet was to have a choki post near it, large enough to accommodate a "necessary guard." For Cautley, this combination of bridge and choki "places the whole machine, viz., of irrigation, inlet, and escape, under the immediate supervision and easy control of the canal establishment."[146] In this view, freeing cultivators from the monsoon required greater control and supervision by engineers. The population of the Ganga-Jamuna Doab—whether drawing water or crossing from one locale to the next—at long last, would be aware of inhabiting a vast irrigation machine, with all of its mechanical and social components. Each rajbuha and choki was a part of the "peaceful conquest" of the Doab.

RAJBUHAS

Proby Cautley, "Diagrams 60–62." **Report of the Ganges Canal Works: from their commencement until the opening of the canal in 1854,** Vol. I. London: Smith, Elder and Co., 1860. (Princeton University).

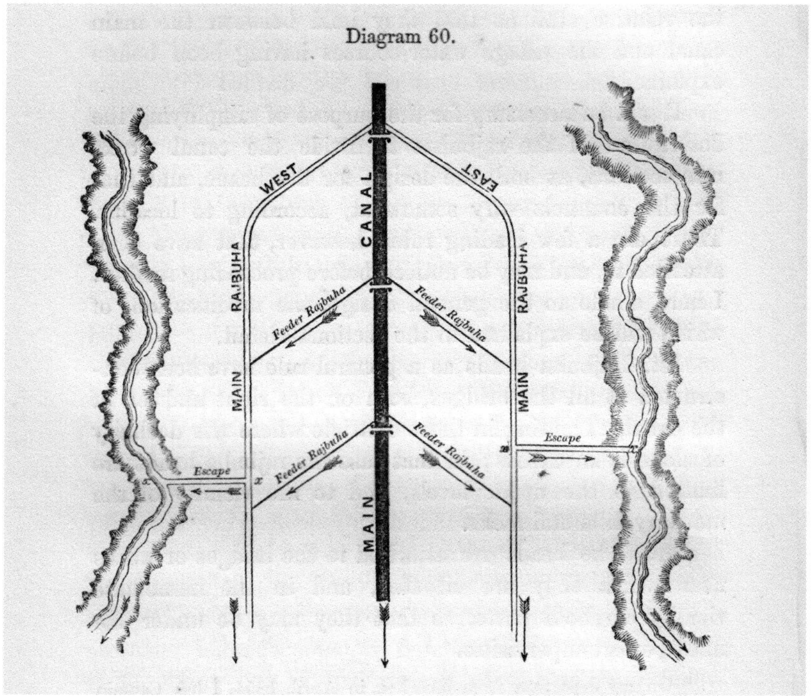

Diagram 60 "explains in skeleton the general design of rajbuhas," where the main trunk of the canal extends two branches or rajbuhas, one branching to the left and another to the right in the shape of a chevron, before then running parallel to the main trunk. A series of feeder rajbuhas, spaced at somewhat regular intervals of four to five miles depending on topography, reconnect the left and right rajbuhas with the main trunk. On the outside edge of the left and right channels, a series of escapes lead to the nearest rivers in the event of an overabundance of water.
Diagram 61 demonstrates how topography deforms the chevron geometries of the system.
Diagram 62, a series of three cross-sections through the three types of channels, defines their required dimensions. The remaining four diagrams depicted how best to construct bridges, aqueducts for the rajbuhas to cross low-lying land, and the dimensions and materials needed to construct the escapes.

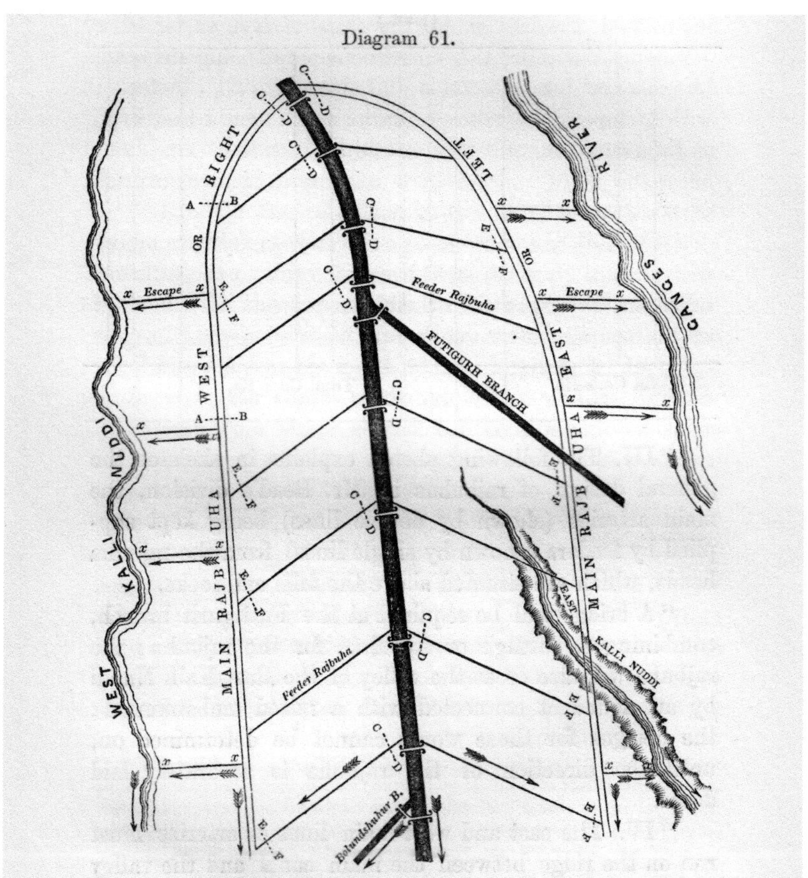

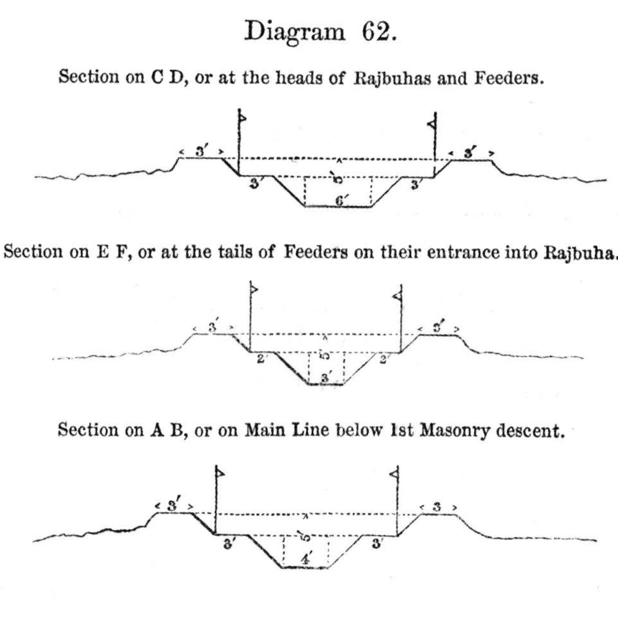

Proby Cautley, "Diagram 66 [Plan and Sections through Rajbuha Distribution System]." **Report of the Ganges Canal Works: from their commencement until the opening of the canal in 1854**, Vol. I. London: Smith, Elder and Co., 1860. (Princeton University).

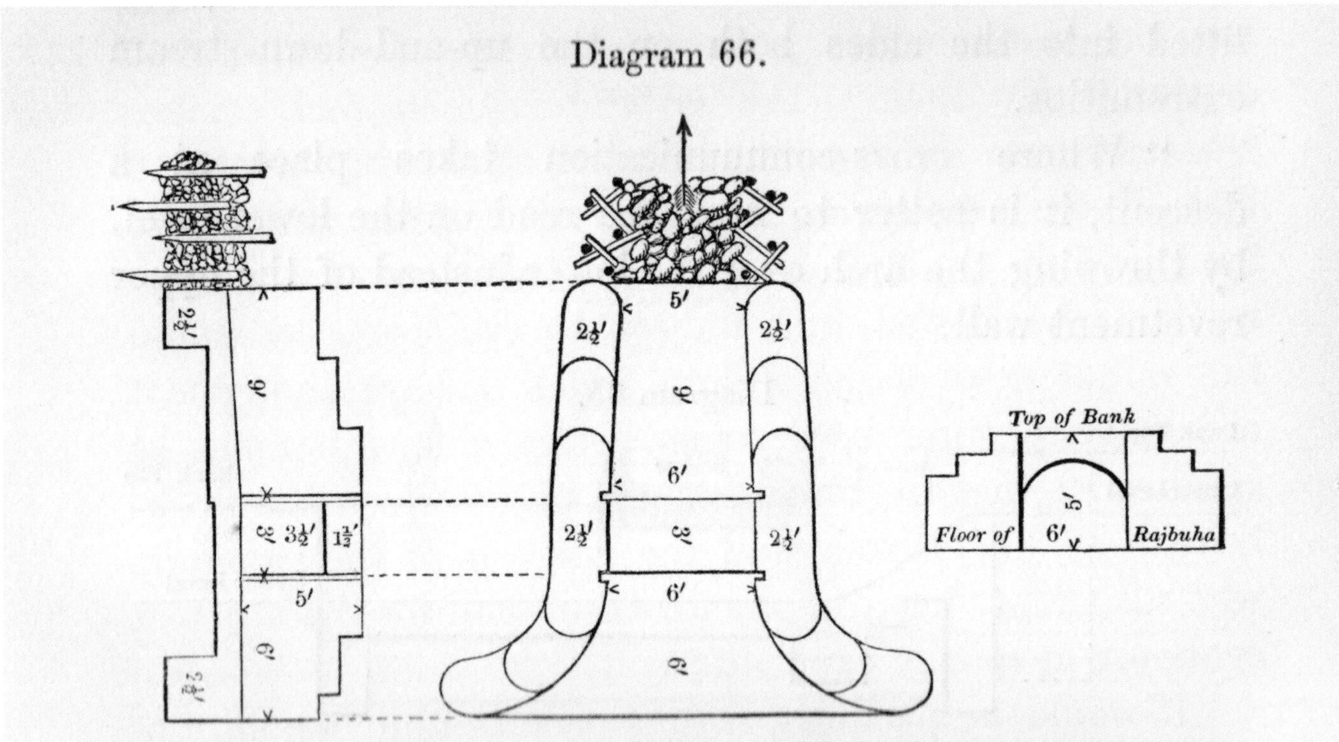

CODIFICATION

By 1871-72, learning to design rajbuhas had become an integral part of engineering education in British India. "What considerations determine the level of the bed, the declivity, the discharge at head, and the depth of Rajbuhas?" asked Captain A. M. Lang in his final examination on Irrigation Works for first year students at the Thomason Engineering College in Roorkee in 1871-72.[147] By 1868, the Ganges Canal was already said to have 3,000 miles (4,828 km) of rajbuha channels—over three times as long as the main trunk.[148] In 1867-68, the government spent a total of Rs. 688,192 on maintenance along the Ganges Canal. It spent Rs. 165,589, almost a quarter of the total budget, on rajbuha clearance alone (55.19 rupees per mile). As the number of rajbuhas multiplied and engineering education was institutionalized in India, engineers were expected to have a technical understanding of how to efficiently design and manage these waterways. Growing the number and length of rajbuhas increased the amount of area to manage. New legislation reinforced the pedagogical importance of technical knowledge and formally expanded the role of the state in irrigation issues by leveraging the expertise of the engineer.

By 1873, with the Northern India Canal and Drainage Act, the government formalized its control of all existing and new watercourses from canals, including rajbuhas.[149] The new law not only replaced simpler legislation on water rights and rents, but also codified the administration of canals in northern India by formally establishing hierarchies of authority within the Irrigation Department. Within this legal structure, the business of the engineer was a purely technical endeavor. Perhaps then, at long last, Cautley's belief in the ability of the engineer, by virtue of "proper adaptation of levels in his artificial channels," was able "to obtain a command of water, which places the whole country under his control..."[150]

If the Solani Aqueduct and the super-passages constitute the hydraulic architecture of the Ganges Canal, then the rajbuhas constitute something akin to its urban-agrarian grid. They frame the boundaries and edges of communities. Like streets and water utilities in most European cities and towns in the nineteenth century, they are regulated and constructed by the state—in this case the colonial state. Like a figure-ground drawing, where solid and void are rendered in black and white, Cautley draws the interior of the rajbuhas as white and the watercourses and channels as solids—they define space. They inscribe a new system, a new space, into the land. The control of water provides the means to govern populations.

Diagram 64.

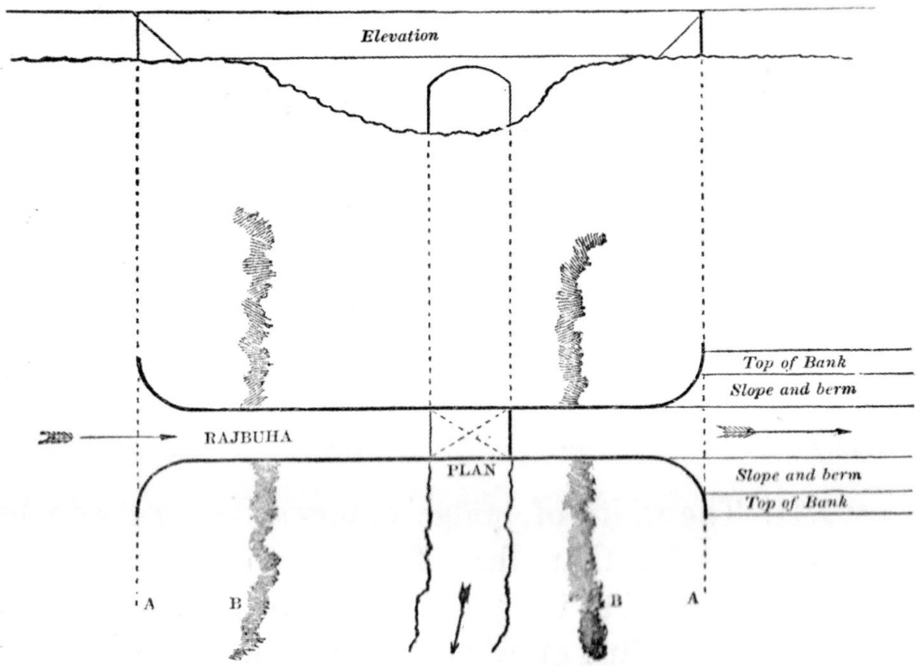

When a rajbuha crosses a drain, brook, or river, miniature aqueducts carry it from one side to the next.

Diagram 70.

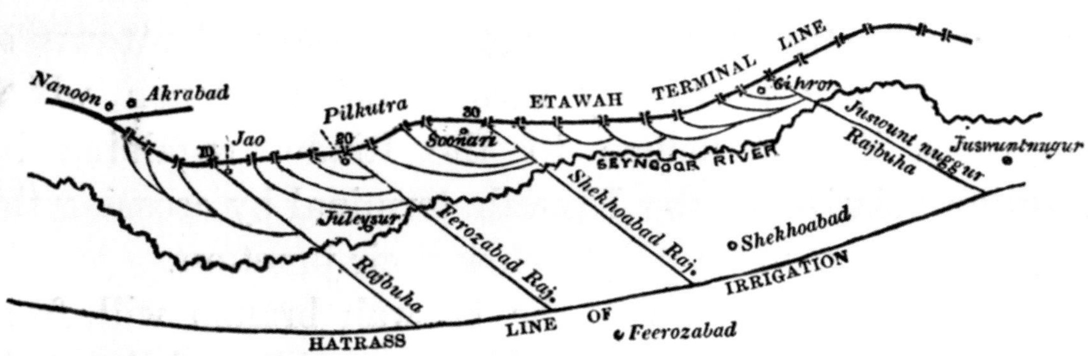

At times, due to topography, the rajbuha system mutates from a series of chevrons into a succession of rings.

Diagram 71.

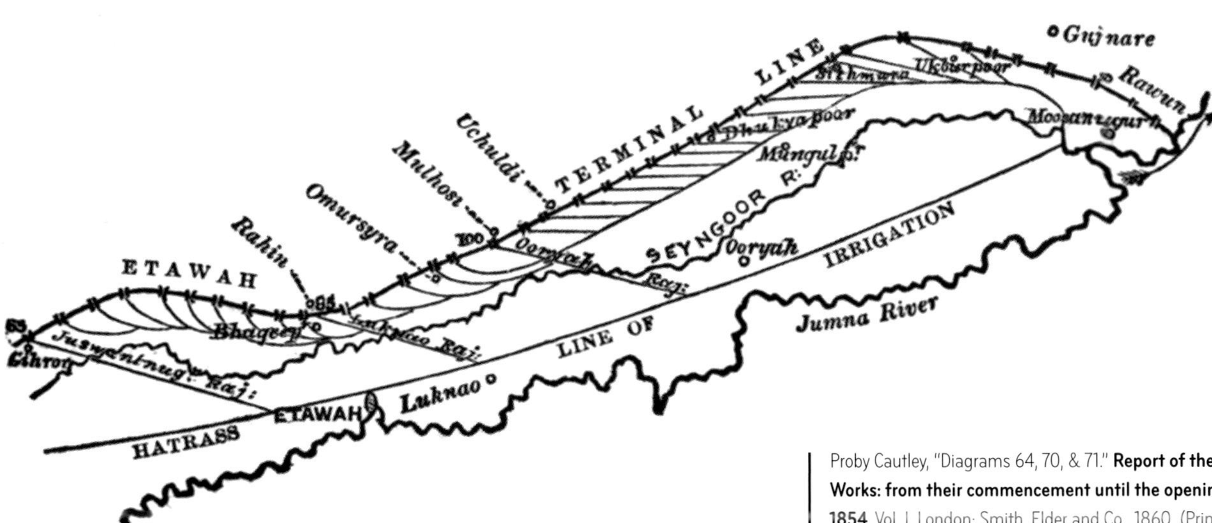

Proby Cautley, "Diagrams 64, 70, & 71." **Report of the Ganges Canal Works: from their commencement until the opening of the canal in 1854**, Vol. I. London: Smith, Elder and Co., 1860. (Princeton University).

SOUTHERN DIVISION

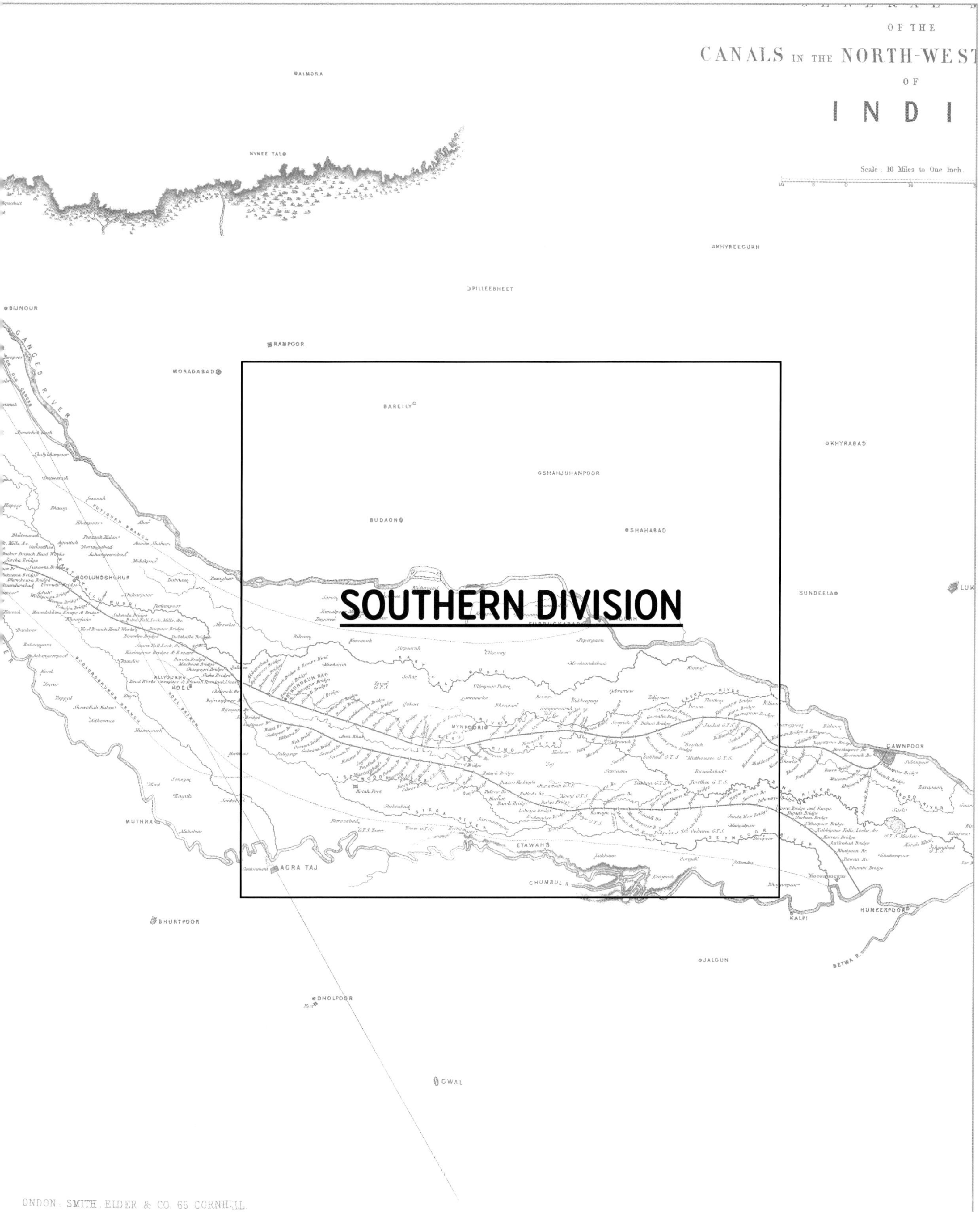
OF THE
CANALS IN THE NORTH-WEST
OF
INDI
Scale : 16 Miles to One Inch.
ALMORA
NYNEE TAL
KHYREEGURH
PILLEEBHEET
BIJNOUR
RAMPOOR
MORADABAD
GANGES RIVER
BAREILY
KHYRABAD
SHAHJUHANPOOR
BUDAON
SHAHABAD
SUNDEELA
BOOLUNDSHUHUR
SOUTHERN DIVISION
ALLYGURH
KOEL
MYNPOORIE
CAWNPOOR
MUTHRA
ETAWAH
AGRA TAJ
CHUMBUL R.
KALPI
HUMEERPOOR
BHURTPOOR
JALOUN
BETWA R.
DHOLPOOR
GWAL
ONDON: SMITH, ELDER & CO. 65 CORNHILL

To Julia A. Stone, the Ganges Canal at Cawnpoore in the 1870s "looked like a new world institution; indeed it appeared like something that I had seen before, with its massive stone locks. I had supposed it was a channel for irrigation only; when lo! it is a highway of inland commerce. Boats laden with clean, golden grain lay at its banks, and I looked around for the great warehouses that Anthony Trolloppe saw, when he told of the 'rivers of corn,' but they were wanting—although this canal in one sense is a river of corn, for it has changed 5,000,000 acres of dry desert land into possible fields of waving grain."[151]

The southern division continues from Nanoon to Cawnpoor (Kanpur) where it drains into the Ganges River, over 500 miles (804.6 km) south of Roorkee. Two massive half-circle flights of stairs (ghats) flank the reunification of the Ganges Canal with its parent stream. Four domed pavilions, "constructed on native models," crown the terminus. It is an attempt to "harmonize with the town and ghat scenery lying in its neighbourhood," according to Cautley. After digging 898 miles (3,054.5 km) worth of trenches and canals to reorder the lives of some six million souls, every attempt was made for the last structure to "meet the habits, and accord with the tastes, of the native community."[152]

Unlike the Solani Aqueduct, where two grandiose lions mark the crossing of the canal into the tableland, here four miniature "pedestals for the reception of sculptured figures emblematical of the great river" sit watch over the waterway.[153] From the drawing (Plate No. XLIV), it would appear that the emblematical creatures are a mixture between a dog, lamb, and cow: a mutant of the pastoral.

Behind this façade were a series of locks, with each "having its passage for cross-communication, and its quays for the landing of goods."[154] The terminal line passes through General Gunj, a neighborhood in Cawnpoore. In order to secure the route of the canal through this area, the Company had to purchase property from local residents.

Cautley only mentions the word property a handful of times in his 471-page report. He casually speaks about payments for ground-rent in his rajbuha scheme. Perhaps the most telling moment comes near the end of the report, when, speaking about the need to develop a machine to adequately measure and distribute water to cultivators, he speaks about the Ganges' water as property: "To know, in fact, the true value of the property that we hold for distribution, and the extent to which that property can be made the most beneficial to the community, are the great desiderata to which all our views ought to be directed."[155] Transforming the bounties of the Ganges River into

"Ghat Cawnpore [along Ganges River]." Publisher Unknown. Circa late-19th or early-20th century (AA).

Proby Cautley, "Plate No. XLIV. Cawnpoor Terminal Line. Terminus Works." **Report of the Ganges Canal Works: from their commencement until the opening of the canal in 1854**, Atlas. London: Smith, Elder and Co., 1860. (Princeton University).

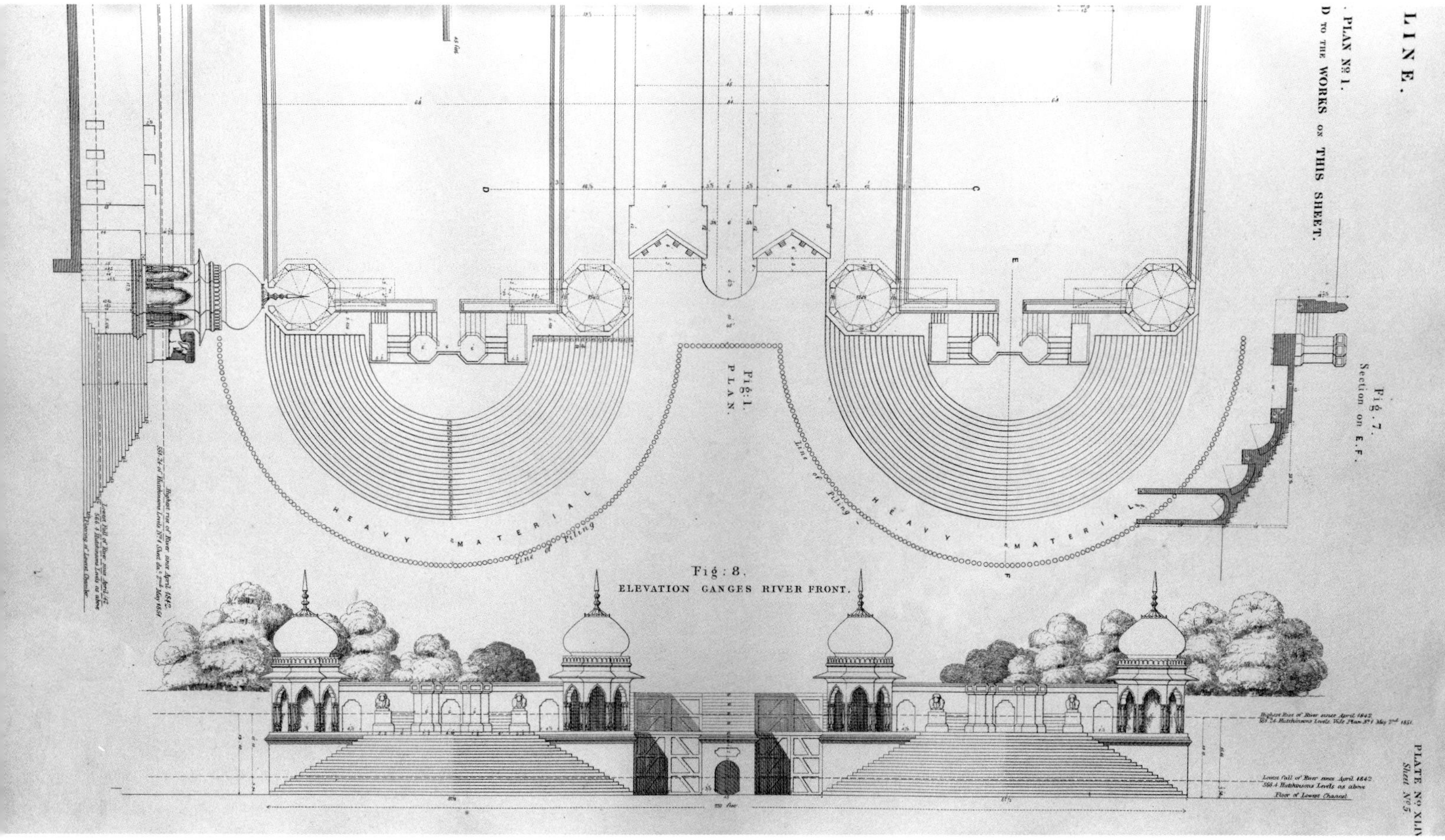

Before reuniting the Ganges canal with its source, it passed through the Cantonment (below).

Once at the terminus, Cautley wished the canal to "harmonize with the town and ghat scenery lying in its neighbourhood." The façade and domes are "constructed on native models" (above).

Diagram 39.

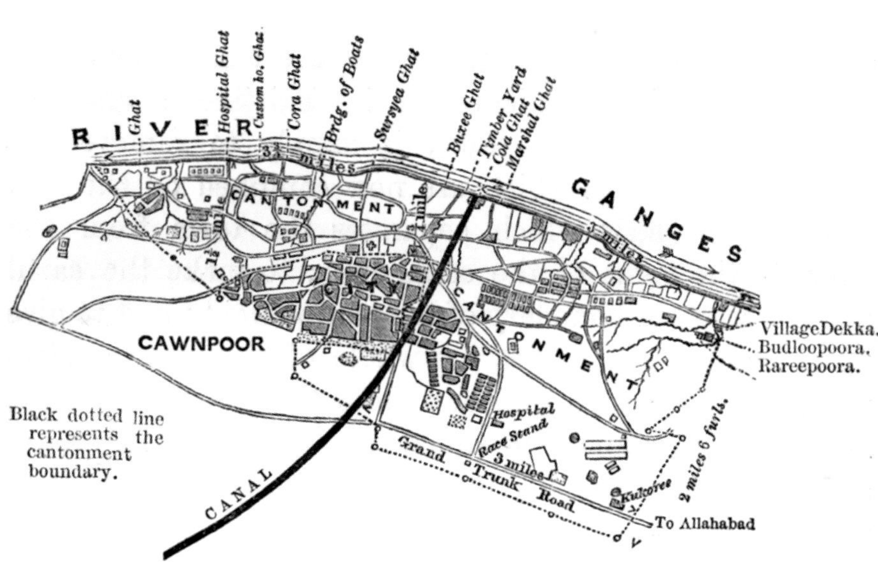

Proby Cautley, "Diagram 39." **Report of the Ganges Canal Works: from their commencement until the opening of the canal in 1854**, Vol. I. London: Smith, Elder and Co, 1860. (Princeton University).

U. S. GEOLOGICAL SURVEY WATER-SUPPLY PAPER NO. 87 PL. VII

GOPALPUR BIFURCATION, GANGES CANAL.

A strip of pasture passes through the city of Kanpur: As the volume of water diminishes, silt and debris build up on its lower courses (right). Herders bring their cows and water buffalo to graze across this provisional strip of land.

"Ganges Canal Cawnpore." Photo by R. J. Divecha for S. B. Varma & Co. Cawnpore. Circa Late-19th century. (AA).

Herbert M. Wilson. "Plate VII. Gopalpur Bifurcation, Ganges Canal." **Irrigation in India**. Washington, D.C.: Government Printing Office, 1892. (AA).

SOUTHERN DIVISION

At Nanoon, the canal bifurcated into the Cawnpoor and Etawah Terminal Lines. From bifurcations like the one at Gopalpoor (Gopalpur) along the Etawah Branch line, engineers could distribute large volumes of water where needed at different times of the year—at least in theory.

Diagram 43.

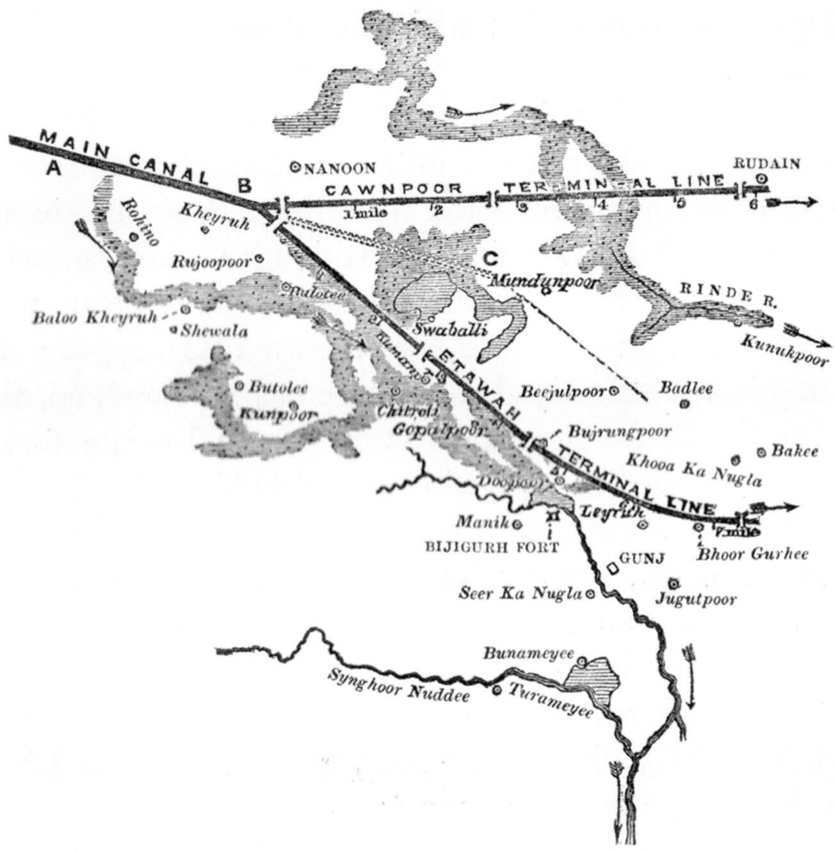

Proby Cautley, "Diagram 43." **Report of the Ganges Canal Works: from their commencement until the opening of the canal in 1854**, Vol. I. London: Smith, Elder and Co., 1860. (Princeton University).

property was easy to imagine. What remained a practical challenge for years to come was assessing the worth of that property—of developing a machine, as Cautley put it, to calculate its value in times of abundance and scarcity. In the meantime, a new division of government was created to manage such newly-acquired properties and structures for the public good.

Cautley departed from India in 1854, leaving behind this question of the value of the Company's newly acquired property. A bust of Cautley was placed in the Town Hall of Calcutta and, as he passed Fort William in that same city en route to catch a steamer back to London, a salute of thirteen guns was fired in his honor at the request of Lord Dalhousie. Certainly Cautley felt a sense of pride in his legacy to India. He had governed over the network of canals passing through the Ganga-Jamuna Doab as an unequalled expert.

1: Panorama in Kanpur, north of Dabauli, at a bifurcation of the Ganges Canal, 2008.

2: Panorama in Kanpur, south of Dabauli, along a branch of the Ganges Canal, 2008.

Canal or Nalah (drain): With parallel roads and bridges crossing the Ganges Canal at regular intervals, the infrastructure layout is ready-made to accommodate agrarian and urban activities alike. In Kanpur and other densely populated areas, its edges become a publicly shared space (above).

3: Panorama in Kanpur along a branch of the Ganges Canal, 2008.

4: Panorama outside of Kanpur on top of a regulating work of the Ganges Canal, 2008.

Begun in the 1990s and completed in the early-2000s, the Ganga Barrage stretches over 1 kilometer across the Ganges River at Kanpur. According to the Irrigation Department of Uttar Pradesh, the barrage is to divert water to ensure the regular flow of water at Kanpur ghats, provide drinking water, increase access across the river with a four-lane bridge, and be an "attractive reservoir."

Ganga Barrage at Kanpur, 2006.

PUBLIC WORKS

Ganges Canal south of Haridwar, 2008.

On July 12, 1854, three months after the Ganges Canal officially opened, Lord Dalhousie, Governor-General of the Company, sent a memorandum to the Court of Directors in London. In this memorandum, he established the office of a Secretary to the Government of India in the Department of Public Works (PWD). Dalhousie stated: "The whole system will thus at once be reduced to order, and Public Works throughout India will become one great separate branch of internal administration, conducted upon settled principles, and with the advantage of the best scientific and professional advice."[156]

Dalhousie transformed Public Works from a decentralized and idiosyncratic department under the superintendence of the Military Board in the three presidencies (Madras, Bombay, and Calcutta) into a new department under the supervision of a single individual. In each of the three presidencies, a chief engineer reported to the secretary. This new hierarchy covered all projects, civil and military: buildings, canals, roads, and railways. Long considered the greatest opprobrium to the Company's administration, the new department was heralded "as one of the greatest improvements even in this age of progression... India will no longer present the appearance it has so long worn of being the neglected estate of a spendthrift landlord."[157] The engineer, by virtue of Public Works, inscribed more than a modern map of India. His labors reshaped the flow of rivers and commerce, the forms of cities and towns, and the nature of colonial governance. Lord Dalhousie single handedly gave engineers such authority.

In the creation of the PWD, Dalhousie empowered a new branch of government to better administer the diverse territories of the Company. And yet, within less than a year of his departure, revolution struck India.

REVOLUTION

On May 10, 1857, *sepoys* (soldiers) employed by the East India Company rebelled in the cantonment of Meerut. This revolt sparked an explosion of diffuse revolts, or "mutinies" as the Company called them, at major sites of British presence across northern India—Delhi, Kanpur, Benares, among other places. The attacks caught the Company off guard, they were grossly outnumbered, the bulk of weapons were in sepoy hands, and it was the hottest time of the year—in sum, the revolt presented a formidable challenge to their rule. But with no central command or common set of goals, the mutiny (today referred to as 1857 War of Independence or 1857 Rebellion) lost a series of major battles with the Company until it was finally suppressed on June 20, 1858, with the fall of Gwalior.

In the years following the engineering feats that combated famine, installed telegraphs and rail lines across the country, and institutionalized a national Public Works department, the Company's sepoys revolted. That is at least how it appeared for decades afterwards to members of the Company: "That great struggle, remember, was in no sense an uprising of the people against our rule, for, if it had been, we could not have held India for an hour," pined George J. Medley in 1872, in his lectures on "India and Indian Engineering" delivered at the Royal Engineer Institute, Chatham. Medley, a veteran of the conflict and principal of Thomason Civil Engineering College in Roorkee, further explained, "But though it was, primarily, a military revolt, caused and aggravated by overweening confidence and bad management, it was secondarily, a struggle of conservatism against the further progress of western innovations; it was a protest by caste and tradition against railways, telegraphs, and national education."[158] Medley, an engineer, served the Company while many of these "innovations" were taking place during the tenure of Lord Dalhousie—along with the annexations of the previous decade, Dalhousie carried a great deal of blame for the revolt.

Just two years after Dalhousie left Calcutta, the Company had lost control of India. On August 2, 1858, by act of British parliament, all the territory of India was vested in Her Majesty the Queen. With this decree, the period of British rule known as the British Raj began, with India standing as the crown jewel of the British Empire until its independence in 1947.

Major administrative changes took place at the start of the British Raj. While a Governor General in Council was stationed in Calcutta (Kolkata), in London the Court of Directors and Board of Control was replaced by a new cabinet position in the government: the Secretary of State for India. Lord Stanley, the first appointee to this position, created a committee of fifteen individuals who met once a week at the India Office in London to deal with all aspects of managing and governing India, conveying instructions to the Governor General in Calcutta.

Under the British Raj, the tools of "peaceful conquest" did not come to an end. They were only sharpened. While the rebellion delayed all work on public works, it only inspired greater commitment and expansion under the new cabinet post. More roads, more railways, and more irrigation were integral to governance.

Sir Proby Cautley was one of the original fifteen members of the council. He was first made a member of the Public Works, Railways, and Telegraphs committee before becoming its chairman as well as a member of the Military Board. He remained a member of the council until 1868.

Between his weekly meetings at the India Office with the Council, Cautley was busy preparing his *Report on the Ganges Canal Works* (1860). Only 500 copies were printed of the four-volume set, which included an elaborate atlas full of construction drawings. Published before all of the lines of the canal were completed, the four volumes are written in a mixture of autobiography and technical prose—a technical travelogue meets how-to manual for building irrigation canals. Cautley's magnum opus describes an ideal state of the canal. The work is also an object of speculation, unable to account for the flaws and imperfect realizations of its construction.

The incomplete state of the canal remained a low priority for the government until 1860–61, when drought and famine struck portions of the North-West Provinces and Punjab. Cautley's former assistant, Richard Baird Smith, submitted a *Report on the Famine of 1860-61* to the Vicreoy and Governor General in Council. He designated portions of the Doab and environs as "Famine Tracts," some of which pass through areas of the still-incomplete Ganges Canal. He placed the blame not on areas where penal fines were leveled after the 1857 Rebellion, nor even on "inferior peoples with no experience in cultivation," but instead on the incompletion of the Ganges Canal, stopped because of the rebellion and its aftermath. The canal's state of incompletion had transformed areas into "Famine Tracts" that were the cause of this strife. In an attempt to persuade the government to finish the canal and create even more dams on the minor rivers of the Doab, Smith redefined the terms of famine: "Our famines are rather famines of work than of food."[159]

When people have work, they are paid, and so they have food, wrote Smith. If there is work to be had elsewhere, then people emigrate to those areas, putting greater stress on those fiscal and edible resources. Mass emigration was not in the interest of the government, which was chiefly concerned with conditions that might produce another rebellion. In other words, maintaining control and supervision of these provinces meant not only providing more irrigation, but also more jobs. Public Works literally meant putting people to work. In his reformulation of the definition, Smith broke from the long-held averted vision of the now defunct Company, placing further social and technical responsibilities in the hands of engineers. Control of the land by means of irrigation now extended to direct control over people.

DEFECTS

In 1863, a major flaw in the canal was aired in public: the slope was too steep. That year, while on the council to the Secretary of State for India, Cautley entered into a protracted public debate with Sir Arthur Cotton, another well decorated irrigation engineer in South India who had undertaken irrigation works across the Cauvery, Godavari, and Krishna rivers in the Madras Presidency.

Cotton leveled nineteen criticisms of the canal at Cautley, five of which he believed were fundamental flaws. Besides claiming that the slope was too great—to which Cautley admitted he was mistaken—Cotton also argued that the head works of the canal were located too far up the river; that the canal was cut too deep in areas; that the canal should have been built using local sub-Himalayan stone rather than brick; that, instead of a single head work, water should have been admitted at numerous points, such that it did not have as far to travel; and that the head works should have been built as entirely permanent structures.

Throughout 1865, *The Times* in London published a series of memorandums with recriminations by both engineers. A few months prior to this public feud, the Public Works Department in India appointed Captain James Crofton to evaluate what parts of the Ganges Canal were to be remodeled and how. Crofton published his report in 1865. Shortly thereafter, the government issued a Dispatch praising its chairman of the Public Works Committee in London, Sir Proby Cautley, for his "eminent services" despite any defects.[160] Cotton remained furious, though the debate eventually fizzled out. In 1903, over thirty years after Cautley died, the Indian Irrigation Commission declared that "the great Ganges Canal—a work which in magnitude and boldness of design has not been surpassed by any irrigation work in India or elsewhere."[161]

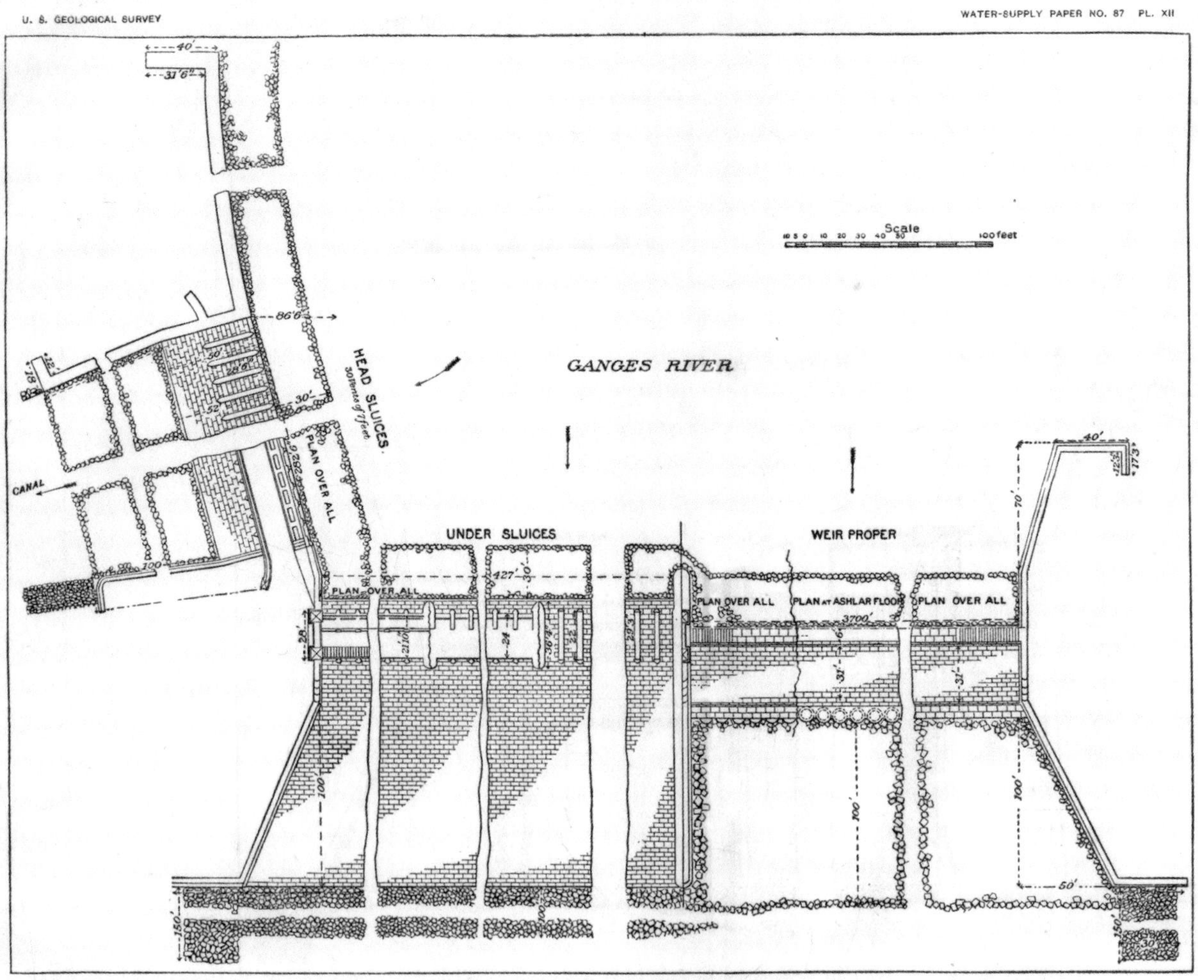

NARORA WEIR, LOWER GANGES CANAL. PLAN OF SUPERSTRUCTURE.

LOWER GANGES CANAL

Following the reorganization of government and the scandal between Cautley and Cotton, the regime decided to extend the Ganges Canal system to the tip of the Doab. The engineers did not choose to construct a modified version of Cautley's head works at Mayapur. Instead, they built a *weir*—a low dam built across a river to raise its level upstream—which was modeled on weirs Cotton threw across the Cauvery River in order to divert water and reduce silt in the first half of the nineteenth century. In this case, in 1872 construction began on a weir across the Ganges at Narora for the Lower Ganges Canal (LGC) system. Completed in 1878, the weir, along with its sluices, head works, and navigation locks would be heralded in 1894 as "one of the finest and most imposing engineering constructions in India."[162]

In 1898, a continuation of the Kanpur branch, known as the Fatephur Branch, began irrigating land all the way to Allahabad District. The Lower Ganges Canal came to command 5,300,000 acres and irrigate 830,000 acres by 1904. A large portion of the canal was navigable (62 miles [100 km] of the main channel and 137 miles [220.5 km] of the branches).[163] With these new main trunks and minor irrigating lines (rajbuhas), the Doab was now "covered with a network of irrigating channels," that conveyed water to millions of acres of cultivated land.[164]

At long last, many years after Lord Ellenborough had died, the Ganges Canal stretched the length of the Doab. With locked channels and raising bridges, boats were able to travel from Kanpur to Roorkee. As it was put in *The Imperial Gazetteer*, the southern portion of the Upper Ganges Canal, "where it meets the Lower Ganges Canal, is kept open chiefly for navigation; and both the Ganges Canals are, in this respect, considered a single system."[165]

Left: "Plate XII: Narora Weir, Lower Ganges Canal. Plan of Surperstructure" and Right: "Plate XI. Narora Weir, Lower Ganges Canal. Plan of Well Foundations." **The Roorkee Treatise on Civil Engineering in India**, Volume 2, compiled by Julius George Medley, third edition enlarged and improved and edited by A. M. Lang. Roorkee: Thomason College Press, 1877. (AA).

U. S. GEOLOGICAL SURVEY WATER-SUPPLY PAPER NO. 87 PL. XI

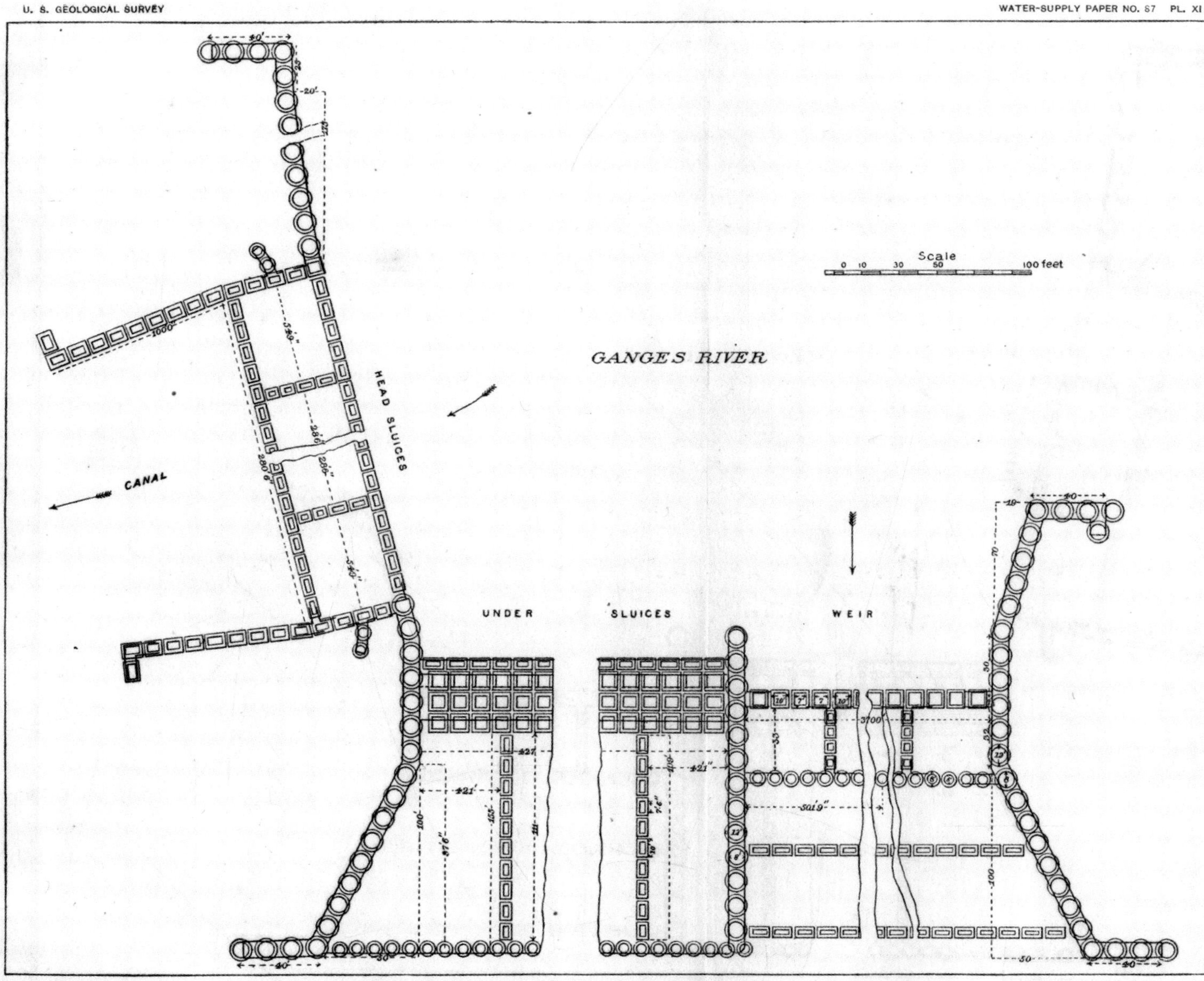

NARORA WEIR, LOWER GANGES CANAL. PLAN OF WELL FOUNDATIONS.

"Ganges River Training Works at Narora. Charts of Ganges Showing Changes Since 1870." **The Roorkee Treatise on Civil Engineering in India**, Volume 2, compiled by Julius George Medley, third edition enlarged and improved and edited by A. M. Lang. Roorkee: Thomason College Press, 1877. (AA).

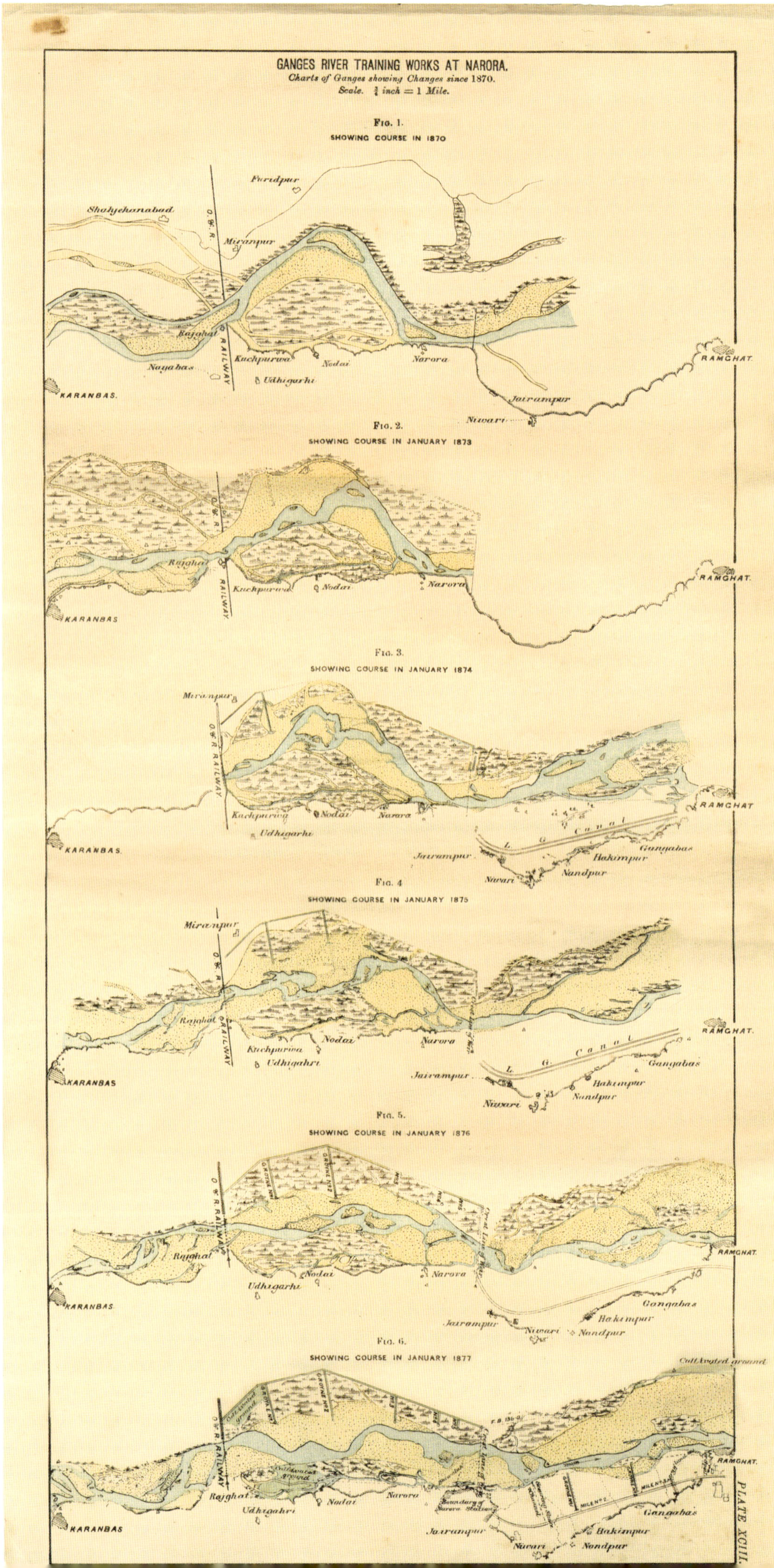

U. S. GEOLOGICAL SURVEY WATER-SUPPLY PAPER NO. 87 PL. XIII

HEAD-WORKS, LOWER GANGES CANAL, NARORA.

LGC

Herbert M. Wilson. "Head-Works Lower Ganges Canal, Narora." **Irrigation in India**. Washington, D.C.: Government Printing Office, 1892. (AA).

STAGES OF SUPER-SURFACE

Development of canal irrigation in the Ganga-Jamuna Doab from the 1830s to the 2000s, 2014.

SUPER-SURFACE

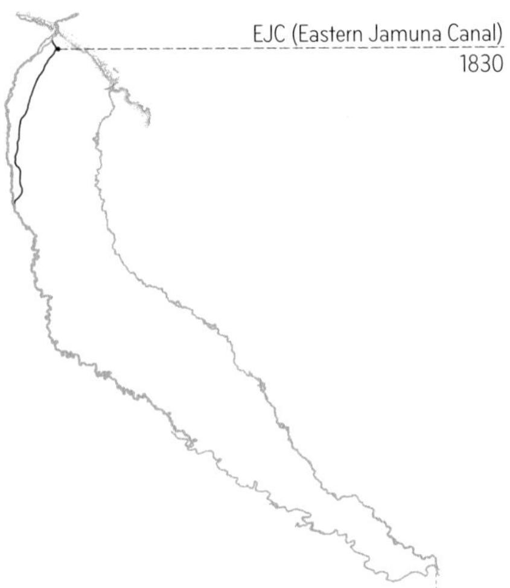

1830

- Doab Canal/Eastern Jamuna Canal (EJC) rebuilt to provide famine relief in the Doab

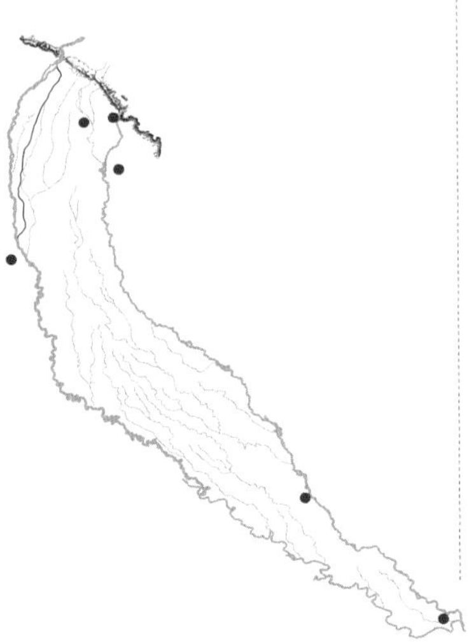

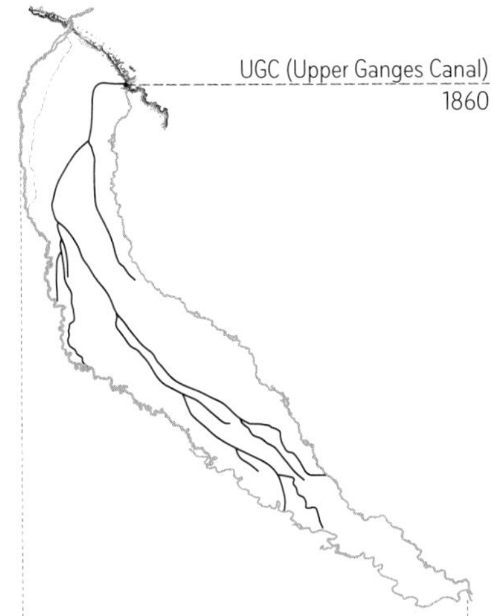

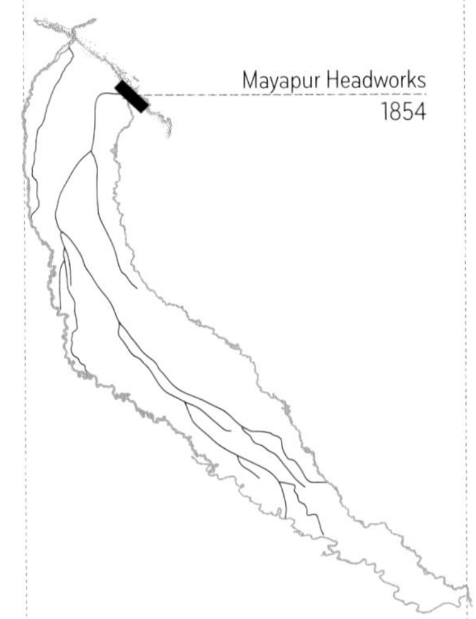

1830 - 1860

- Later known as the Upper Ganges Canal (UGC), work commenced in 1847 on the Ganges Canal--designed to provide famine relief and increase agricultural production--it was the longest canal system in the world

- Mayapur Headworks proved a difficult structure to maintain due to design flaws in terms of managing silt and debris

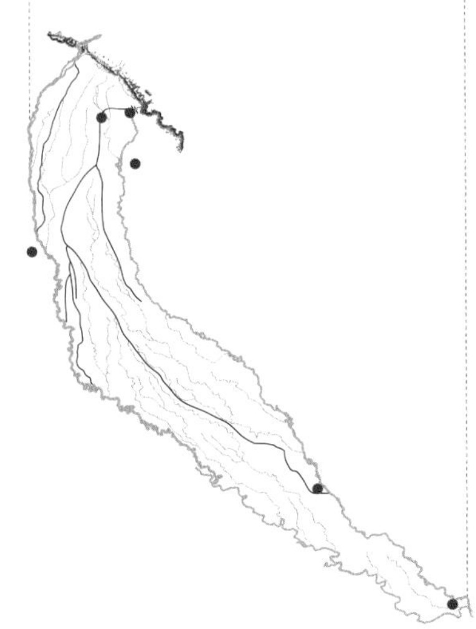

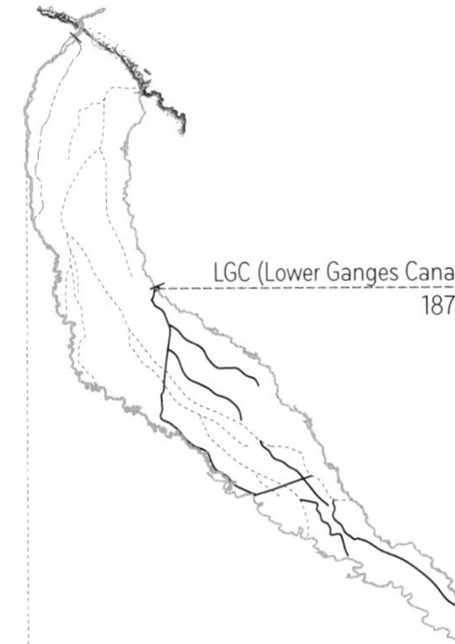

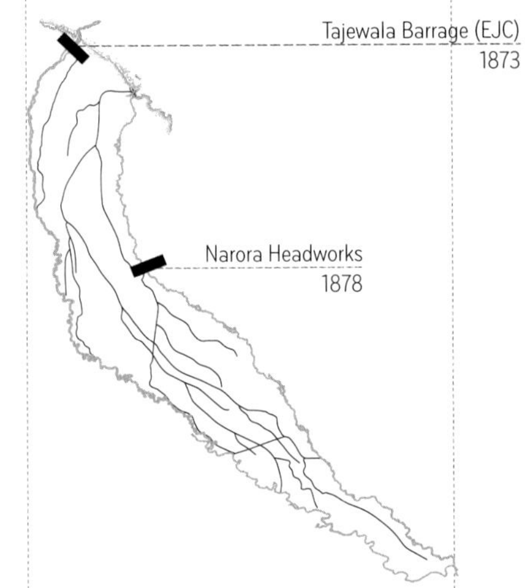

1860 - 1880

- Commenced in 1872, Lower Ganges Canal (LGC) was built to increase irrigation and navigation across the Doab

- Narora Headworks uses a weir system similar to Sir Arthur Cotton's weirs in Madras in the mid-19th century

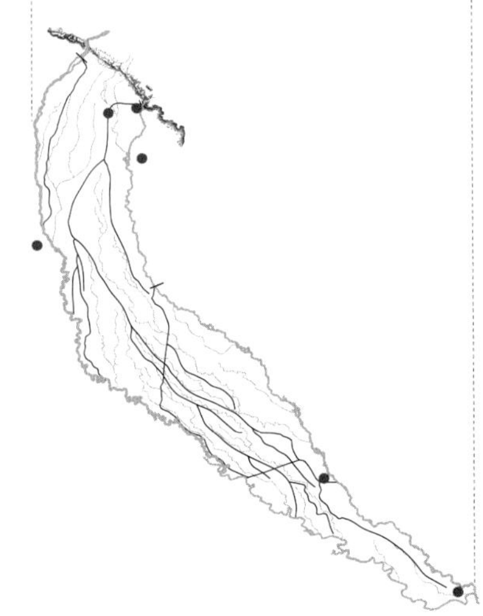

SUPER-SURFACE

NEW CANALS

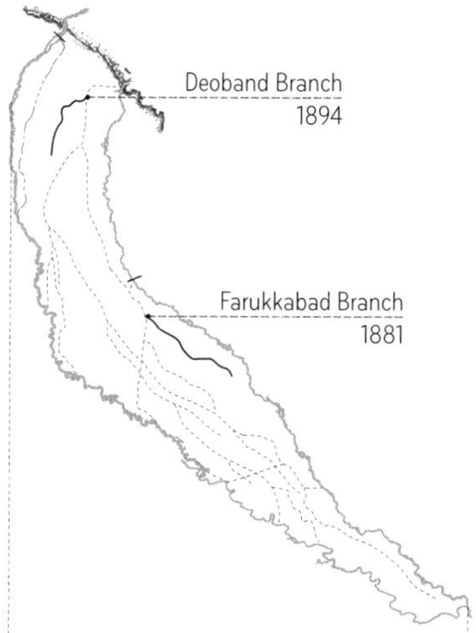

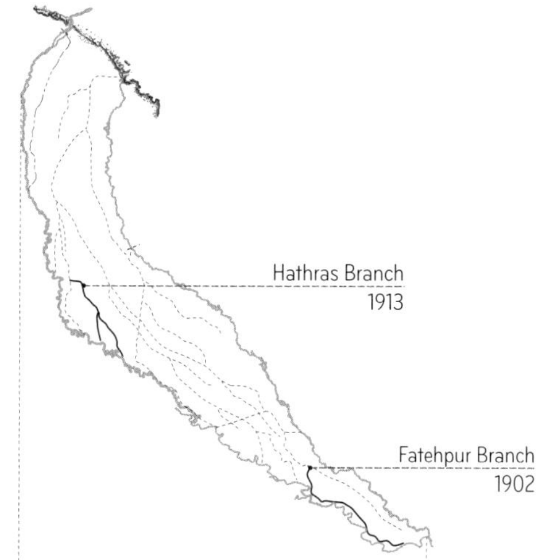

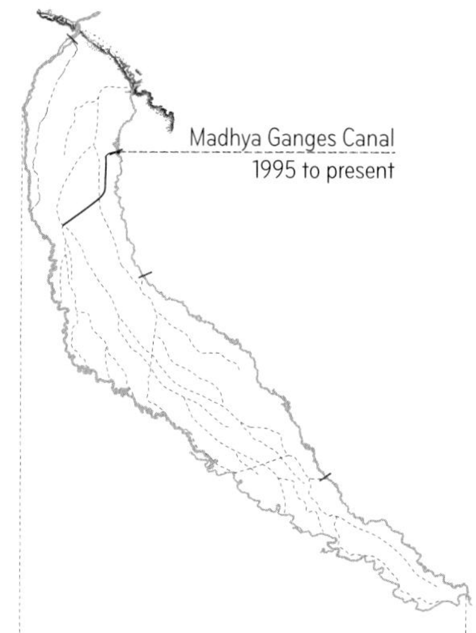

CANALS & BARRAGES

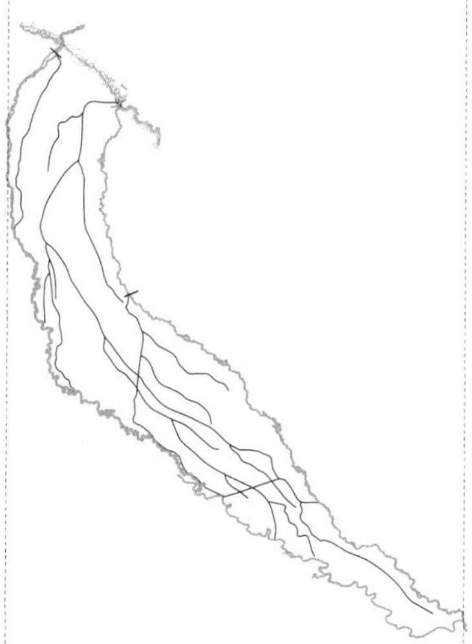

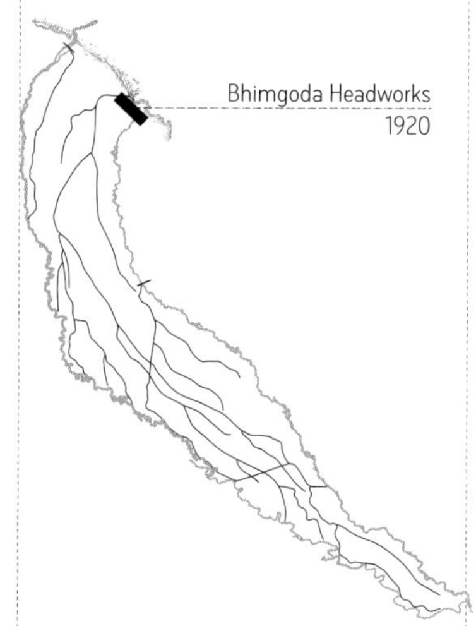

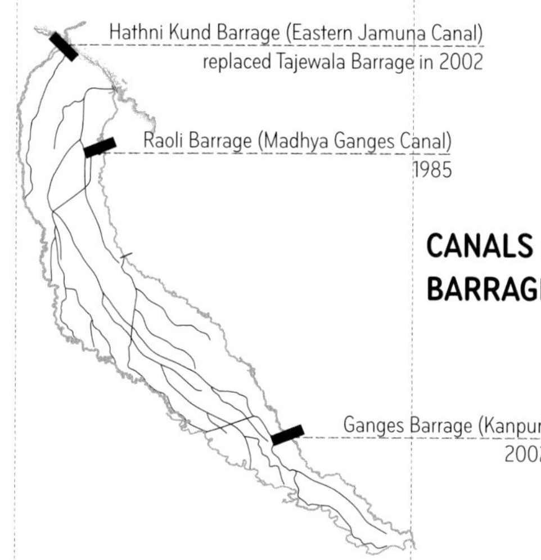

1880 - 1900

- Deoband Branch was added to the UGC to provide outlet for water no longer needed to feed the tail ends of the Etawah and Kanpur sections

- Narora Headworks uses a weir system similar to Sir Arthur Cotton's weirs in Madras in the mid-19th century

1900 - 1950

- Bhimgoda Head Works (head discharge of 190 m3/s) replaced temporary embankments built every year during the southwest monsoon and diverts water to ghats in Haridwar

- Hathras Branch extends UGC system

- Fatehpur Branch extends LGC system

1950 - 2014

- Hathni Kund Barrage (2002) diverts water to the EJC and WJC (Western Jamuna Canal)

- Raoli Barrage (1994) diverts water during the monsoon to Madhya Ganges Canal

- Madhya Ganges Canal, a monsoon canal, recharges aquifers and waters paddy rice

- Ganges Barrage ensures water for drinking, the ghats, control erosion, improve ecology

SUMMATION

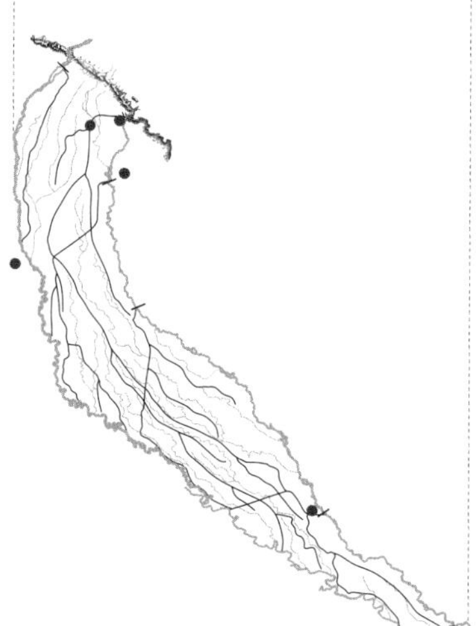

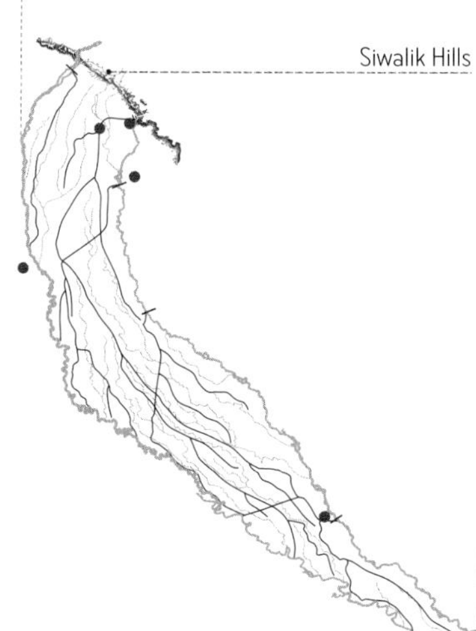

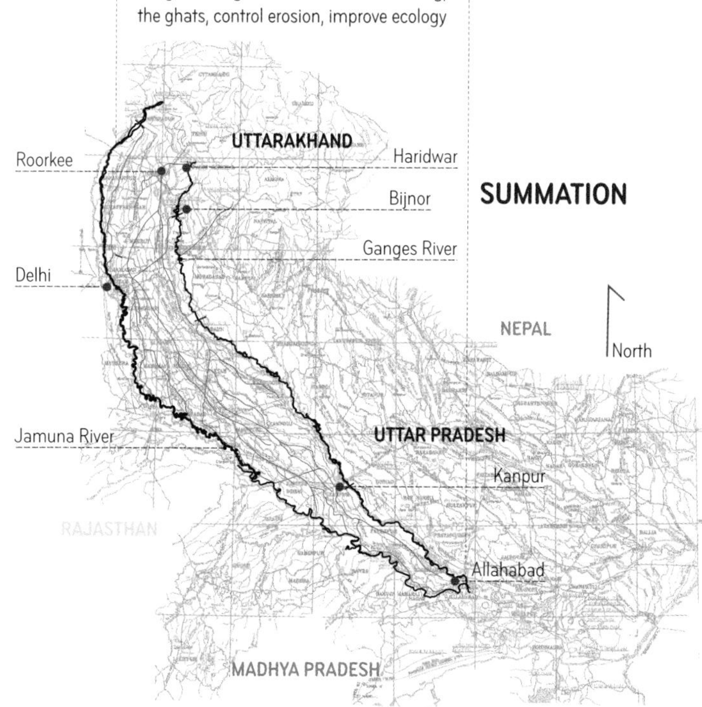

By the end of the long nineteenth century, as hydraulic engineering was professionalized throughout the British Empire and around the globe, a number of reports, articles, and books cataloguing major irrigation works appeared, and few were as universally praised as *Irrigation Canals and Other Irrigation Works* (1892). Written by Patrick J. Flynn, a former executive engineer for Public Works in Punjab before serving in the San Francisco and Los Angeles municipalities, the book profiled canal works from around the world. Flynn lauded the canals of India as "the greatest number of canals that can, in many respects, be quoted as good examples..." Unlike in the United States, India's canals "are permanent, that is, that very little wood or other perishable material enters into their construction."[166] For Flynn, permanence, along with "length, cross-sectional dimensions, discharging capacity, number and aggregate mileage," all rendered India's canals "the greatest in the world."[167] Although he identified four types of irrigation works in India—"Perennial canals, Inundation canals, Tanks or Reservoirs, and wells"—the canal systems received the greatest attention. To Flynn, these were heroic projects, and there was no doubt that Sir Proby Cautley, despite his miscalculation in fixing the slope of the Ganges Canal was, "the greatest canal engineer that ever existed."[168]

Flynn was not alone in his heroic descriptions. Engineers and government administrators, from Australia to the United States, journeyed to India as pilgrims in search of the most modern irrigation engineering in the world. Their articles, books, and government reports profiled Indian engineering with awe.

Among the awestruck was Herbert M. Wilson, a geographer sent to India to investigate canals for the U.S. Department of the Interior and Geological Survey. Wilson toured the great works for technologies to import to the American West, and in the 1890s he published several articles and reports on his findings. He compiled descriptions and drawings of technologies powered by people and livestock, like the *peacottah* (well sweep), *latha* (swinging water by basket), Persian Water Wheel driven by bullocks, and bullocks lifting water by bag (see illustrations).[169]

Impressed though he was with the technologies, he was struck most by the system of water governance in India: "The powers of control over the waters for irrigation are entirely centralized" in the hands of the government of India, controlling the flow of water "to make it the greatest benefit to the community at large."[170] The government of India owned the rights to water and land in perpetuity. Public Works Department engineers routinely patrolled the canals from their stations (chokies) to control the flow and extraction of water. They made sure cows did not trample ditches and cultivators did not damage the channel of the canal with illegal structures. Profits were small with low sums in terms of direct returns to the government. But, "the indirect return to the government is always large in enhanced revenue from the rental of land, [and] in immunity from famine."[171] In contrast, wrote Wilson, in the U.S., "the laws relating to the ownership of irrigable lands and works, and rights of way have become so voluminous and differ so greatly in the various states as to create one of the most serious impediments to the inception of irrigation enterprises."[172] Because the government did not patrol and police the canals of the U.S., "Gophers are permitted to burrow at discretion in the canal banks, and small leaks go unattended to, as their dangers are unappreciated, while flumes and weirs and other perishable works are permitted to remain unrepaired until they are past repairing and must be renewed."[173] In the face of animal intrusions and systemic decay, the engineer was not entirely powerless in the U.S., but his acumen and policing powers were limited when compared to those of his British counterpart in India.

Another traveler who chronicled these developments was Alfred Deakin, the former Prime Minister of Australia. Following his tour of India and Ceylon in the 1890s, he saw the British administration of irrigation in India in slightly different terms. For him, British administration was "benevolent tyranny, leaning a little towards unnecessary officiousness." From time to time this tyrant did thrust "advantages" on his subjects. But by Deakin's estimation, this was rarer and rarer because "India can never be British in its conditions, and will never be British except in rule and principle."[174] In the 1890s, few if any of Deakin's contemporaries would have argued with this point. But then how did the thousands of miles of new irrigation and rail lines fit into British rule? For, after all, according to Deakin, "We found a despotism, and we preserve it, striving not to alter or weaken it, but to make it sympathetic and just."[175] At the tail end of the nineteenth century, public works remained the

HEROISM

means to sublimate the juxtaposition of preservation and obliteration produced by British rule in India. Deakin's words dwell less on the contradictions of imperial rule—in spite of everything, the British Empire never approached the misconduct of the Romans, according to him—but instead he conveys a sense of melancholy about the heroism of a sympathetic and just tyrant. He looks at these works with both awe and trepidation at a time when others are casting serious doubts on the imperial project.

Herbert M. Wilson. "Mot," "A. Persian Wheel," "B. Paecottah." **Irrigation in India**. Washington, D.C.: Government Printing Office, 1892. (AA).

U. S. GEOLOGICAL SURVEY WATER-SUPPLY PAPER NO. 87 PL. IV

A. PERSIAN WHEEL.

B. PAECOTTAH.

Obsessed with lift irrigation, Herbert M. Wilson hoped that a study of the Mot, Persian Wheel, and Paecottah [well sweep] would inspire engineers to develop new, mechanized lift irrigation for the American West.

. S. GEOLOGICAL SURVEY WATER-SUPPLY PAPER NO. 87 PL. III

MOT.

DRAIN

Not everyone saw India benefiting from large capital investments in public works. Just before the great famine in 1876, Dadabhai Naoroji read a series of searing indictments of imperial rule, "Poverty of India" (printed as a book in 1878) to the Bombay Branch of the East India Association. One of the founding members of the Indian National Congress and mentor to Mohandas K. Gandhi, Naoroji showed how—by turning official facts and figures of the Raj against their contradictory policies—all large-scale investment in the country, from railways to irrigation, were draining wealth from India and investing very little (if any) of that wealth back into the country. As he put it, "even the most well-intentioned acts of Government become disadvantageous."[176]

For instance, after reviewing all of the maps and tables in government reports, he noticed a pattern of high prices in good seasons occurring in districts serviced by lines of public works while those with the lowest prices were the furthest away from the lines.[177] Next, he showed how increases in prices for commodities like cotton and wheat, long thought to indicate growing prosperity, only occurred in India because of scarcity and famine. Because these cash crops sold at higher prices in the market than most foodstuffs, cultivators paying high rents for canal irrigation water chose to plant cotton and wheat to offset the cost of canal irrigation. There was something inherently destructive to public works in India—not in terms of their creation—but in their administration under British Rule. If facts and figures were not enough, he quoted official government reports at length, which either called British fiscal policy "oppressive and depressing in its influence" or showed how perverse it was to call Indians "miserably poor" for decades, while still insisting on levying taxes on them.[178]

Naoroji was not alone in his criticism of imperial rule during the Victorian Age. British liberals, like social reformer Florence Nightingale and journalist William Digby, as well as Cautley's nemesis—Sir Arthur Cotton—all blamed, in one way or another, British policy in India as the root cause of famine.[179] In 1878, drawing on thousands of reports issued by the government as well as documents never made public, William Digby published *The Famine Campaign in Southern India*. In this two-volume set, he chronicled the mismanagement and abuses of the government in its campaign. He made special note of the fact that not only did grain made available by railways drive prices up everywhere, but that the government was aware of it.[180] When viewed in totality, Digby presented a government that was at least an accomplice, if not the sole perpetrator, of the 1876 famine.

In response, the government formed a commission in 1878 to examine the causes of the famine. Richard Strachey, who was not only an engineer and former Inspector-General of Irrigation (1866-69), but also the brother of John Strachey (budgetary advisor to Governor-General and Viceroy Lytton), led the commission. After its examination, the commission did everything except blame the government for the crisis. The commission, echoing Lord Lytton, proposed protecting the country against famine through what the Strachey's would call the "instrumentality of public works": railways and irrigation.[181] They proposed extending railways from 10,000 miles of track to 20,000 (16,093 km to 32,187 km) to meet the need to shuttle surpluses. Irrigation was to be extended as well, although on a far smaller scale.[182] Furthermore, the commission sought to precisely quantify the answer to the question, "How much food does a human need to live and work?"[183] All social problems were to be usurped by the technical "instrumentality of public works."

Years later, in 1901, after the great famines of 1876-77, 1896-97, and 1899-1900, Naoroji noted: "how strange it is that the British rulers do not see that after all they themselves are the main cause of the destruction that ensues from droughts; that is the drain of India's wealth by them that lays at their own door the dreadful results of misery, starvation, and deaths of millions..."[184] By analyzing the finances of public works and noting its drain on the economy of India, Naoroji showed the economic and physical violence perpetrated by "peaceful conquest." Just as the Company and the Raj constructed the Ganges Canal and railways, so too did they construct a poverty-stricken society in India.

Herbert M. Wilson. "Map of India." **Irrigation in India**. Washington, D.C.: Government Printing Office, 1892. (AA).

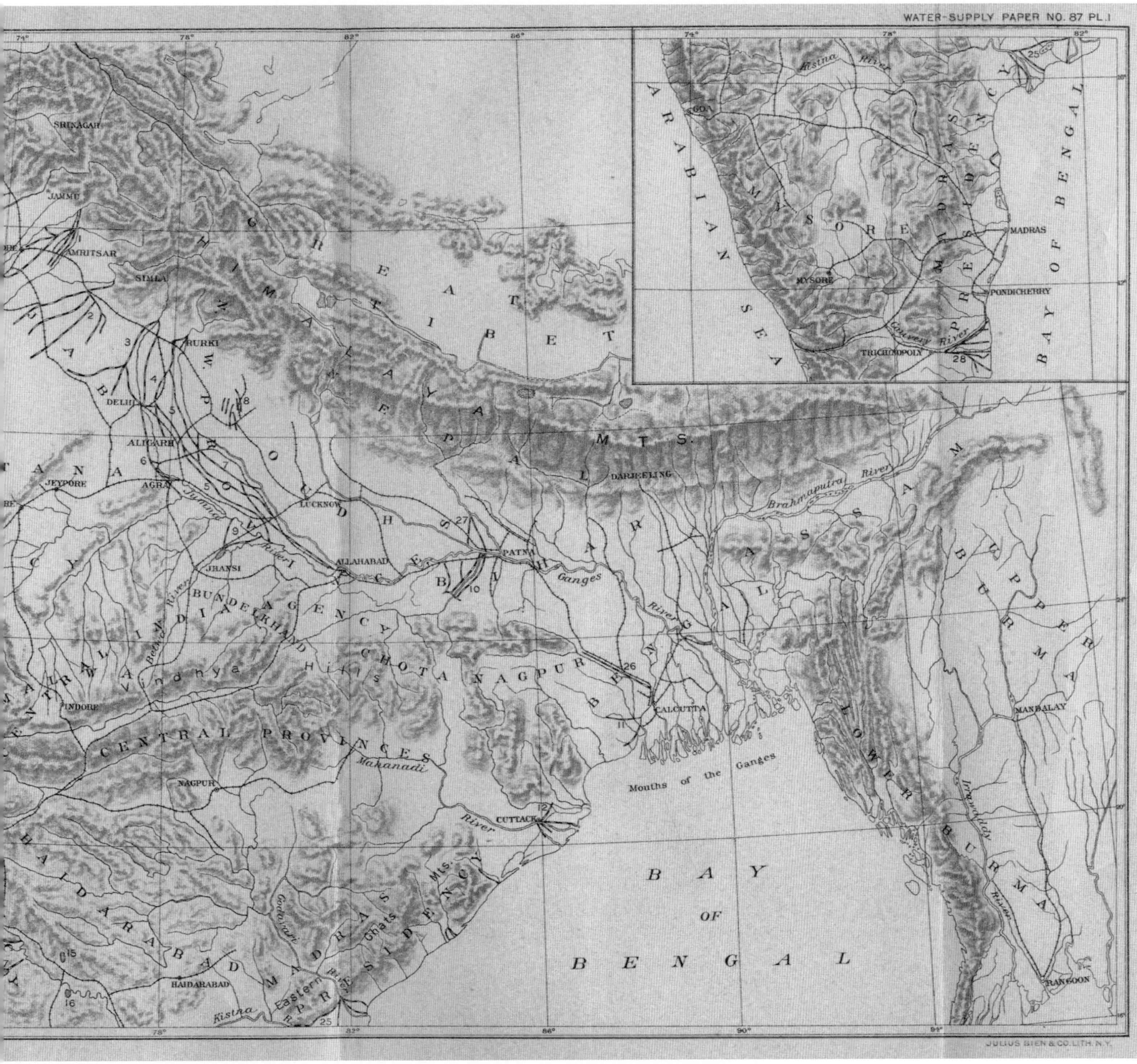

DRAIN

By the turn of the century, others began to question the techno-scientific principles of "peaceful conquest" as well. Senior scholars such as Romesh Dutt, a professor of history at University College London and a member of the Bengal Legislative Council, criticized both the execution of public works and the financing of public debt. Building on Naoroji's theory of economic drain, Dutt specifically compared expenditures and functions of two notable public works in India: irrigation and railways. Whereas irrigation proved remunerative and averted famines in a largely agricultural country like India, "railways led to a permanent loss to the Government year after year."[185] And yet, railway investment outpaced irrigation investment by a two-to-one margin. Unlike canals, for which the government provided almost no guarantees of remuneration to investors, railways enjoyed a guaranteed profit of five percent. Presuming that private investors would not want to take the risk of investing in Indian railways without a guaranteed return, if any railway fell short of five percent, the government made up the difference. Such fixed guarantees lead to a building boom in areas with low-profitability coupled with over-spending on construction. Rail lines requiring a subsidy from the government in 1900 accounted for seventy percent of all the lines.[186]

Adding to the favoritism of railway construction over canal irrigation was yet another change in the definition of famine. By the mid-1870s, railways were deemed instruments of "famine relief" while irrigation works were "preventative."[187] With the most remunerative large-scale irrigation works completed by this time, the government did not encourage incurring greater debt from the construction of new canals. This coincided with a shift in thinking about the causes of famine. Still a spatial problem, famine was produced by unequal distribution of food as opposed to water. Railways functioned as networks for re-distributing agricultural surpluses of one area to make up for scarcity in another. They were also excellent means of transporting commodities for export abroad and for dispatching troops.

These reasons explain why railways became the favored infrastructural project for Britain, but the case was much more difficult to make for India. Within nineteen years of the Queen of England assuming the title of Empress of India in 1877, the Crown had doubled the national debt of India incurred by the previous century of Company rule: from 69.5 millions to 139 millions sterling.[188] The majority of this debt came from the financing of railways, a British priority, as well as through the annual tribute paid by India. For Dutt, like Naoroji, it was quite clear: If India had representation in British Parliament, then irrigation, rather than railways, would figure into averting famine. Even a despotic government like the Crown should have consulted Indians on such a matter, Dutt observed, "But Englishmen in their own country were more familiar with railroads than with canals; and they made the mistake of judging the needs of India accordingly."[189]

Railways, unlike canals in India, were built with British steel and British machinery. Canals required very little importation of building materials or machinery. By the 1940s, less than 700 locomotives were produced in India compared to the 14,000 imported from Britain since the 1850s.[190] Thus, not only did railway subsidies benefit British investors, but also directly benefitted British industries and labor in ways that artificial irrigation ever could. Such stark imbalances led Indian nationalists to criticize these practices as the "deindustrialisation" of India.

These policies led to a burgeoning national nebt in the "peaceful" years after the 1857 revolt. For Dutt, "Modern European nations create National Debts mainly to extend their conquests and colonies, and to maintain their position among rival nations. India seeks no conquests; she has no rivals in Asia; her position under a strong and good government is invulnerable."[191] So why create a national debt for India, asked Dutt? In large part, debt had accrued to finance the railways, which did perversely little to help India's domestic economy.

What railways did, by virtue of their role in famine relief, was create a second phase in the life of the Ganges Machine. They extended the reach of the canal's original mission: famine relief. They were an integral part of the internal administration of the Public Works Department.

With no real direct representation of Indian interests in London, criticizing the financing and administration of public works was a way to expose British despotism and assert the political demands of Indians. Just as the management of public works was used to justify British rule in India, its mismanagement could also be used to undermine British authority. The systemic mismanagement of infrastructure would become a principal means of asserting greater Indian representation and eventual independence in the twentieth century. Mohandas K. Gandhi, following Naoroji and Dutt's lead, reiterated how railways both depleted the countryside of food and destroyed village handicrafts, all contributing to the causes of famine and poverty.[192]

DECENTRAL-IZATION

After over five decades of focusing only on works of major irrigation, by the beginning of the twentieth century the Government of India could no longer ignore minor works of irrigation as well. Wells and water tanks, still used by the vast majority of Indian cultivators to irrigate their fields, never garnered much attention from the Company or the Raj. In the wake of recurring famines in India, an Indian Irrigation commission was established in 1901 to investigate, yet again, how to prevent famine. For the most part, the commission advocated more of the same: extend railways and canals. It included formulas to calculate the "direct protective value" of irrigation and admitted that certain expenditures might place a burden on the population that "may be even greater than that of famine itself."[193] Buried within the otherwise typical report, however, was a radical departure from the large-scale centralization works of the government: a proposal for small-scale irrigation works and the decentralization of irrigation control.

Sir Colin C. Scott-Moncrieff, an engineer and president of the commission, was no radical engineer. "He had none of those irresistible flashes of genius which lift a man suddenly above his fellows," read his obituary in *The Spectator* in 1917.[194] Still, this "ordinary" engineer was a long supporter of large-scale irrigation works—in the 1870s, as head of irrigation works for the whole of Egypt, he oversaw the redesign and repairs to the Nile Barrage, a weir built across the Nile River to store water for irrigation. A new focus on small-scale, minor works was thus a departure from his previous experience. In his 1903 Report, he advocated for the "provincialization" or decentralization of all existing and proposed imperial irrigation works in India.[195] He also supported the introduction of private development of productive irrigation works. His report subdivided such works into three major categories: productive, protective, and minor. Productive works were directly remunerative, protective works were indirectly remunerative, and minor works included "all works which have not been classed as productive or protective."[196] In the era of large-scale, monumental irrigation works, minor works were precisely what the canals were supposed to replace. But Scott-Moncrieff argued that the increasing rate of famine, coupled with an expanding population with limited irrigable land, demanded the systematic inclusion of minor works into the larger irrigation machine.

Large-scale irrigation works like the Ganges Canal were still to be built and extended, but they were to be augmented as well. The Ganges Canal in particular was seen as a success in many ways, but also a failure; it provided insurance against famine in many places, but not all; it raised productivity but also produced waterlogging and salinization of the soil, as well as causing an incalculable number of deaths from malaria; and it would go on to provide inspiration for such monumental, centralized projects throughout the world, even as its length and bureaucratic centralization made it unwieldy to manage.[197]

In rethinking the three classes of irrigation, the Commission not only redefined the parts of irrigation, but also redefined the whole. This was its major innovation.[198] By linking canals, railways, wells, and tanks into a larger system, engineers were forced to extend their technical expertise to managing surface and sub-surface water regimes. Accustomed to thinking in layers—both geologically and geographically—groundwater had always posed problems to areas around canals in terms of percolation and waterlogging. Now engineers needed to find new ways to make better use of groundwater for irrigation.

The 1903 Report was not only a set of findings and suggestions made by the government; it was also a design project. It thickened the intersection between agriculture, infrastructure, and topography. It suggested new sites of small, decentralized interventions. In a climate where the risk of famine appeared increasingly certain, the three-dimensional nature of water distribution provided a new way to think about human settlement.

Colin Campbell Scott-Moncrieff, "Diagram Comparing the Volumes of Rainfall, Surface-Flow, and Portion of Surface-Flow Utilized for Irrigation, in the River-Basins of India, Excluding Burma, Assam and Eastern Bengal." **Report of the Indian Irrigation Commission, 1901-03**. Calcutta: Office of the Superintendent of Government Printing,1903. (AA).

If the Ganges Canal simultaneously multiplied water-rents and territories through irrigation, then the 1903 Report theoretically rationed the Ganga-Jamuna Doab (and other irrigable land in India) into smaller and smaller parcels. Much like Cautley's system of rajbuhas, thrown off the main trunk of the Ganges Canal, the proposed harvesting of groundwater resources, coupled with provincialization, promised an almost infinite growth of water extraction without loss of legibility or control.

Decentralization also represented a political advantage: by putting people "in charge of their own future," it displaced blame away from the central government and more in the hands of provincial and local authorities. After years of progressive centralization by the Irrigation Department, Scott-Moncrieff suggested a system of decentralization akin to the years before Dalhousie centralized all public works throughout India into a single department.

The provincialization of irrigation control and renewed focus on small-scale wells was also portrayed as a benefit to individual farmers. Just two years after publishing the 1903 Report, Scott-Moncrieff gave an "Opening Address" to the British Association on "living, life-giving water." He noted that there was one practical advantage to wells over canals: "It is in the farmer's own hands. He can work his pump and flood his lands when he thinks best. He is independent of his neighbours, and can have no disputes with them as to when he may be able to get water and when it may be denied to him." Individual wells also insulated the farmer from giving bribes because "corruption is rife among the lower subordinates of government."[199] Wells accounted for thirty percent of all annual irrigation in India (13 million acres). If Indian farmers had access to lift machinery, like pumps, then a well-fed farm could function like an artificial island. A farmer held the irrigation captive, but the farmer was also captive to its proper functioning.

But how? India had few Artesian wells to draw from, and Indian farmers typically did not have enough funds to purchase expensive, imported equipment like pumps. Borrowing a strategy that Cautley had employed to finance rajbuhas, Scott-Moncrieff suggested that the government might either advance cultivators' funds to make such purchases or "joint stock companies might be formed with the same object."[200] However, with no government-sponsored program in place, groundwater utilization remained limited to technologies like the Persian Wheel or the mot. Gravity-flow canals remained the most feasible means of irrigation.

"Tanks and wells ought not to be suffered to fall into decay," remarked professor of Sanskrit at Oxford University, Monier Williams in 1879. "It might even be possible, say some," he continued, "by means of anicuts to intercept the onward flow of streams, and diffuse every particle of liquid by a network of small channels and feeders over every tract of arid country." But then, rather sarcastically, he noted that even those "admirable theorists forget that the dry regions of India are often on table-land, and that no engineer can make water flow up hill."[201]

Two decades later, Scott-Moncrieff offered a similarly integrated vision, but it was still not financially feasible. It would take another two decades before motorized groundwater utilization could be realized and another four before it was adequately mapped. Initial tests began in the Indus River Basin, but like so many other water technologies developed in India, the Ganga-Jamuna Doab proved the best testing ground for these new technologies. "Peaceful conquest" was about to unleash a whole new culture of decentralization across the Ganges River Basin.

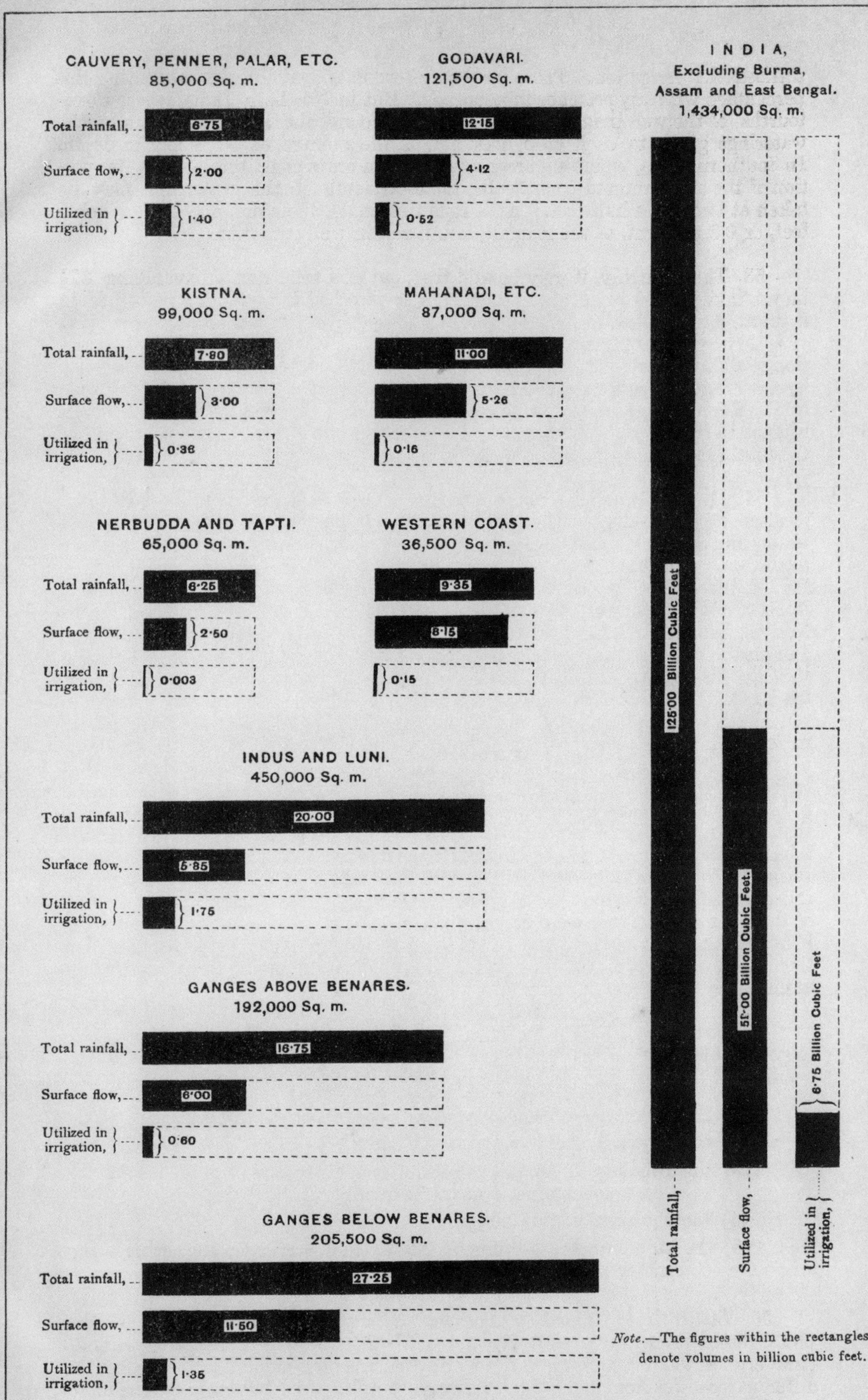

DIAGRAM COMPARING THE VOLUMES OF RAINFALL, SURFACE-FLOW, AND PORTION OF SURFACE-FLOW UTILIZED FOR IRRIGATION, IN THE RIVER-BASINS OF INDIA, EXCLUDING BURMA, ASSAM AND EASTERN BENGAL.

A common technological fixture throughout the Gangetic Plains, hand pumps can be found in cities and on farms, like the one below, outside Kalpi (U.P.), 2008.

PART 3

Himalayan Dreams...Super-Surface...Survey...Crisis in Slow-Motion... Liberation...Poets...Tubewells...Stations...Explosions...1966... Family Planning...Compact Oasis...K/G Ratio...HYVs...Implementation... Schemes

DECEMBER 17, 1946.

A few months before Indian independence, an irrigation adviser to the Government of India gave a lecture to the East India Association at Caxton Hall, Westminster. There, he shared a dream that previously he had only told to a handful of his friends. It was a "Himalayan Dream," one built on "East and West in Co-operative Action." An irrigation engineer in northern India, Sir William Leonard Stampe prefaced his fantasy with what he called a "Death Diagram": a Malthusian vision of how India's population growth would outpace its food supply. Unless groundwater irrigation could be developed for large tracts of land not serviced by canals, Stampe predicted that thousands would perish from hunger and millions more would suffer from malnutrition. Standing beside his graph, a statistical *memento mori*, Stampe offered an engineering solution to this cataclysmic disparity: the tubewell.

The tubewell, still in use today, is composed of perforated steel or plastic pipe (or occasionally bamboo) that is bored between 16 and 200 feet (5 and 60 meters) below the ground to extract water from an aquifer, driven by either an electric or oil-powered engine. After it is pumped from the ground, a series of distribution channels convey water to surrounding fields. If the water must travel great distances, constructing and maintaining the channels often costs more than all the other components. This is particularly true for publicly-owned tubewells, which typically irrigate more land than privately-owned. While centralized canals dispersed the water from the Himalayan and Vindhya rivers, the tubewell pulls water from "the vast underground river that is now proved to exist some 20 feet [6 meters] or so below the arid surface of the great alluvial basin of the Ganges and to some extent below the Indus also."[1] For over a decade, Stampe had been hard at work attempting to puncture this "vast underground river," or what he called the "silent Saraswati (the hidden river of Hindu mythology)."[2] Stampe's "Himalayan Dream" described a way to link these two systems, transforming the Ganges River Basin into a vast super-surface of canals and tubewells, roads and railways, cities and farms. The tubewell's triumph was in its liberalization, its potential to disperse and decentralize the control of water. Unlike the topographic limits placed on canals, nearly every farm in the Ganges and Indus river basins could potentially be outfitted with a tubewell. The pace at which this technological amenity proliferated throughout the basin testified to how urgently it was needed.

"Driven Wells." **Appletons' Cyclopaedia of Applied Mechanics: A Dictionary**, Vol. II, edited by Park Benjamin. New York: D. Appleton and Company, 1880. (Princeton University).

The resulting super-surface, dense with these technologies of water extraction and distribution, is an embodiment not only of its long and fraught history, but also a peculiar mixing of ideologies—the centralized character and management of the canals versus the diffuse, expansive distribution of thousands of tubewells. This mixing of ideologies anticipated, in technological and environmental terms, the politics of post-colonial development that came to agonize the second half of the twentieth century. Stampe's Malthusian dilemma, solved by science and technology, was a dress rehearsal of the arguments governments and international development experts later deployed to "liberate" agrarian communities in India and around the globe: a green revolution.

Stampe's audience at Caxton Hall no doubt knew of the Upper and Lower Ganges Canals. They may have even known that the Saraswati River was believed to surface from underground to join the Ganges and Jamuna rivers at the tip of the Doab, the place known as Triveni Sangam (Allahabad/Prayag). But what of this underground river? Before telling them how exactly he proposed to utilize this subterranean resource, Stampe performed a psychoanalysis of the topography: the river "flows on through the dark sands, unconscious equally of the administrative boundaries above it, as of racial creeds, political differences and salutes, only hoping, perhaps, as we do, to 'surface' with the aid of co-operative energy, and thus see the light of a brighter economic day."[3] Stampe's rhetoric of impartiality allowed him to transcend political concerns like "administrative boundaries" and "racial creeds"; the engineer is once again the great revealer of "natural truths."

In his speech, Stampe had to be pithy—he had only twenty minutes to convey his vision. He therefore couldn't give a thorough account of the tubewell's complicated history. He couldn't waste time recounting how the tubewell was a transcontinental, military invention, initially used by British engineers in the Abyssinian War of the late-1860s to quickly pump drinking water from the Ethiopian desert. He neither had time to explain how the first experiments with tubewells in India began in 1911 in and around the city of Amritsar in Punjab. These experiments were not only designed to irrigate land, but to reduce percolation and waterlogging from the Upper Bari Doab Canal in urban areas. Running three miles outside the city of Amritsar (with a population of 150,000 in 1920), the Upper Bari Doab Canal artificially raised the subsoil water level during the dry season from 60 feet to 10 feet (18 to 3 meters). During the southwest monsoon, mosquitoes

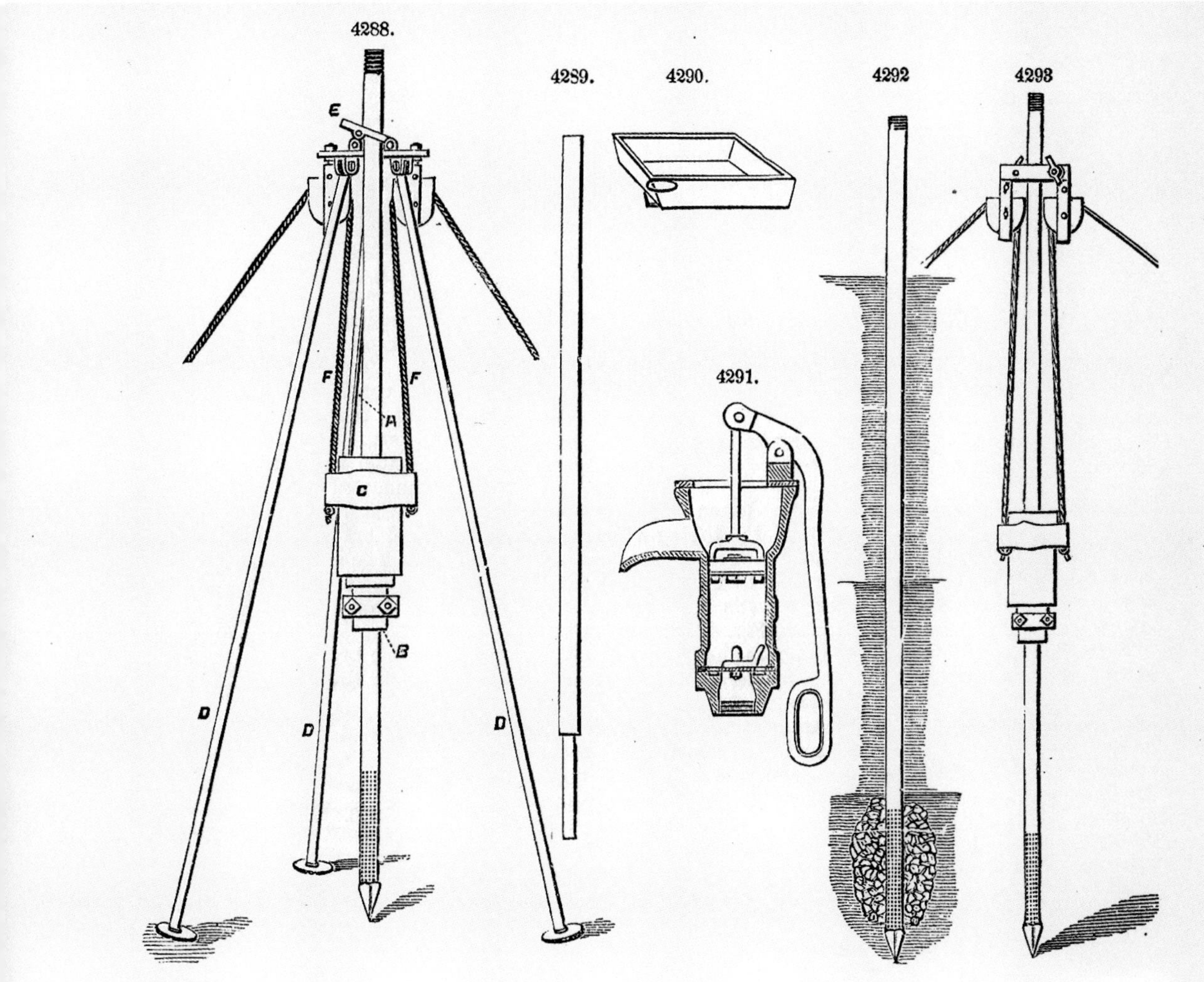

"DRIVEN WELLS—The process of driving tube-wells resembles pile-driving, but this distinction, that, while piles received the blows of the monkey on their heads, the tubes are not struck at all, the blow being communicated by the clamp, which receives the blow near the ground. The tube-well, as ordinarily used, is not intended for piercing rock or solid stone formations, but it is quite capable of penetrating very hard and compact soils... The first or pioneer tube A, shown in Fig. 4288, is furnished with a steel point of bulbous form, and perforated with holes varying from one-eighth of an into to one-quarter of an inch... When water is struck, and stands several feet in the tube, the pump, Fig. 4291, can be applied. Fig. 4292 shows a well-tube that has been driven, and the soil disintegrated around the perforations. Fig. 4293 shows the mode of driving the tube without use of supporting standards..." (Appletons', pages 928-929).

During the second half of the 19th century, encyclopedias dedicated to applied mechanics showcased tools and methods for drilling and boring into soil and rock—often for artesian wells and later oil. Such instruments and techniques became instrumental, not only for the design of the Abyssinian well of the 1860s, but also later for the tubewell.

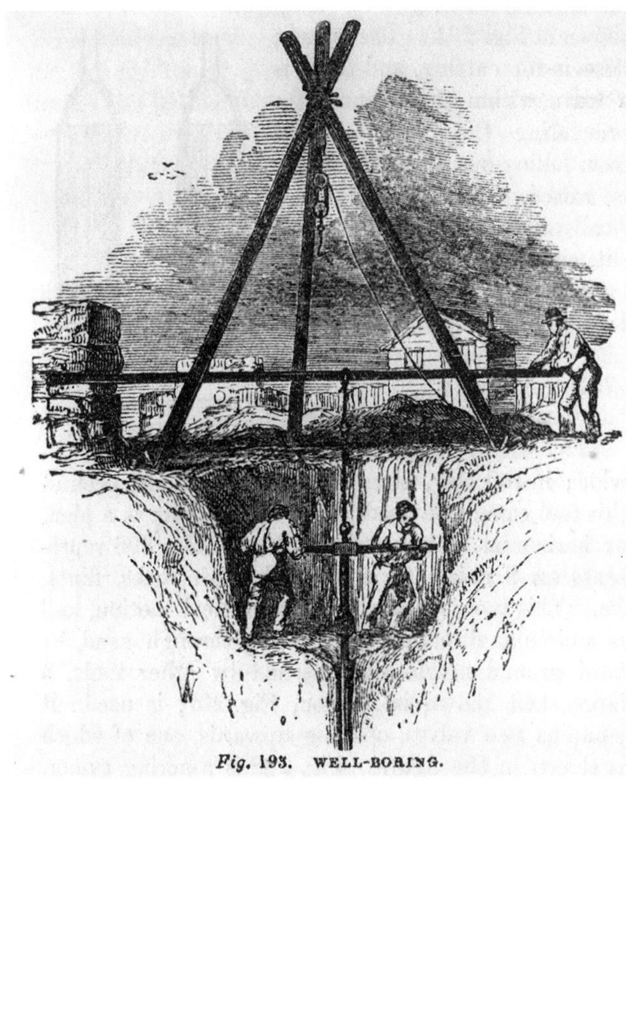

Fig. 193. WELL-BORING.

"Fig. 193. Well-Boring." **Cyclopaedia of Useful Arts, Mechanical and Chemical,** Vol. I, edited by Charles Tomlinson. London: Virtue and Co., 1866. (Princeton University).

bred "with great rapidity in this stagnant water, and infest the city, carrying with them the germs of disease."[4] Neither did Stampe have time to tell about how he already led "The Ganges Canal Hydro-Electric Scheme with Its Associated State Tube-Well Irrigation Project" in northern India in the 1930s. He couldn't relay how, by 1937, ten of the thirteen engineered falls along the Upper Ganges Canal were outfitted with "rotating machinery, which an electric age has since imposed upon them," electrifying 88 villages and powering a network of some 1,500 tubewells capable of irrigating 600,000 acres.[5] Stampe saw this use of the Ganges Canal as in line with Sir Proby Cautley's vision of the canal serving a "double purpose, viz., that of direct irrigation in the adjacent fields and also in wide tracts hundreds of miles beyond the command of the canal waters themselves."[6] Extending over and beyond Cautley's rajbuhas, Stampe's "hydro-electric power grid" transmitted the energy of the Ganges to over 1,000 tubewells dotting the Doab. Producing electricity from the canal, for Stampe, paid honor to Cautley's ghost by widening the command area of this vast irrigating machine.

But he didn't have time to say a word of any of this. Instead, he told the audience how, after inspecting the tubewell experiments in 1934, the Governor of the United Provinces concluded his inspection with the statement: "'Whilst appreciating the technical and economic advantages of the project, what impresses me most about the scheme is the prospect that it offers for better relations between the official and the village community.'" For Stampe, this epitomized the ways in which Indians might become "owners and custodians of their destinies," while the British, "possessing longer mechanical experience as well as greater factory resources," would then foster "[post-independent] India's agricultural economy."[7] Despotism was to become mechanical paternalism—"peaceful conquest" could continue.

Tubewells recast scale and power relations in the Doab. Stampe's introduction of the tubewell defines a territory that could theoretically be shaped and controlled by a single individual, unfettered by the need for centralized infrastructure (like the Ganges Canal). It is a device to insulate the individual farmer from the caprices of both the monsoon and water bureaucracy. It has the potential to create an artificial, compact "oasis." Plot by plot, Stampe's idea succeeded in transforming more small patches of earth than canals ever had. Those patches were turned into oases that could increase productivity, supply the freedom to choose what to cultivate (basing decisions on higher market prices rather than the availability of water), and outlive droughts. He reimagined agrarian space in vertical sections as opposed to the horizontal plan: a decisive victory against the realities of topography. Within this new schema, the farm became a refuge and the tubewell became its technological savior. Every farmer could lead his own peaceful conquest of land.

Stampe surveyed these developments from the air. After three years of inspecting thousands of miles of the Doab from an airplane, he computed an approximate "Khaki-Green ratio" (K/G). Unlike the canals, with bands of green hugging their edges, the areas serviced by the Ganges hydro-electric tubewell scheme formed a mosaic of green patches.[8] Over the course of his aerial surveys, he found that outside the canal command zones of northern India, the "Khaki-Green ratio" was as high as 10 or even 11 to 1. "It was the main object of the 'Himalayan Dream' to reduce this ratio from, say 90:10 to perhaps 50:50."[9] This was not to be a change in degree, but a change in kind—a spatial and social transformation. The tubewell was the fulcrum upon which the farm could be raised.

Seen from this distance, tubewell technology was liberating. Barely legible from the air were the other tiny, quickly disappearing lifelines: electricity. As self-sufficient as they seemed, tubewells required energy inputs. Energy to electrify village wells was supplied by the Government of U.P. Electrification Board, which was in turn produced from the falls of the Ganges Canal. Initially, in the 1930s, oil-powered motors were too expensive for most small landholders, but by the late-1960s and '70s, over 90 percent of all private tubewells would be powered by diesel (only government tubewells remained electric-powered). Because electrified tubewells were tethered to state-pro-

Tubewell with Submerged Pump and Covered Diesel-Powered engine near Fatephur (U.P.), 2008.

Photo-etching Roorkee College.

BAHADURABAD FALLS, GANGES CANAL.

"Bahadurabad Falls, Ganges Canal." **Roorkee Treatise on Civil Engineering: Irrigation Work in India**, edited by J. Clibborn. Roorkee: Thomason Civil Engineering College Press, 1901. (New York Public Library).

duced electricity, it was logical for state governments and economists to calculate energy consumption in regard to agricultural production. The oil tubewell, on the other hand, was a far less transparent technology, particularly from the point of view of state governments and economists. After it was paid for, it was up to the purchaser to decide how and where it was used. However, where there was no electric grid yet in place, Stampe and others supported the use of oil-powered tubewells until rural electrification was completed.

Stampe proposed three major initiatives in his scheme for tubewell dispersion. First, the creation of a Central Ground-Water Section (C.G.W.S.) to facilitate demonstrations on the "efficacy, economy, and speed of the mechanical drilling of tube-wells."[10] Second, sending Indians to the recently-opened Central Drilling School in Roorkee, the headquarters of the Upper-Ganges Canal, to be trained as drillers by "expert drilling engineers." Lastly, in cooperation with the Geological Survey of India and other agencies of the government, the creation of a "ground-water contour map of the whole of alluvial India, which will readily indicate the prospects of irrigation from the Saraswati and its branches."[11] Stampe estimated that such a contour map would help site "many thousand tube-wells" across the country to produce millions of additional tons of food. To make good on his carefully calibrated transformation of agriculture, the government would need to control the allocation of energy. Electricity was their preferred power source, but there was no chance of completing hydroelectric dams in time to meet the growing food demands of an exploding population. Unlike electricity, which was metered at or near the site of its consumption and tethered to power lines, once oil left the pump, it was mobile and could be stored indefinitely. A diesel-powered tubewell was a wild card, a wily technology compared to the accountable, electrified tubewell. Nevertheless, Stampe supported the use of oil-powered tubewells, in the interim, until electrification was complete.

By 1937, ten of the thirteen engineered falls along the Upper Ganges Canal were outfitted with "rotating machinery, which an electric age has since imposed upon them," electrifying 88 villages and powering a network of some 1,500 tubewells capable of irrigating 600,000 acres.

Built in 1930, Bahaderabad Power House was one of the original powerhouses of William Stampe's Ganges Canal Hydro-Electric Supply. Today, the interior has largely been gutted and is a popular hangout with lovers and quadrupeds, 2008.

Although the state made ambitious plans to build hydroelectric dams to meet certain agricultural demands, their other smaller initiatives were more immediately successful at increasing government oversight of agricultural production. Electricity and water extraction were not the only methods for engineers and bureaucrats to monitor production: fertilizer was another important input. After experimenting with mixing ammonium and water at the tubewells in various proportions to produce a "fertilized water"—an effort that ultimately failed to work—fertilizers were sold to farmers at tubewell demonstration centers. Field staff from the Department of Agriculture and the Department of Irrigation trained farmers to apply fertilizer in order to make the most use of tubewell irrigation.[12] Training centers were placed in key areas throughout the tubewell area; they served as advertisements for a new mode of living and working.

Long before he shared his "Himalayan Dream" in 1946, Stampe's scheme in western U.P. inspired others to dream, as well. After touring Stampe's experiments in 1937, the American-born missionary and founder of Allahabad Agricultural Institute, Sam Higginbottom, wrote in *The Allahabad Farmer* "This tube-well area is one of the outstanding pieces of irrigation work anywhere in the world." Besides its practical and economic uses, Higginbottom was impressed by how the tubewell could reshape the social life of the village. "Because there is electric power at every tube-well, this could easily become the social centre of the village, a radio receiving set and adult school, demonstrations of better crop varieties and methods."[13] The tubewell's technical assistance to irrigation was its entry point into village life; from there it could reshape the life of the community. The tubewell was anything but mono-functional—it was a Trojan horse.

By the late-1940s, the future of the Ganges River Basin was anyone's guess. Few would disagree with the amazing potential of the tubewell, but no one really knew if it was capable of meeting Higginbottom's vision or Stampe's "Himalayan Dream." Stampe, an avid aviator, crashed his aircraft and died outside Devon, England in 1951. His obituary, appearing in *Civil Engineering and Public Works Review*, remembered him "as a far sighted irrigation engineer in Upper India, and in particular for the principle of supplementing great canal systems by feeder tube wells"—a fitting final line for a man who had extended the work begun by Sir Proby Cautley with the Ganges Canal.[14] These two technologies co-existed, for a time, but they would soon mutate into an ungovernable groundwater revolution.

FORMATION OF SUPER-SURFACE

Canals and tubewells, along with roads and railways, cities and farms produce a layered super-surface. Dense with these technologies of water extraction and distribution, it is an embodiment not only of a long and fraught history, but also a peculiar mixing of ideologies—the centralized character and management of the canals versus the diffuse, expansive distribution of millions of tubewells. From the days of the Company and Raj through to Independence and the Soviet-inspired planned economic growth, the tubewell came to dominate (unofficially) irrigation in U.P. Unlike the canal, its spatial pattern of development is not linear, but diffuse.
2010.

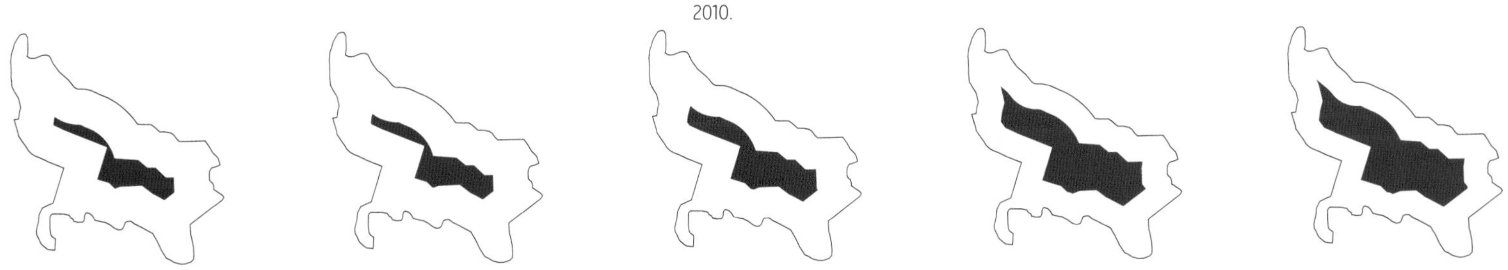

PRE-PLAN (Before 1950) 25.53 | FIRST PLAN (1951-56) 28.83 | FOURTH PLAN (1969-74) 40.80 | SEVENTH PLAN (1975-80) 66.58 | ANNUAL PLAN (1996-97) 70.28

DEVELOPMENT OF IRRIGATION IN U.P (PRE-1950 TO 1997)

IRRIGATED AREA (x 100,000 ha)

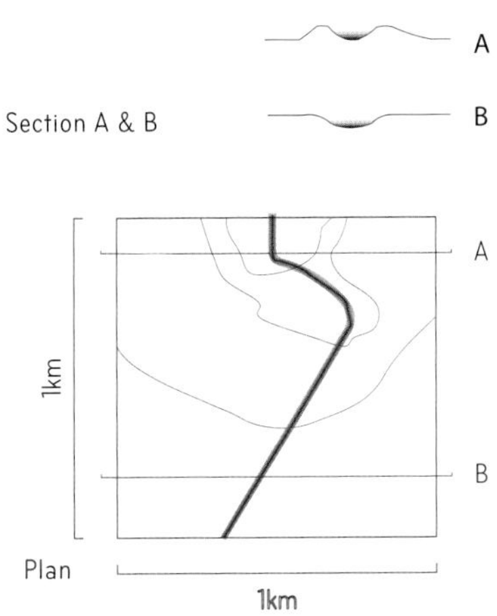

CANAL

	IRRIGATED AREA (X 100,000 ha)	TOTAL LENGTH OF CANALS (km)	TOTAL NUMBER OF PUBLIC TUBEWELLS
CANAL	51.70	74, 762 km*	
TUBE WELL	11.91*		28, 636**

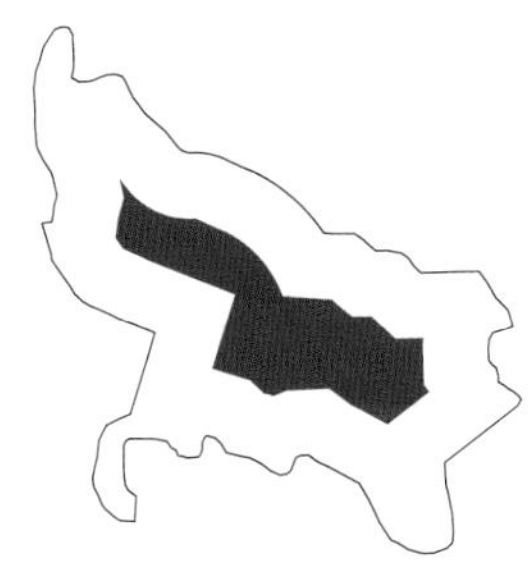

* official number as of 1996-97 from Government of U.P.

TUBE WELL

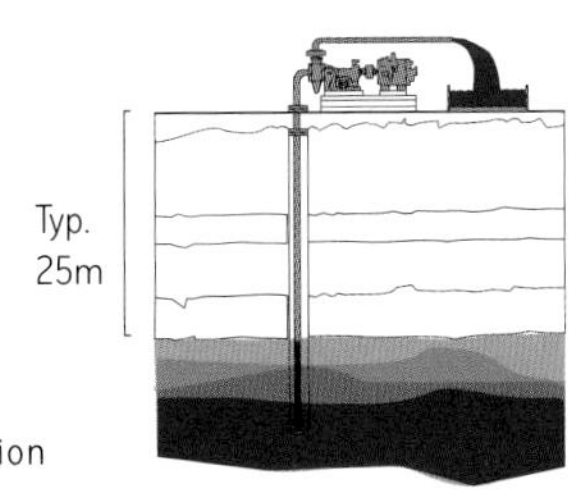

Section

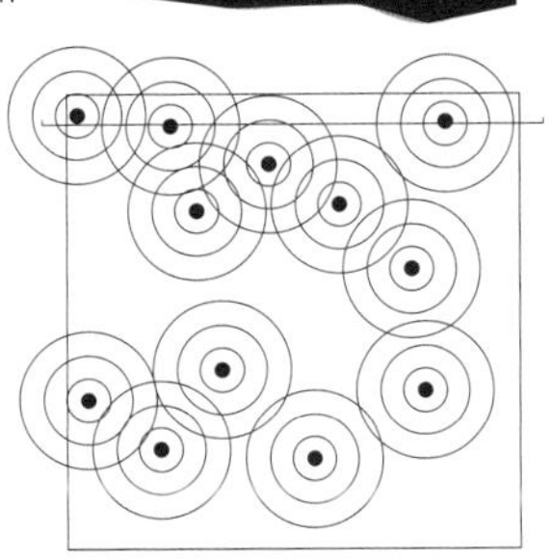

Plan

*official number as of 1996-97 from Government of U.P. (In actuality, 11.91 is lower due to the uncounted number of tubewells)

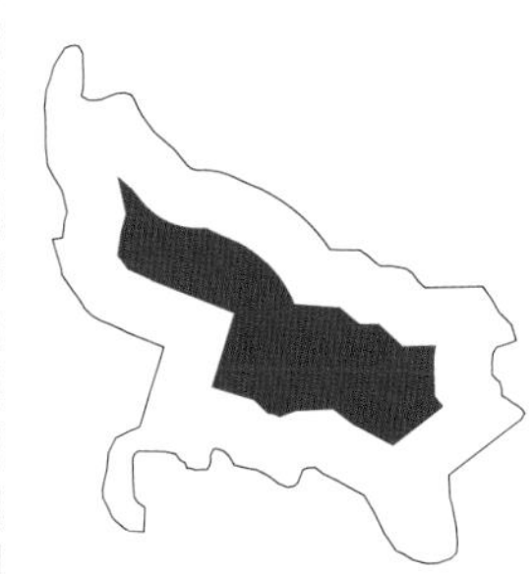

TOTAL IRRIGATION
7,028,000 hectares**

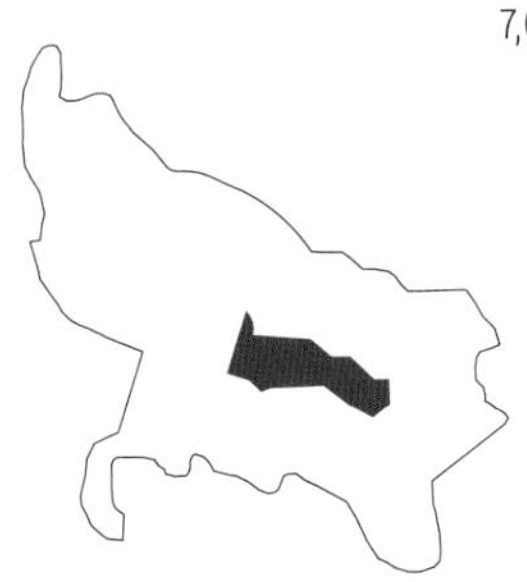

**official area as of 1996-97 from Government of U.P., which includes present day Uttarakhand. (Because it is estimated that there are over 3.5 million private tubewells, this area is much greater)

SUPER-SURFACE

SURVEY

Aboard an airplane en route to Varanasi from Delhi, one looks down to a sparsely forested, diffusely populated mosaic; the landscape crisscrossed by irregular grids of beige cropland. On a cloudless spring day, during what is known as the rabi season, when the air and the land are both dry, the Ganges River contracts markedly. Large sandbars and shoals develop in the bed of the river and reduce its flow in parts to that of a brook. It is at this time of year when the Ganges Canal is most visible. Its edges are green, its water is flowing, and its rajbuhas weave in and out of fields and cities. A highway like the Grand Trunk Road, which stretches from Delhi to Kolkata (Calcutta), pierces through the Doab and crosses the Ganges at Allahabad. Villages, often clustered around an irregularly shaped algae-colored water tank, speckle this agricultural stretch of U.P. (Uttar Pradesh).

A return trip in the summer renders the landscape below unrecognizable. What were once dry and reticulated divisions of agricultural plots now dissolve into stains and smudges of soupy soil, dully reflecting clouds up above. Within weeks, chartreuse shoots of basmati rice sprout above this thickened ground, inaugurating the kharif season. Farmers tiptoe through fields, tending to the shoots, clearing them of all manner of debris. The airplane provides a certain perspective, an often-criticized god-like view. Questioned though it may be, looking from above reveals interrelated figures and marks made in the landscape, especially if one can compare those same figures and marks from another perspective.

Aboard a bus on another spring day, en route to Varanasi from Delhi, a rider looks across a bustling highway of people dodging cars and horse drawn carts and livestock. The air is dry and the heat is intense. Countless canals and nalahs (drains) sneak beneath the road. They cut clear linear vistas deep into the horizon. If one is not looking for these drains, one still hears them; rolling over a canal bridge produces a slight hiccup in the bus's cabin. All the while, like a series of jump cuts in a film, villages appear along the side of the road and then abruptly end, only to be replaced by another in the distance. Homes and shops, built out of brick and elevated, or the rare mud home capped with a thatch roof, are all tethered to the nearest electrified power line (at least, when electricity is available). Elsewhere, near a house or out in a field, sometimes covered by a tin roof, but mostly exposed to the elements, sits a 10-15 horsepower engine. Nowhere is an electrical line drawn to it. These private pumps are powered by diesel. If one hops off the bus and explores the villages or the outskirts of a town, such diesel-powered engines can be seen in every direction. At certain hours of the day and year, a careful listener can walk across any field and hear these pumps running, gurgling, and chugging water drawn from Stampe's subterranean Saraswati. And late at night or early in the morning, one hears the pumps filling water tanks for a day's supply atop homes and businesses. This is not just in villages and small towns; it also occurs in sprawling metropolitan areas. Where there are people, there is water pumping.

According to a 2010 report from the World Bank, India extracts nearly 230 cubic kilometers of water per year, making it the single largest user of ground water in the world.[15] Some eighty-five percent of drinking water and sixty percent of irrigation for agriculture is derived from groundwater. In the Indo-Gangetic plains, much of this groundwater extraction is done by means of private tubewells. U.P. is no exception. Private tubewell users extract groundwater in villages and cities as well as towns and industrial estates at alarming rates. The shift from the centralized water distribution of government-owned canals to "democratic" forms of water extraction by means of privately owned tubewells exposes a landscape borne out of the discontinuities of politics, overlapping bureaucracies, and cultural practices. The only limitation placed on the number of tubewells and the hours they ran has been personal wealth. State governments have not required any registration of tubewell ownership. For decades owners drilled as many wells as they saw fit.

As with any open access system in which users may partake as much as they wish of a natural resource, whether it be groundwater or grazing grounds or ocean fisheries, a tragedy of the commons has emerged. In the case of the Ganges River Basin, since the mid-1960s, small and marginal farms, with little available cash and credit, have found it impossible to compete with medium and large farms equipped with private tubewells. Mining groundwater, a practice whereby water is extracted faster than an aquifer can recharge, has annually lowered the water table level in areas of western U.P. and Punjab. This depletion has lead farmers and companies to sink tubewells at even greater depths, producing "tubewell wars." Economists have noted that the mining of a common resource like groundwater gives "rise to technical diseconomies," particularly for those with no access to pump irrigation at all.[16] Over the last five decades, as the population continues to grow in states like U.P. and private tubewells are sunk to meet the competing demands of irrigation, industrial manufacturing, and drinking water, these inter-personal and inter-regional imbalances grow more stark and tragic.

A tal (tank) and a set of fields along Grand Trunk Road (NH2), less than forty kilometers southeast of Allahabad: between June 14 and July 26, 2006, with the arrival of the south-west monsoon, the khaki to green ratio is inverted.

GROUNDWATER POTENTIAL

Ganges River Basin contains the greatest groundwater potential in India: a shallow sub-surface volume—Stampe's Saraswati. The tubewell connects the surface with the underground, 2011.

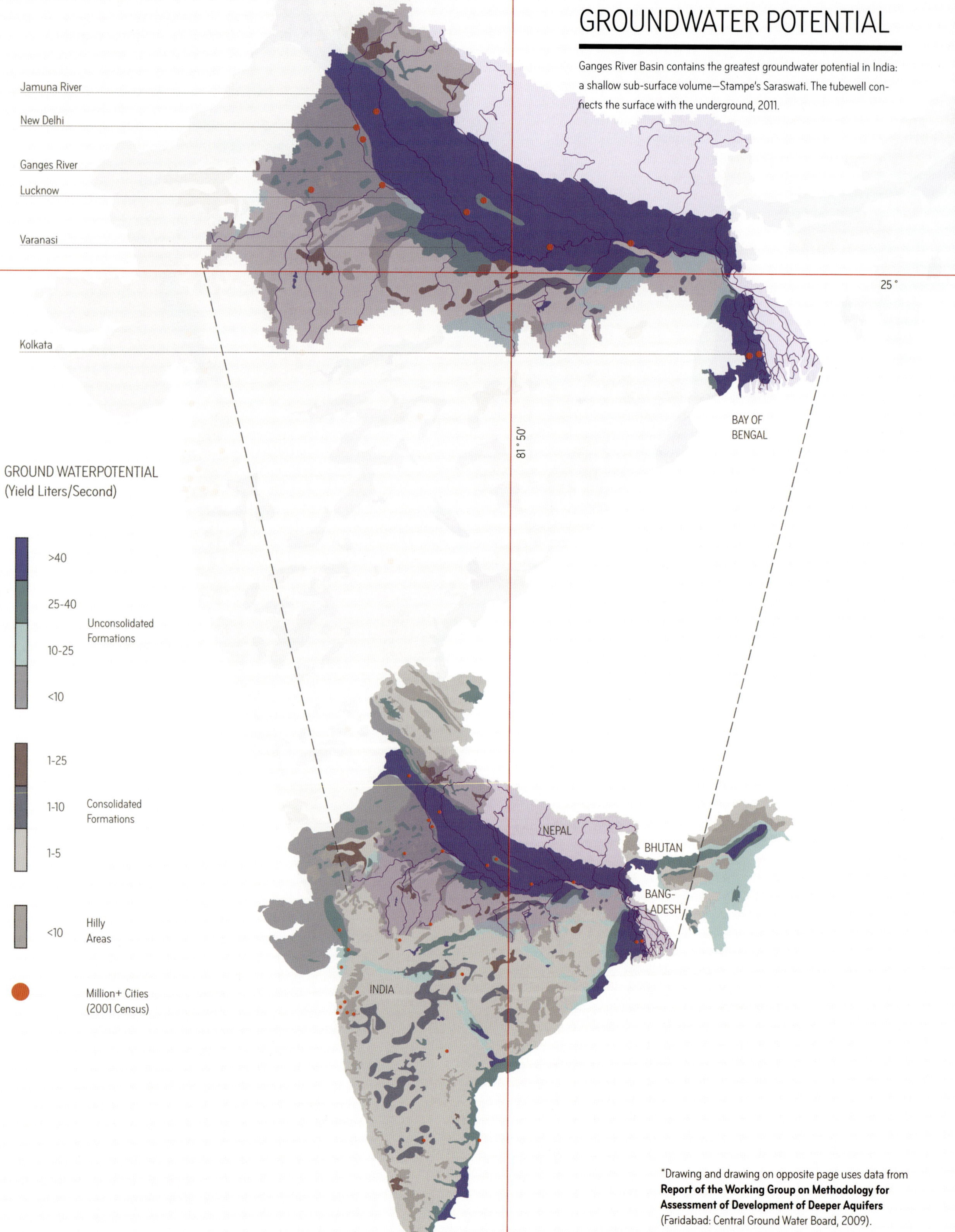

*Drawing and drawing on opposite page uses data from **Report of the Working Group on Methodology for Assessment of Development of Deeper Aquifers** (Faridabad: Central Ground Water Board, 2009).

AQUIFER SCHEMATIZATION IN U.P.

Districts in and around the Doab remain some of the best irrigated in the state, while those in Eastern U.P., long neglected by the state, have been increasing their share since the 1980s.
2011.

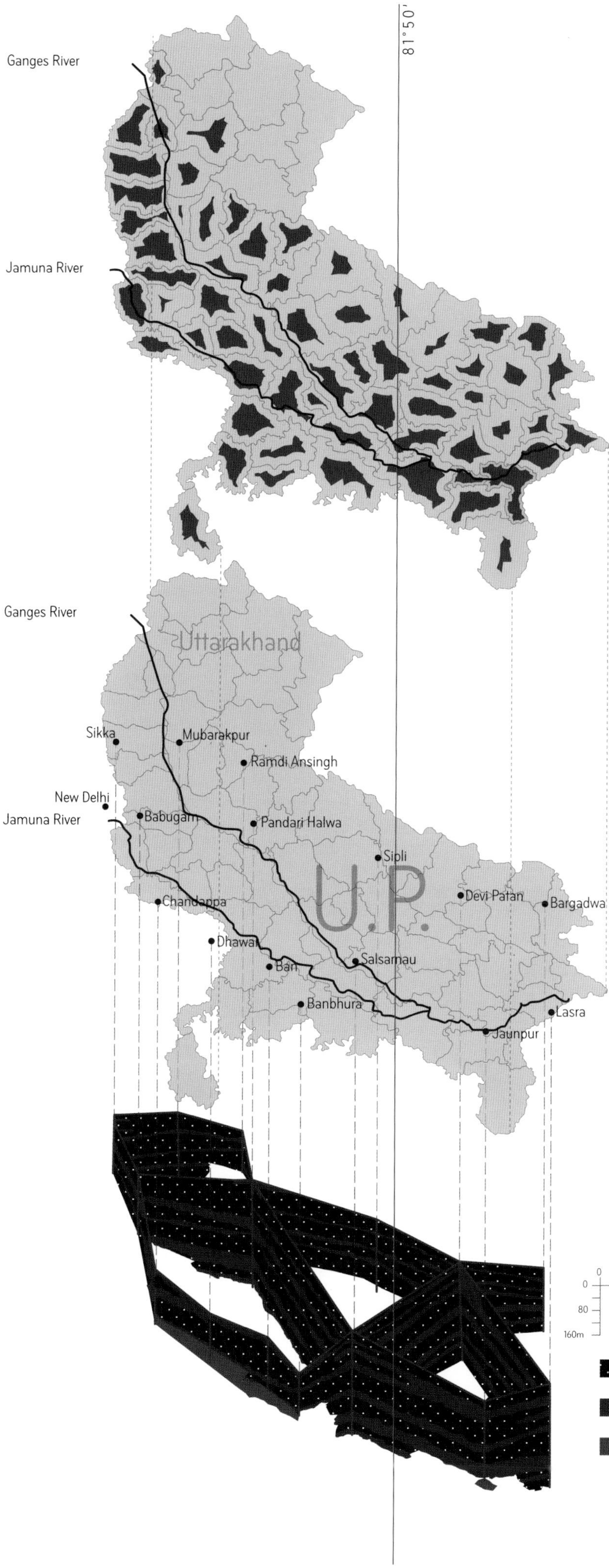

SURFACE IRRIGATED/DISTRICT IN U.P.*

SURFACE WITH DISTRICTS OF U.P.

AXONOMETRIC OF SUB-SURFACE SOIL CONFIGURIZATION

0 60 120 km

0 80 160m

Sand, Clay, Kankar, Gravel, Boulders

Clay & Kankar

Sandstone, Shale, Granite

*Note that this is the net irrigated area/district and not the actual spatial configuration of each district—each district is far more fractured.

Outside Fatehpur (U.P.), 2008.

Many crises, like hurricanes, tornadoes or earthquakes, occur instantly or happen in a few days or weeks. By contrast, the proliferation of new forms of water pumping technologies produce a crisis in slow motion—rather than a collision of "countryside versus city," political and material conflicts arise over a shared set of infrastructures for water extraction. Roadways, railways, canals, rivers, agriculture, and human settlement further layer this already stratified condition. It is the water in aquifers, stored in the voids and fractures of soil, sand, and rock that occur beneath the surface of the land, which set off a set of unforeseen cultural, political, and spatial mutations.

It is worth considering how privately owned pump technologies became so ubiquitous. How did the state of U.P. go from 3,000 private tubewells in 1951 to over 3.5 million in the last ten years?[17] Who owns these 3.5 million private tubewells in a state inhabited by just over 200.5 million people? Who can afford these technologies and who has access to them? How does today's reality of private tubewells and water pumps compare with Stampe's publicly electrified "Himalayan Dream" of surface and sub-surface water integration? Finally, what is the relationship between the monumental surface water projects like the Ganges Canal and the diffuse tubewell irrigation across the state?

Pumping water from a canal running parallel to National Highway 76 to irrigate a field, outside Manpur (U.P.), 2006.

LIBERATION

The canal and the tubewell came to represent the limits of the Ganges River Basin's infrastructural vocabulary. They each, in their own particular way, defined the extremes of the basin's territorial development. The canal is the most monumental, most expansive structure to shape space. With its hierarchy of trunks and branches, it creates legible corridors and two-dimensional order. The tubewell is an inversion of the canal—the thinnest, most inconspicuous structure to mark a site. With a minimal footprint and maximum draft of water, it creates undreamed of independence and three-dimensional chaos.

A shallow tubewell may be bored or drilled in under an hour by a few industrious souls. A canal requires countless laborers, at least one surveyor and engineer, and takes weeks, months, even years to complete. Tubewells produce an ever-expanding constellation of sites, each capable of creating their own gravitational pull. The tubewell is the great liberator of the topographic surface and symbol of economic liberalization. It is a fantastic technology, precisely because of its differences to a canal: it can be sunk almost anywhere, it is portable, it provides water on demand, it is managed independently, and it is not a monument. It is, in fact, the antithesis of a monument. Unlike the canal, no one writes poems to praise it or to herald its heroes.[18]

Despite (or perhaps because of) this lack of poetry, the motorized tubewell is the final act in the peaceful conquest of the Ganges River Basin. If surface water operates in a shallow X-Y coordinate system, then the tubewell introduces a deep Z-coordinate, breaking the limitations of the flat alluvial plains. Managing the caprices of climate, weather, and precipitation are within the reach of the farmer through the thickening of this multi-layered super-surface.

STATIONS

"Since the Indo-Gangetic plains are better known for their spirit of individual instead of cooperative endeavour," noted B. D. Dhawan in 1982, "the development of tubewell irrigation on an extensive scale depended primarily on the direct initiative of the state in the pre-plan days."[19] Following Indian Independence, the promise of pump technology led to government-run stations of tubewell irrigation. These sites were set up as experiments to determine the best practices for tubewell use in regard to agricultural productivity and equitable water distribution. Images of tubewells circulated in newspapers, academic reports, and journals as advertisements for the wonders of mobile and compact modern machines. Lifting water from the dusty soil, in such images the tubewells appeared to instantly alter reality.

As part of the planned economic development of the successive Soviet-inspired 'Five Year Plans,' the first Five Year Plan (1951-56) aimed to expand both major and minor works of irrigation. Throughout the 1950s, canal building and tubewell drilling grew in tandem in all the states across the country, but in U.P. above all others. Since Independence, U.P. has been the most populous state in the nation. It has grown from 60 million inhabitants in 1951 to over 200 million in 2011, occupying an area the same size as Great Britain. If U.P. were a country, it would be the sixth most populous in the world. Less than one-quarter of its inhabitants—44.5 million—are said to live in "urban regions." The remaining 175.5 million live in some 106,000 villages or "rural regions."[20] These pressures led to a concentrated need for irrigation in the state and so U.P. became the incubator of irrigation in the Ganges River Basin.

At these early stages of tubewell development, there was a dearth of expertise when it came to tubewell irrigation (in contrast to the long history of canal engineering in India). Developing a decentralized technology involved different methods of distribution, financing, training, and oversight. The Indian government came to rely on technical assistance from foreign donors, like the U.S. government and the World Bank, to help exploit the potential for groundwater irrigation. Focusing primarily on developing India's productivity for self-sufficiency and industrial development, these funding models and policies drew India into larger geopolitical debates that emerged around this time about the nature of hunger, industrialization, and economic security.

Though introduced in Amritsar to counteract the rising water-table level of the Upper Bari Doab Canal (Punjab), the tubewell and the canal have long lived separate lives—only occasionally reunited (at least officially).

Canals and Pumps between Kanpur and Varanasi (U.P.), 2006-08.

WORLD POPULATION TO LAND AREA: A TICKING TIME BOMB?

If U.P. were a nation, it would be the sixth most populated country in the world, with a land area the same size of Great Britain, 2014.

1,354,040,000 (2013)

PRC 9,596,960 km2

1,210,193,422 (2011)

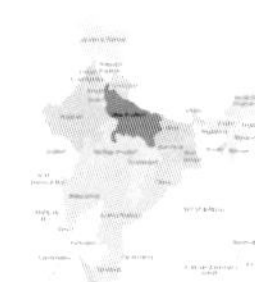

INDIA 3,287,590 km2

507,069,424 (2014)

E.U. 4,271,600 km2

315,183,801 (2013)

USA 9,631,420 km2

237,424,363 (2010)

INDONESIA 1,919,440 km2

199,581,477 (2011)

U.P. 238,566 km2

190,732,694 (2010)

BRAZIL 8,514,877 km2

But the first step was to determine just how much groundwater potential India, or even just the Gangetic plains, actually had. How was the state to make the most of this vast underground resource? What areas contained confined aquifers (covered by impermeable materials like rock formations) and what areas contained unconfined aquifers (where the top of the aquifer is covered by a permeable material like alluvium)? Finding answers to these questions would require a comprehensive subterranean map.

In June of 1950, Charles C. Taylor, Jr. arrived in India to create one. On loan from the USGS (United States Geological Survey) to the GSI (Geological Survey of India) until 1955, Taylor, a geologist and hydrogeologist, came to "design a long term program of technical assistance to the GSI in groundwater geology and hydrology."[21] Funded by U.S. President Harry S. Truman's Point Four Technical Assistance Program to Developing Countries, Taylor's first order of business was to train the GSI's hydrogeologists. He divided their time between the office and the field. Hydrogeologists, in teams of three to four men, would accompany him on four to six month long "reconnaissance ground-water investigations" on which they conducted "hydrogeologic mapping, well inventories, some aquifer tests, evaluation of exploratory drilling by cooperating agencies, and groundwater sampling and mapping." This data was then collated and organized at the offices of the GSI in Calcutta (Kolkata).[22]

In those five years, Taylor came to see some additional 2,200 productive public tubewells sunk in the Gangetic plains of India, mostly in U.P. and Bihar. He also helped establish institutions that would support this new infrastructure, like the Exploratory Tubewell Organization (ETO), formed in 1953 and placed in the bureaucratic structure of the GSI and the Ministry of Food and Agriculture (with USGS technical assistance).[23] Over the course of his work, Taylor identified eight "ground-water provinces of India." His 1957 'Map Showing the Areal Extent of the Ground-Water Provinces of India,' based on a GSI map from 1949, designated the Ganges-Brahmaputra Province as the single largest and most productive of the eight. What was clear, both from his map and his descriptions, was that the Ganges-Brahmaputra basin was to be the center of tubewell development.

Political, administrative, and infrastructural lines already demarcated this vast surface, but they were not the only forces at play. How to disperse tubewell technology equitably and productively remained a challenge for several reasons. Outside the new stations and select areas, India was facing a serious food shortage. Taylor's sponsored collaboration with branches of the Indian government overlapped with a request, made by the Government of India, for two million tons of wheat from the U.S. Just three years after Independence, in 1950, India's food supply was threatened by floods followed by drought in Bihar and Madras. In the aftermath of WWII, the U.S. was the only country capable of supplying this aid. The U.S. was experiencing unprecedented agricultural surpluses, surpluses which led American lawmakers, particularly in agrarian states, to see food as a persuasive tool in foreign policy and international development.

After a series of political and economic quarrels between Prime Minister Jawaharlal Nehru and American legislators, two million tons of American wheat arrived in the port of Bombay in August of 1951.[24] In less than three years time, U.S. President Dwight D. Eisenhower signed into law the Agricultural Trade Development and Assistance Act of 1954, commonly known as PL-480 or Food for Peace. The law established the comprehensive foundation for U.S. foreign food aid, while also liquidating large surpluses of wheat and grain in the U.S—that is, agricultural aid would be used to bolster the U.S. international image, while also propping up its domestic agricultural industry. With the passage of the law and the perceived threat of communism in Asia, American politicians and policy experts saw raising the standard of living in the world's largest democratic republic—India—as a crucial counterpoint to Mao Zedong's Great Leap Forward in China. American food aid to India also provided the necessary cover for implementing long-term technical assistance to programs like groundwater extraction.

EXPLOSION

By the end of the 1950s, the U.S. government, western economists and academics, philanthropies, and research institutes began to look at India's food aid in relation to the perceived "ticking time bomb" of agricultural scarcity combined with a burgeoning population. Policymakers began to question the tenets of centralized planning by the Indian state: detailed regulation of private industry, government investment in large-scale industrial projects, and widespread control of foreign trade. Reinforcing this view was Karl Wittfogel's *Oriental Despotism: A Comparative Study of Total Power* (1957), which introduced Americans to the idea that Oriental cultures create "hydraulic societies" with deep connections to irrigation, bureaucracy, and despotism.[25] Within this set of emerging geopolitical positions, the ideals of open markets and decentralization of power promised greater freedom. The tubewell, with its decentralized control of resources and its ability to spur the creation of local markets—in water, energy, food, and land—fit easily into this worldview.

No document did more to solidify this point of view than the distressingly titled, *Report on India's Food Crisis and Steps to Meet It*. Published in April 1959 and sponsored by the Ford Foundation, the 269-page report was the result of a ten week investigation conducted by a team of a dozen American agricultural specialists. Written with a sense of dire urgency, the team focused on 1966 as a decisive year. By that time, the report estimated, India would reach 480 million people. If the country failed to triple its agricultural output, it would face a grain shortfall of 28 million tons. To cope with this new Malthusian dilemma, the specialists proposed turning away from "heavy industry or very large irrigation projects" and investing in minor irrigation, fertilization, and mechanization to reduce the threat of mass hunger.[26] Furthermore, shifting the focus of the national economy to agricultural production was not enough for the Ford Foundation team—reducing birth rates and developing effective family planning initiatives were also needed to curb rampant population growth. The Report's focus on small-scale technological inputs and social programs shaped the policies and mechanisms for dealing with food and population growth for decades to come.

In response to the Report, in August 1960 the Indian government launched the Intensive Agricultural District Programme (IADP). The program was aimed at raising "farm productivity in well endowed districts by applying a package of inputs consisting mainly of high quality seeds and chemical fertilizers."[27] For its size and scope, the Government of India considered the program a success; however, it was unable to meet India's successive agricultural shortfalls. This focus on agricultural productivity and population control only intensified during the tenure of President John F. Kennedy, before reaching a critical point in 1965 during President Lyndon B. Johnson's administration.

Two major blows to the Indian economy in the early 1960s further exacerbated economic growth and food security. First, following the 1962 Sino-Indian War, defense spending increased dramatically. Second, major failures from the monsoon crop in 1965 compromised food production.[28] Both of these, combined with international pressure from development experts and World Bank, led to the launch of a new strategy for agricultural development: the High Yield Variety Programme (HYVP). The new program was a comprehensive "technological package" of new inputs to raise productivity, most notably "miracle seeds": high-yield varieties of wheat, rice, maize, sorghum, and millet. These were developed by American biologist Norman Borlaug and a team of agronomists breeding fertilizer-responsive seeds in Mexico in the 1940s. By 1955, the U.S. based Rockefeller Foundation was funding and promoting Borlaug's research internationally to increase agricultural yields.

With the impending population boom and food gap in mind, the government of India introduced these seeds to farmers across the country. Funding was allocated not only for new seeds, but also for credit markets, agricultural research, education services, and, most importantly, fertilizer and irrigation. Without fertilizer and a controlled water source, the "miracle seeds" would not stand a chance. Tubewells became the preferred water controlled technology in this new campaign. During this period—eventually christened the Green Revolution in 1968—a private "tubewell explosion" took place across India, particularly in Punjab, Haryana, and western U.P. This burst moved from west to east in northern India, coming to a halt in central U.P. in the mid-1970s, before the widespread "dieselization" of eastern U.P.'s tubewells in the 1980s.[29]

U.P. POPULATION DENSITY DIAGRAMS

Between 2001 and 2011, the population of U.P. increased by 20% (from 166 million to 199 million). Estimates run as high as 270 million people by the year 2020.
According to the official U.P. State Water Policy, the state is said to contain 20.0 mha of land, with 17.4 mha under cultivation (as of 2011). By 2020, the state has assessed a food grain requirement of 63 million tons. With the exception of a few major regional capitals like Kanpur, Allahabad, Varanasi, Agra, and Lucknow, the Agropolis is the predominant form of human settlement in the state.

At present, cultivated and non-cultivated land is spread out like a mosaic across the state. To give some sense of how agriculturally productive the state is and how densely populated it is, imagine if each non-cultivated land was consolidated in four different ways:
1) northern tip
2) center
3) rectangle
4) outer edge of the state
2011-2014.

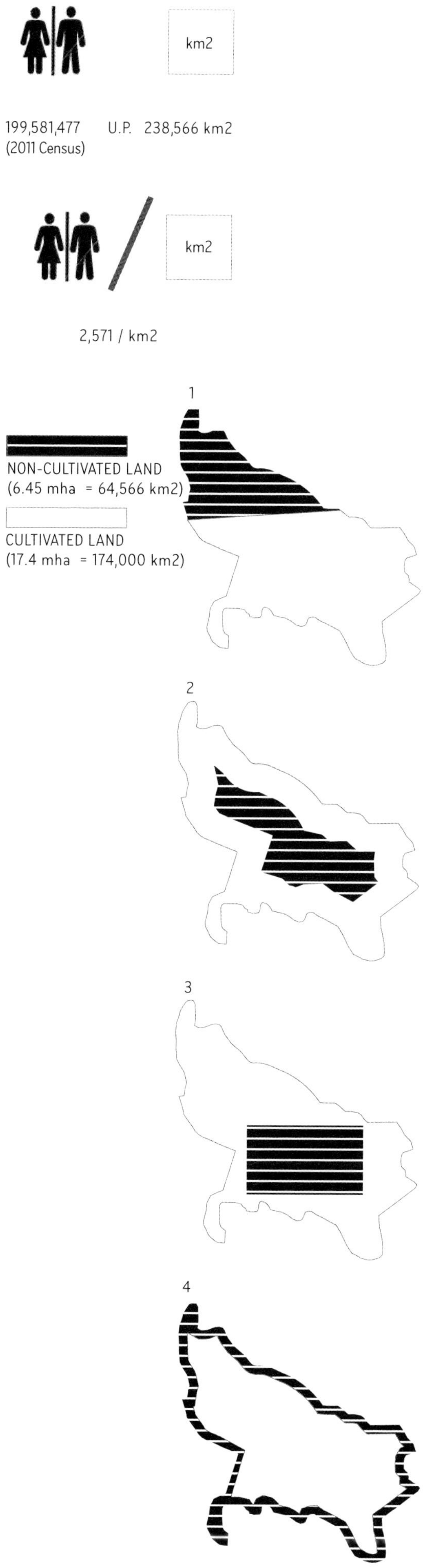

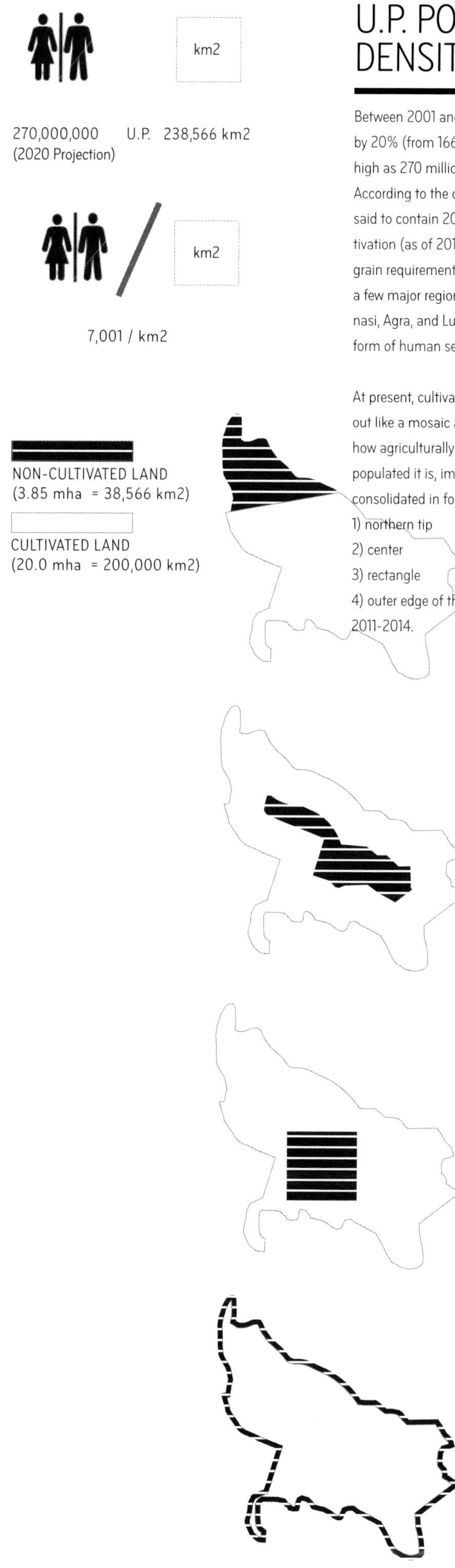

Just as inorganic fertilizer, which was produced using petroleum and was subject to waves of global surplus and scarcity, the private, diesel-powered tubewell explosion was imbricated in the global petroleum market too. Government and World Bank efforts to install public electric tubewells in U.P. almost always failed, largely as a result of inadequate or erratic electricity supplies from the U.P. State Electric Board (SEB).[30]

The introduction of purchased inputs, namely petroleum products, was a way to industrialize farming and open Indian agriculture to foreign investment. To impose these new inputs, the Johnson administration developed a 'short tether' policy, where all food deliveries from the U.S. were contingent on the adoption of a new package of agricultural policies.[31] This led economist Andre Gunder Frank to conclude that the Green Revolution was "business as usual" in terms of capitalist development. Writing in the same year as the energy crisis of 1972-73, an event that raised petroleum prices and stopped India's oil-fired and diesel-powered tubewells, Frank's criticism of the Green Revolution was prescient.[32] Fertilizer and tubewells, two of the essential "technological inputs" to increase food security, were thus tethered to global commodity fluctuations.

For Frank, the Green Revolution was in fact "counter-revolutionary"; it was a way to "contain revolution through 'peace.'"[33] The conflicted, perhaps contradictory, structure of development was made manifest when Norman Borlaug was awarded the Nobel Peace Prize in 1970. His HYVs had produced more food, but also exposed developing markets to larger global forces. The very term, Green Revolution—first spoken by David Gaud, Director of U.S. Agency for International Development (USAID)—contrasted these successes in agricultural production to the communist (red) revolutions in Southeast Asia. In addition, these larger geopolitical and market forces intersected with longstanding internal social reforms that had constrained agricultural productivity since Indian Independence. U.P. was the testing ground for these historical developments, and the diffusion and adoption of tubewell technology was the first step in this radical transformation.

IMPLEM-ENTATION

The seeding of a state as densely populated and geographically large as U.P. with public and private tubewells had posed difficult economic and spatial challenges since the first introduction of tubewells to northern India in the 1930s. Land reform, scarcity of electricity, as well as the lack of rural banks were all impediments to the spread of tubewell technology in U.P. Prior to the passage of the U.P. Tenancy Act of 1939, the majority of farmers were tenants-at-will. They did not own land and paid taxes to local *zamindars*, a hereditary group of rent collectors who paid taxes to the Company and later the Crown. With the passage of the 1939 Act, followed by the U.P. Abolition and Land Reform Act of 1951, millions of farmers became *bhumidars* (land owners holding the title to their land) and *sirdars* (people with hereditable rights to land, but unable to build on or sell the land). While the new laws abolished the longstanding zamindari system, they also produced a more equitable, if fractured topography. The size and configuration of these new farms—many were small landholdings and discontinuous—made it difficult for farmers to irrigate all of their land with a single tubewell.[34] Many landholdings were small and they were often discontinuous, with plots sometimes scattered several kilometers from one another. This distribution and size of land was compounded by inadequate access to credit markets to finance privately owned tubewells as well as an irregular supply of electricity, all of which diminished the state's ability to keep up with demand for groundwater irrigation.

One corrective to this problem was to reorganize fragmented farms. Land consolidation, a process whereby each farmer exchanged discontinuous plots of land for contiguous plots of equal value, was introduced to raise the economic viability of farms and increase social equity. Still, U.P. lagged behind the other states in land consolidation, as well as in electrification and rural credit. For instance, in 1956, 54% of all of the farms in the neighboring Punjab and Haryana were consolidated, while only 1% in U.P. were. By 1974, both Punjab and Haryana completely consolidated agricultural lands while U.P. was estimated to reach 70%.[35] (Admittedly, Punjab and Haryana did not face the same scale of population as U.P.) Because tubewell technology was capital intensive, without subsidies from the government to small-farmers in places like eastern U.P. and Bihar, only large farms could afford a private tubewell. Furthermore, a farmer's access to loans to finance a tubewell was based on their farm size. As a result, a small farm would pay a higher interest rate than a large farm in the same market. This same disadvantage was duplicated when small farms sold their produce at markets. For the small and marginal farmer, as economist B. D. Dhawan observed, "there is hardly anything in either the product market or the factor markets or the institutional framework, that is loaded in his favour."[36]

Despite these structural challenges, small farmers did find ways to access tubewells to irrigate their farms. Primarily, this occurred through the secondary groundwater markets, where small farmers could purchase water without possessing a tubewell of their own. These emerged from increased electricity prices in the 1970s, when the costs of metering electricity had become higher than the profits state-run electricity producers received from selling it. In response, the U.P. State Electricity Board (SEB) introduced a flat tariff on electricity. As economist Tushaar Shah has noted, this arrangement encouraged electric tubewell owners to recover the high investment costs "by supplying irrigation service to other farmers," often selling water to small and marginal farmers.[37] Other government policies continued to drive this phenomenon, as state subsidies encouraged electric tubewells (with their corresponding secondary market) to be developed at a greater pace than diesel, increasing the market size and reducing the rates for groundwater as competition grew.

However, because the SEB did not anticipate increased usage of electricity, and because state politicians were reluctant to raise tariffs any higher, the SEB began what Shah has called "unobtrusive acts of omission." Instead of directly cutting off power to certain communities, the SEB would simply not repair transformers and power lines. This led to a process of "de-electrification" in eastern U.P. between 1985-1995, as electricity supply diminished due to the lack of network repairs. In the face of this, secondary groundwater markets continued to exist through diesel-powered tubewells, which again began to be preferred over electric pumps. Ultimately, private markets for groundwater irrigation proved adaptive, responsive, and cheaper to provide than public initiatives. As far as Shah was concerned, the "groundwater revolution" acted as "something of a safety valve," democratizing access to irrigation and keeping the peace.[38]

At the same time, Shah noted the downsides to diesel pumps. Besides being a costlier energy source than electricity, diesel-powered pumps cost more to maintain and repair than electrified tubewells. With every increase in diesel prices the cost of access increased and "resource-poor farmers" were forced to economize on water. "Despite all these," writes Shah, "it would be appropriate to say that nothing else has produced as much welfare for the small and marginal farmers of eastern UP as diesel-pump-driven shallow tube wells."[39]

Buying and selling in these groundwater markets created new networks of water distribution. They also introduced a new sense of timing to local farmers as water was sold by the hour. It was precisely this new sense of timing and on-demand services that state electricity boards struggled to keep up with. A new sense of local time, contingent on global supply and demand of petroleum, altered the way people managed space and labor. In the last twenty years, smaller, mobile pumps began to augment tubewell irrigation. These pumps, often diesel powered 5 horsepower engines (or smaller), allow owners to siphon water from water tanks, rivers, canals, drains, and large puddles. Cultivated land is now measured in liters to fuel engines, ounces to fertilize crops, and hours to water.

SCHEMES

Another reason for the surge in diesel power was the creation of a free boring scheme (FBS) by the state of U.P. In the 1980s, the average farm holding of two-thirds of farmers was less than one hectare in U.P., 50% below the national average. Private tubewells were out of reach for most small farms. The state, realizing that there was a great disparity in access to groundwater irrigation, initiated FBS for marginal farmers. Under this scheme, the state would cover 50% of the costs of sinking and installing a tubewell. By 2002, the Minor Irrigation Department estimated that 2.5 million tubewells had been bored under FBS.[40]

Like so many of the government programs in this area, the FBS occurred without a clear plan. As recently as 2012, the scheme was suspended in select districts due to a rapidly dropping water table level. While the state has never enforced a system of licensing tubewells, the FBS at least provided a record of where and at what depth millions of tubewells were sunk. But a running record is not enough; without some sense of how tubewells are spatially distributed, it has been impossible for government at all levels—national government and state government, all the way to the district and village panchayats—to monitor the nature of this common resource.

Pump irrigation does not necessarily result in the creation of freestanding oases. Instead, pump irrigation creates an archipelago of interconnected islands, with uneven power relations between the different stakeholders. This arrangement, however, has proven robust in the face of discordant agricultural policies by the government and lending institutions. What is more, the tubewell has completely changed the calculus of scarcity in the Ganges River basin: historically, the scarcest resource in this region was water; but today, with the aftereffects of the tubewell explosion, combined with a growing population, the scarcest resource is land.[41] Agricultural intensification—made possible by the tubewell, fertilizer, and other "technological inputs"—has allowed farmers to produce more with the same amount of land. Meanwhile, heavy population pressure from small and marginal farmers has further increased the need for diversification and intensification of agriculture.

From 1975, tubewell irrigation has been the primary source of irrigation for these developments. However, as of 1996, nearly one-fourth of U.P.'s groundwater recharge is "contributed by canal irrigation, according to the estimates by the State Groundwater Department." And, because water loss by seepage runs as high as 75% in the thousands of kilometers of unlined canals, it is most likely even higher.[42] Thus, despite their vastly different spatial and bureaucratic implications and despite a common narrative of infrastructural succession—canals, then tubewells—groundwater irrigation and canal irrigation are deeply intertwined in the Ganges River Basin.

Integrating these two systems—what is known as conjunctive irrigation—was a major component of the definition of the Ganges Water Machine by Roger Revelle and V. Lakshminarayana in their 1975 essay. Since then, numerous calls to integrate the networks of canals and the "atomistic" logic of the tubewell explosion have been made. Economists, engineers, and policy experts have exposed how the tubewell is imbricated in larger networks of trade and politics. They have created equations and graphs to elucidate the interconnections and need for greater integration. They have deftly acknowledged the layering of these spaces. But they have yet to visualize their current and potential integration into a larger Ganges Machine.

With the introduction of the tubewell, a visible change has taken place in the Gangetic plains. The narrative arc of this change points to liberalization, but its overall structure is more complicated. Such liberalization is not possible without the canal water supply to raise water table levels and the FBS to subsidize the costs of tubewell installation. These layers come together temporally and spatially, producing new sites and conditions across this ever-thickening super-surface.

The existence, construction, and interaction of these layers require a different kind of map. There is a deep rhythm—a choreography—that must be discerned visually. These layers contain not only the farmers and tubewells and the bureaucracy that administers canals, but also a collective way of managing space. These different kinds of management encounter one another spatially and temporally; their observable intersections render these layers intelligible and interconnected. What follows is an exploration of these layers as they have been developed to the east of the Ganga-Jamuna Doab.

SURFACE TO SECTION

The canal and the tubewell came to represent the limits of the Ganges River Basin's infrastructural vocabulary. They each, in their own particular way, define the extremes of the basin's territorial development. If surface water operates in a shallow X-Y coordinate system, then the tubewell introduces a deep Z-coordinate, breaking the limitations of the flat alluvial plains.
2010.

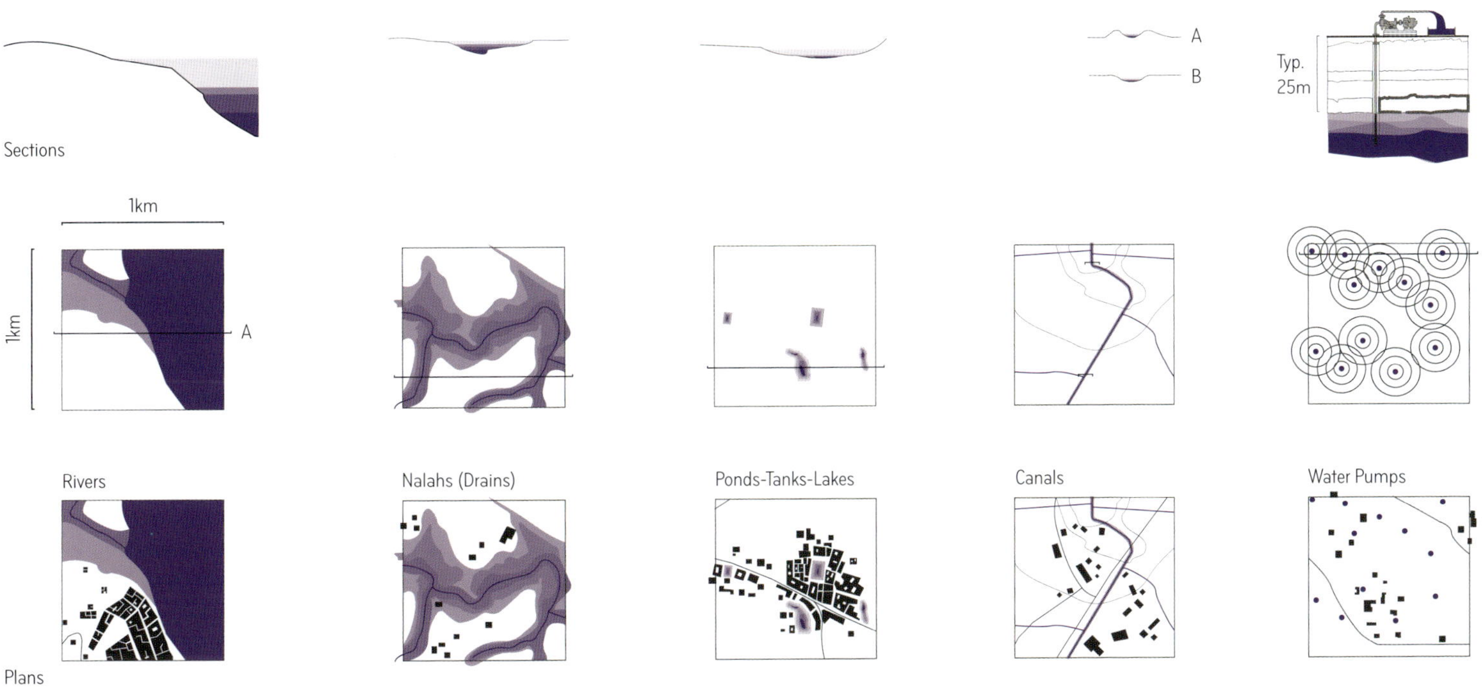

The introduction of a deep Z-coordinate reorients how people develop fields, homes, towns, and cities. No longer reliant on the proximity of tanks and stepwells or state sponsored water distribution systems, wealthy urbanites and farmers develop "compact oases" off the "grid." By and large these compact oases produce diffuse archipelagos of agriculture and estates rather than greater urban density. As groundwater becomes more available and land more scarce, water tables will continue to drop due to the reduced rate of recharge by means of soil and tanks. Over time, as land scarcity continues to increase, the tubewell may go the way of other vertically oriented technologies—most notably the elevator—and provide the scaffolding for greater urban density, intensive agriculture, and larger surfaces to replenish aquifers.

If tubewells and other lift irrigation are to be used as fulcrums to shape space and define settlement, then they must be seen as one layer of a larger spatial whole. Integrating these layers requires thinking and drawing in vertical section and horizontal plan—both above and belowground. Choreographing how and where surface and subsurface water come into contact with one another demands an iterative set of drawings: a dynamic atlas.

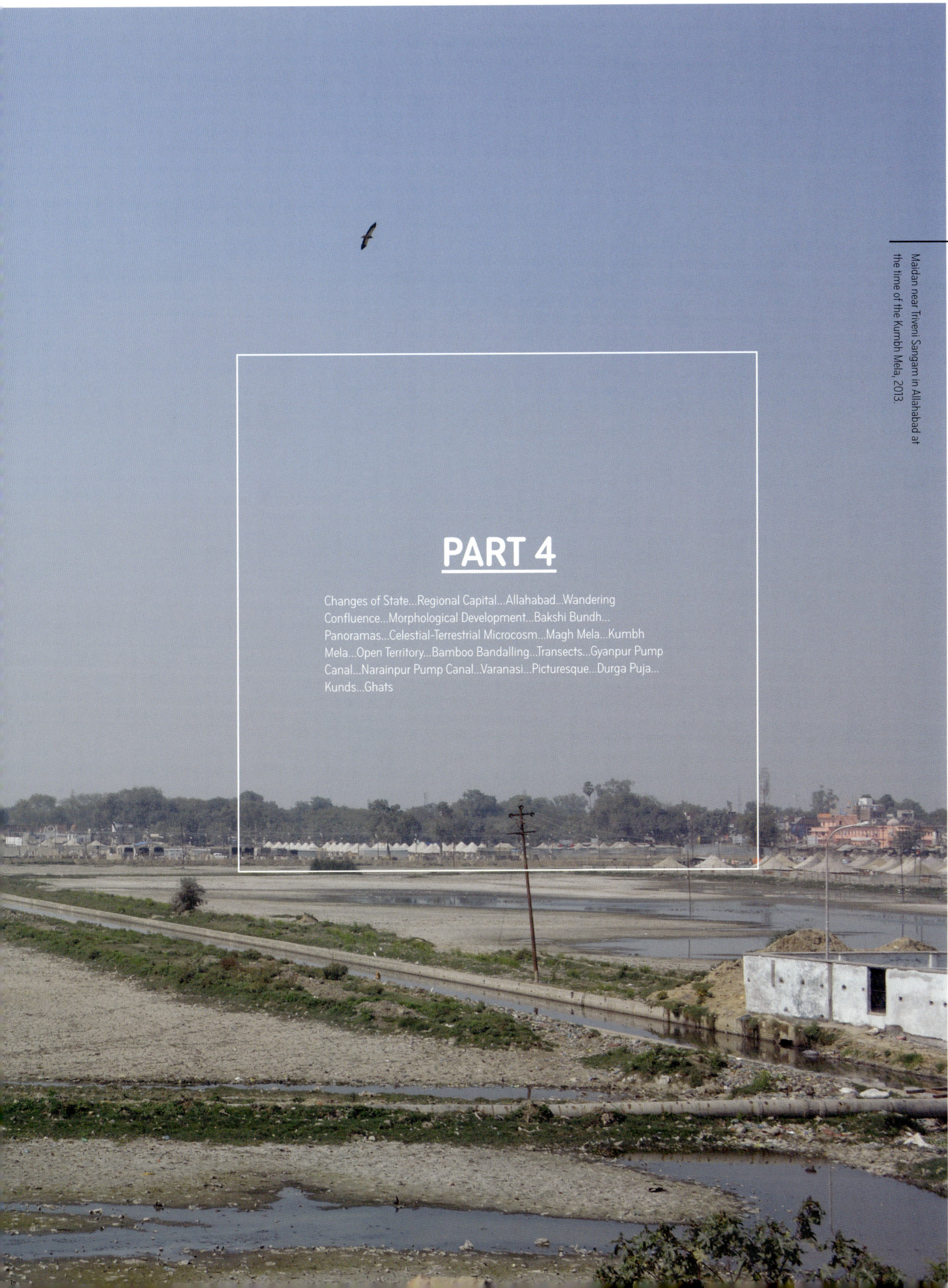

PART 4

Changes of State...Regional Capital...Allahabad...Wandering Confluence...Morphological Development...Bakshi Bundh... Panoramas...Celestial-Terrestrial Microcosm...Magh Mela...Kumbh Mela...Open Territory...Bamboo Bandalling...Transects...Gyanpur Pump Canal...Narainpur Pump Canal...Varanasi...Picturesque...Durga Puja... Kunds...Ghats

Maidan near Triveni Sangam in Allahabad at the time of the Kumbh Mela, 2013.

CHANGES OF STATE: ALLAHABAD TO VARANASI

With its vociferous advocates and unwitting practitioners spreading the culture of peaceful conquest to all corners of the Ganges River Basin by means of hydrological infrastructure, the insatiable Ganges Water Machine is constantly subdividing existing territory into new territories and adding new territory by extending the reach of its super-surface. The omnipresence of this process—of this surface and its spatial configurations—makes traditional terminology an insufficient descriptor for the changing geography.

The goal of this book—a dynamic atlas—has been to reveal the narratives and developing spatial logics that generate this vast mega-machine and its urbanism. Because of the mixture of actors, the scale of inhabitation, and the widely varying techniques of interventions, this landscape of infrastructure requires a different kind of map. Now, over 150 years after the opening of Proby Cautley's Ganges Canal, the time is ripe to integrate the fantastic and tragic heroism of the canal systems and the tubewell revolution.

After the mutation of William Stampe's Himalayan Dream into an ungovernable tubewell revolution, the following set of drawings should be considered an elucidation of the elements of the Ganges Water Machine. This will be communicated in part through words and historical documents and artifacts, but more significantly through a series of visual surveys and provocations made from the tip of the Ganga-Jamuna Doab at Allahabad (Prayag) to the city of Varanasi (Kashi). These drawings, which are the outcome of nearly a decade-long analysis of the Ganges Water Machine by foot, boat, and car, explore the speeds of cyclical change produced by the monsoon and agriculture, and the evolution of urban settlement.

The tract of land between Allahabad and Varanasi exemplifies the overlapping states of urban, rural, riparian, and agrarian spaces permeating this vast alluvial basin. The seats of former empires, principalities, and satrapies, Allahabad and Varanasi have always served as regional capitals and continue to do so today. At each of their historic cores, they are consolidated cities: dense, compact settlements. Both are revered as important religious sites within a larger sacred geography, and each city, at different times of the year, hosts a large festival. The open territory between these two consolidated, regional capital cities is less dense, but equally complex in terms of its smaller, decentralized enclaves of villages and towns within a tight patchwork of agriculture.

This particular belt in the Gangetic plains occupies a number of soluble states: it is at once water, mud, dirt, dust, deity, and atmosphere. The term "state" not only refers to the coincidence of physical changes of state (molecular transformations from solid to liquid to gas)—but also political and cultural changes of state (the village, the town, the city, the nation as well as the religious festival and winter and spring harvests). In other words, the impact of weather and climate are intertwined in the matrix of everyday life in all of its manifestations. Physical changes of state induced by weather and climate often are synchronous with urban events and festivals—most notably Magh Mela in Allahabad and Durga Puja in Varanasi.

CHANGES OF STATE

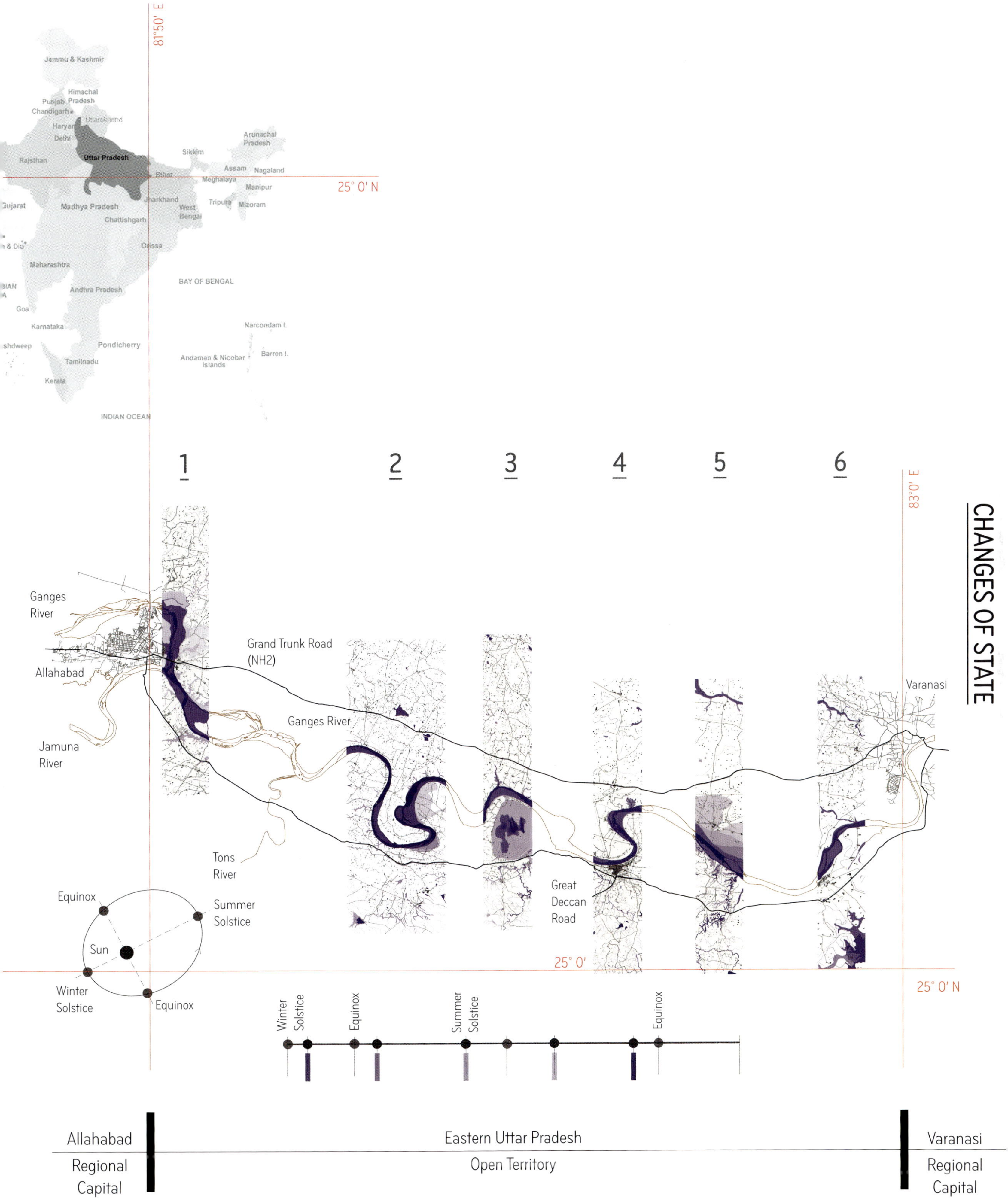

ALLAHABAD

Allahabad, a fortified town of Hindostan, in the province of Allahabad, most favourably situated in a dry and healthy soil...It has been occasionally the residence of royalty, and still contains some fine ruins: but it never appears to have been a great and magnificent city, and has now a desolate and ruinous appearance, having obtained among the natives the name Fakeerabad, or Beggar abode.–The Encyclopædia Britannica *(1842)*

"View of Allahabad, Showing the Fort," Unsigned. Circa 1850s (AA).

Lieutenant-Colonel Forrest. "Raj Ghaut and Fortress of Allahabad, at the Confluence of the Ganges and Jumna Rivers." **A Picturesque Tour Along the Rivers Ganges and Jumna in India**. London: R. Ackermann, 1842. (Princeton University).

Allahabad Fort looking north along the Jamuna River. 2005.

(left) Khusro Bagh (17th century). 2006.
(right) All Saints' Cathedral (1871-29), designed by William Emerson in the Gothic Revival style. 2006.

ALLAHABAD

Muir College (1872-76), designed by William Emerson, the building is a mixture of styles, sometimes referred to as tropical gothic. 2010.

(left) Allahabad Fort looking northeast from confluence of Ganges and Jamuna Rivers. 2005.
(right) Victoria Memorial (late-nineteenth century) in Alfred Park. 2006.

ALLAHABAD: WANDERING CONFLUENCE

Dangling at the tip of the Ganga-Jamuna Doab, where the Lower Ganges Canal system terminates, the city of Allahabad overlooks the confluence of the Ganges and Jamuna rivers. While the Jamuna, to the south of the city, runs deep and narrow, the Ganges, to the north and east of the city, runs shallow and wide. Where these two rivers meet (and a third mythical river, the Saraswati), is known as the Triveni or Sangam, the most sacred site within Hinduism. Crowning the Sangam is Allahabad Fort, a large and imposing fortress built of sandstone in 1584. The scale and position of the fortress attests to the strategic importance of the site in terms of communication, trade, and security.

Historically, Allahabad was important as it occupied the furthest point upriver where cargo boats could dock year-round, however by the end of the nineteenth century railways and Grand Trunk Road (NH2) had replaced barge transportation. Allahabad is retaking some of its former prominence due to its position on the Grand Trunk Road—one quarter of the recently completed Golden Quadrilateral, a 3,633 mile (5,486km) long highway system which links Mumbai, New Delhi, Kolkata, and Chennai—which will assure that the city continues to be the site of major urban and infrastructural development. Today the riverfront is primarily used for agriculture and hosting religious festivals. However, since 1982, when the Ganges from Allahabad to Haldia (West Bengal) was declared National Waterway No.1, there have been plans to re-introduce barge transportation all the way to Allahabad.

Allahabad city is the district headquarters of Allahabad district. The district contains seven *tehsils* (sub-districts). Throughout its storied history, Allahabad has been an important religious and administrative center. When Prince Salim (later known as Jahangir) was governor of Allahabad (1600-03), he built Khusro Bagh just north of Allahabad Fort. The gardens contain a number of tombs, most notably Prince Khusro. The Company acquired Allahabad in 1801, but it was not until the 1857 Rebellion that Allahabad became an important colonial outpost. Sir Richard Strachey, a former assistant to Sir Proby Cautley, designed the layout of a new British enclave called Canning Town (Civil Lines). A number of villages were razed to make way for Strachey's gridiron plan, which included spaces to house administrative offices, bungalows, parks, and a church for British officers and denizens. Rather than using a system of blocks like those found in the old city, Strachey used a system of streets to define the urban space of Canning Town. Large lawns separated bungalows and offices, creating little density in each of the blocks. Over time these blocks have been filled in with new buildings and residences. Allahabad Junction was placed as a buffer in between the historic core and the new British enclave, and Alfred Park was a buffer to the east of Canning Town. Since the implementation of Canning Town, Muir College and the High Court were established.

Today the city has expanded significantly, with growth to the east and west. Transportation corridors, especially highways like NH2, drive most of this growth. The Allahabad Development Authority consistently produces Master Plans with little or no fidelity to conditions on the ground. Instead, the city grows in a highly erratic, piece meal fashion. Once a regional capital, Allahabad has all of the administrative elements of a major capital city, but no longer a seat of government worthy of its layout.

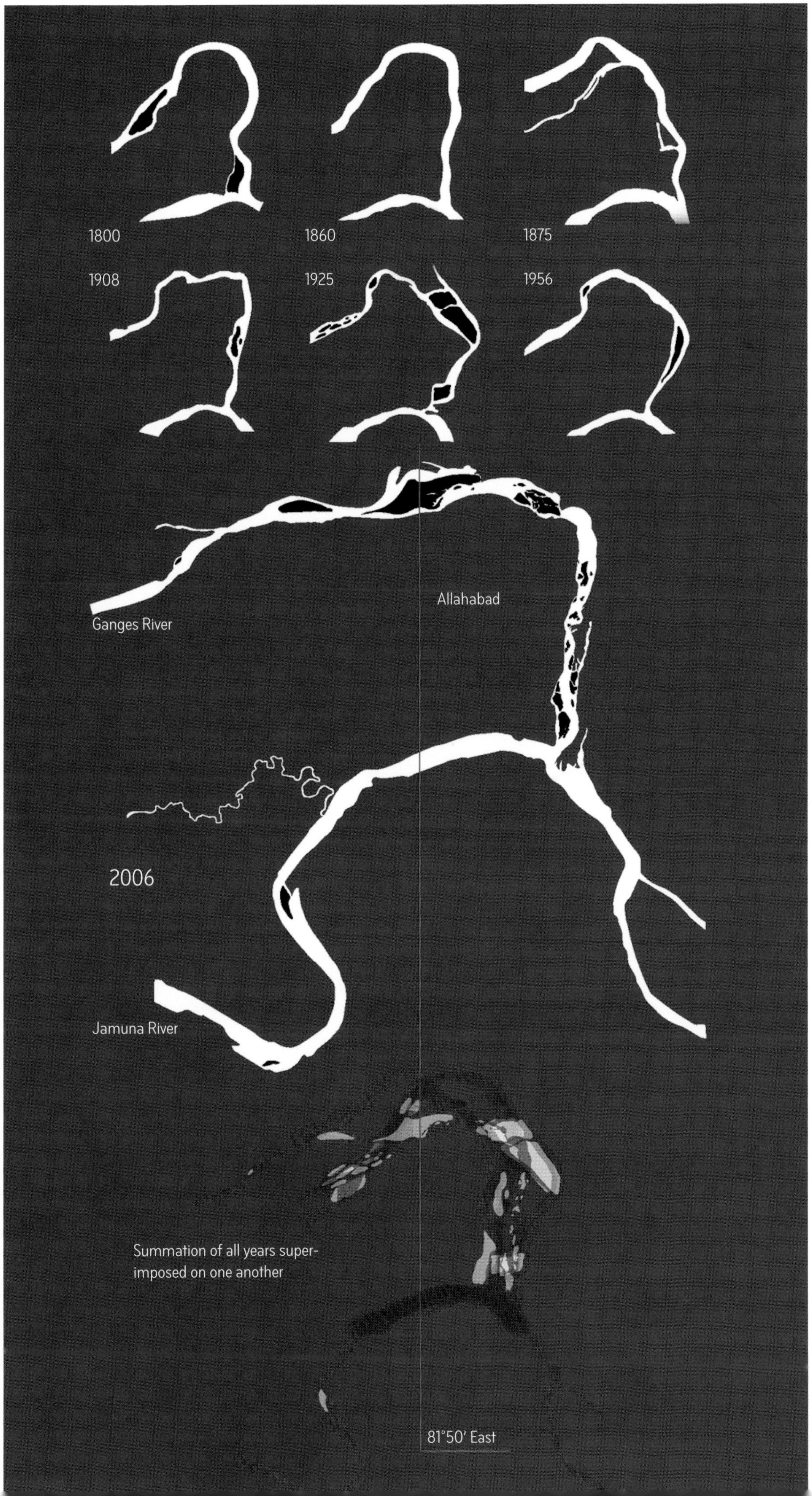

ALLAHABAD

Wandering Confluence: Allahabad, 1800-2006.2006.
A mapping of the changes in the course of the Ganges and Jamuna rivers at Allahabad.

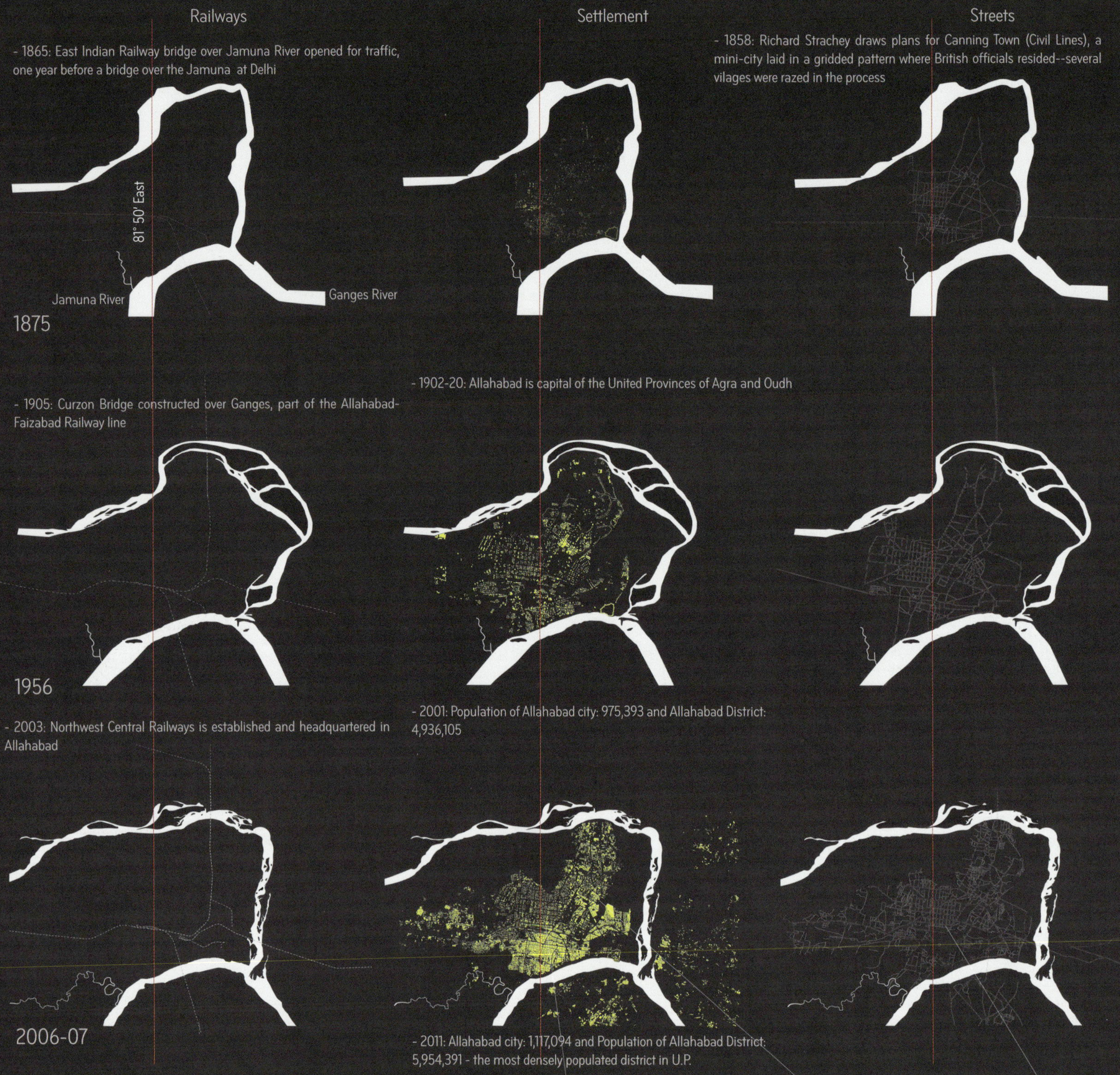

Since the late-sixteenth century, Allahabad (Prayag) has been an important administrative center. Following the implementation of Richard Strachey's plan for Canning Town (Civil Lines) in 1858—a gridded town with large thoroughfares—Allahabad has steadily grown from a town to regional capital. Until the establishment of the new imperial capital of New Delhi in 1911, Canning Town was the largest urban planning scheme implemented in India under the Raj. A few smaller, planned enclaves were built in the following decades, but by and large the city grew along major thoroughfares like the Grand Trunk Road. Agriculture, due to the rich alluvium and high-water table level, has traditionally followed the edges of the Ganges River. Since the dissemination of lift irrigation (most notably tubewells), agriculture and diffuse urban growth has expanded to the north, east, and west of the city.

Note: 1875 and 1956 drawings based on maps in Ujagir Singh, **Allahabad: A Study in Urban Geography**. Banaras: Banaras Hindu University, 1958.

Canning Town (grid) Historic Core (below) — Near Church Lane — Near Allahabad Junction

ALLAHABAD — MORPHOLOGICAL DEVELOPMENT

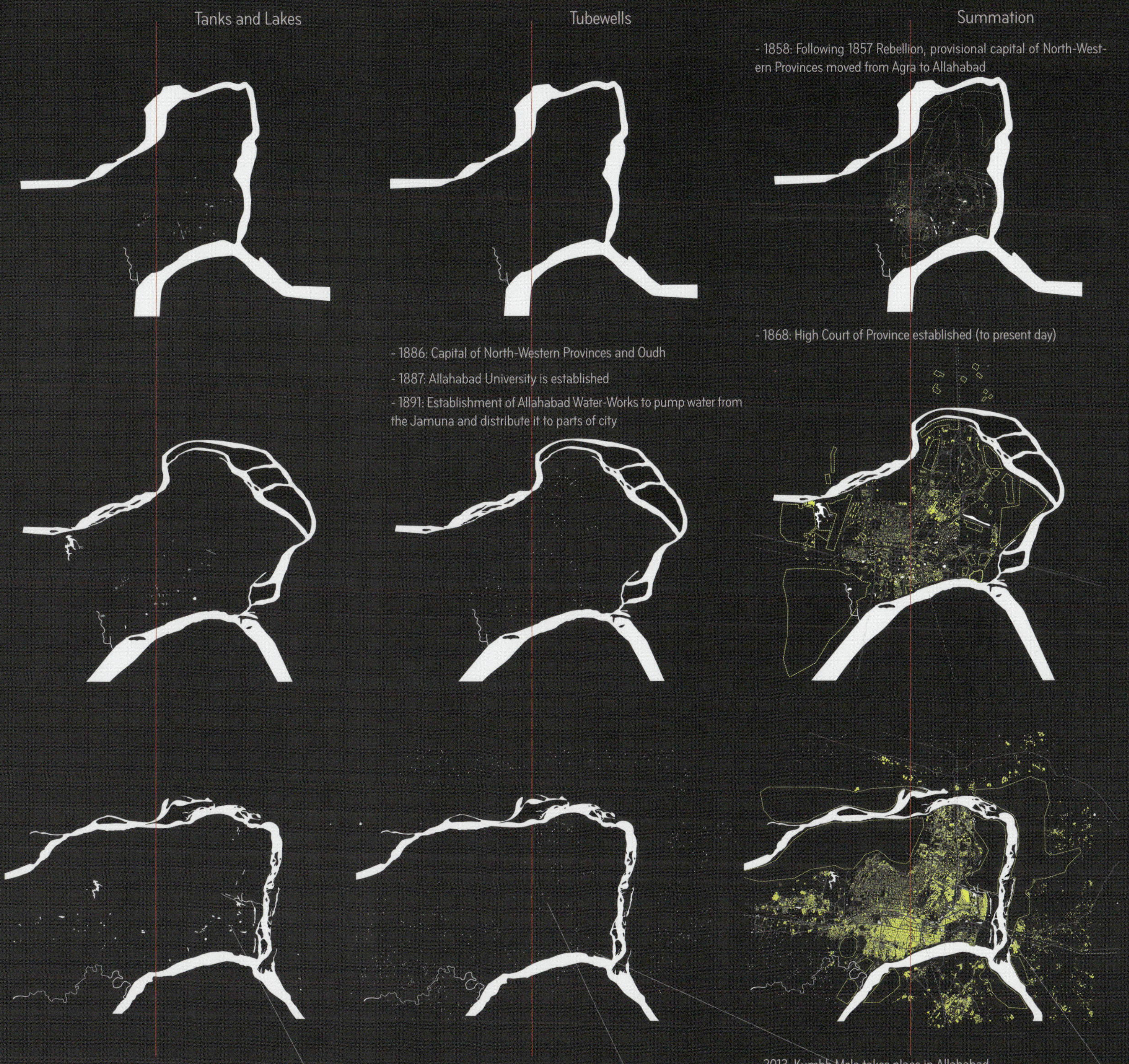

A natural depression (jheel)

Tanks and Jheels

Tubewells and Hand Pumps

Pumping Station along Jamuna River

Mechanical Sluice North of Triveni Sangam

Near Ganges River, June (2006).

Atop Hotel Kanha Shyam in Civil Lines, July (2006).

ALLAHABAD

Atop Railway Embankment, Allahabad, June (2006).

Near Daraganj, June (2006).

North of Triveni Sangam, May (2006).

Panorama 1 from east to west atop Bakshi Bundh, July (2006).

Panorama 2 from east to west atop Bakshi Bundh, July (2006).

Southside of Bakshi Bundh with pumps, April (2006).

Northern side of Bakshi Bundh where nallah exits city, July (2006).

Farming and grazing take place on the northern side of the bundh.

BAKSHI BUNDH

Since it was built during the reign of the Mughal Emperor Akbar during the sixteenth century, Bakshi Bundh (embankment) has been extended and rebuilt numerous times to divert the flood waters of the Ganges from the low-lying areas of Allahabad. Today the embankment runs 1.1 miles (1.8km) and protects most of the eastern quarter of the city. At the base, municipal diesel-powered pumps lift water over the opposite side during heavy rains to prevent excessive flooding and waterlogging. A paved road crowns the embankment, with elevated rail lines crisscrossing the road and embankment.

Panorama 3 from east to west atop bundh, Allahabad, July (2006).

SANGAM: CELESTIAL-TERRESTRIAL MICROCOSM

Every twelfth year, the sleepy university town of Allahabad is transformed into a colossal tent city populated by millions of pilgrims for the *Kumbh Mela* (literally Pitcher Celebration). And it all seems to happen so fast. The waters of the Ganges and Jamuna slowly recede after the monsoon. A city grid is tattooed into the banks and shoals at the Sangam. Tents and temples pop up in October. Pontoon bridges stretch from one bank of the Ganges to the other and pilgrims begin to arrive in January. Then come reporters and camera crews from all over the world, who come to document the life of what must at first appear to be the world's largest Instant-Mega-City: an ephemeral tent city with the major infrastructure of a metropolis.

Yet how can an instant city possibly provide a constant flow of running water and electricity, a coherent layout of avenues and roads defining 14 sectors, 18 pontoon bridges, 38 hospitals, 30 police stations, 30 fire stations and 35,000 toilets? Given the months or years of planning involved to make the Kumbh happen, it is anything but an instant city or a permanent metropolis. It is impossible to standardize the plan of the city as the waters of the Ganges recede differently each year, depositing a different configuration of sandbars and edges. It is a methodically planned ephemeral city. To make this happen, government officials collaborate with public health workers and *akharas* (religious orders, sects) to organize the physical layout of the temporary city. While at first glance it might have the ephemeral qualities of a pop-up city, its deep structure is that of a carefully planned metropolis in an ephemeral and dynamic landscape. It is ephemeral and dynamic, but it is not instant.

A similar cyclical process takes place every year with the bathing festival *Magh Mela* at the Sangam, which, like the Kumbh, is held during the month of Magh (January-February). There, too, a gridded city plan is laid out. Plumbing is buried in the sandbars dotting the Ganges. And an electrical grid provides power for millions of pilgrims and news crews. The Kumbh and Magh Melas are so regularly timed that they just seem part of the annual cycle of the Ganges to residents of Allahabad.

While the exact date of the Kumbh Mela at Allahabad is debatable, the sanctity of the Sangam and the occasion of Magh Mela date back well over ten centuries. Hieun Tsiang, a Buddhist monk from China traveling throughout north India during the seventh century CE, witnessed a yearly festival at the Sangam. He not only recorded bathing at the confluence in great numbers and alms giving to Brahmans, but also the Akshyabat tree, noted in the *Matsya Purana* for its profound sanctity and its timeless state of green. By the late-sixteenth century, the site of the Akshyabat tree was enclosed within Allahabad Fort. A portal with controlled access to the tree and temple was constructed for pilgrims.

Since Hieun Tsiang, numerous travelers visited Allahabad and recorded their experiences of Magh Mela. Touring the British Empire in 1895, Mark Twain visited Allahabad and described the festival: "Two millions of natives arrive at this fair every year. How many start, and die on the road, from age and fatigue and disease and scanty nourishment, and how many die on the return, from the same causes, no one knows; but the tale is great, one may say enormous. Every twelfth year is held to be a year of peculiar grace; a greatly augmented volume of pilgrims results then." The year of grace Twain described is the modern day Kumbh Mela. By the late-nineteenth century, the Kumbh rotated every four years between Allahabad, Haridwar, Nasik, and Ujain.

A decade later, standing atop the ramparts of Allahabad Fort in 1906, Sidney Low described a sprawling human settlement of two million souls at the Sangam. For Low, "the camp, indeed, should not be called by that name. It is a town—a temporary town, it is true, but while it lasts one of the great cities of the world, more populous than Peking or Vienna." Describing the avenues and crossroads laid out by the government, Low observed trade and commerce taking place like in any city. Low's account is similar to many foreign travelers who were amazed and appalled by the deluge of pomp and color of the gleaming palanquins, beggars, sadhus (holy men), bathers, and banners of silk and gold rustling above the crowds. Since the time of the Raj, the government has played a marshal role in the layout and design of the temporary city.

But what is more spectacular than the reliable delivery of water and electricity, and the planning for millions of pilgrims—no doubt feats unto themselves—is the inherent ephemeral nature of the Sangam and its place in the larger system of the Ganges once the bathing festivals end in March. After the tents and temples are dismantled and the cameras are gone, the gridded city remains etched into the banks of the Ganges. Footprints of tents remain in the grey silt. Farmers use these residual lines from the sprawling metropolis to cultivate wheat and rice. Herds of cows and goats graze while water buffalo plunge into the river. The ephemeral metropolis is transformed into an agropolis. This process occurs every year: transitioning from a pastoral landscape to a densely populated tent city to islands of wheat and rice paddies to a monsoonal deluge of gelatinous silt. This process is not instantaneous; it happens in slow motion.

Triveni Sangam: Celestial-Terrestrial Microcosm, 2006.

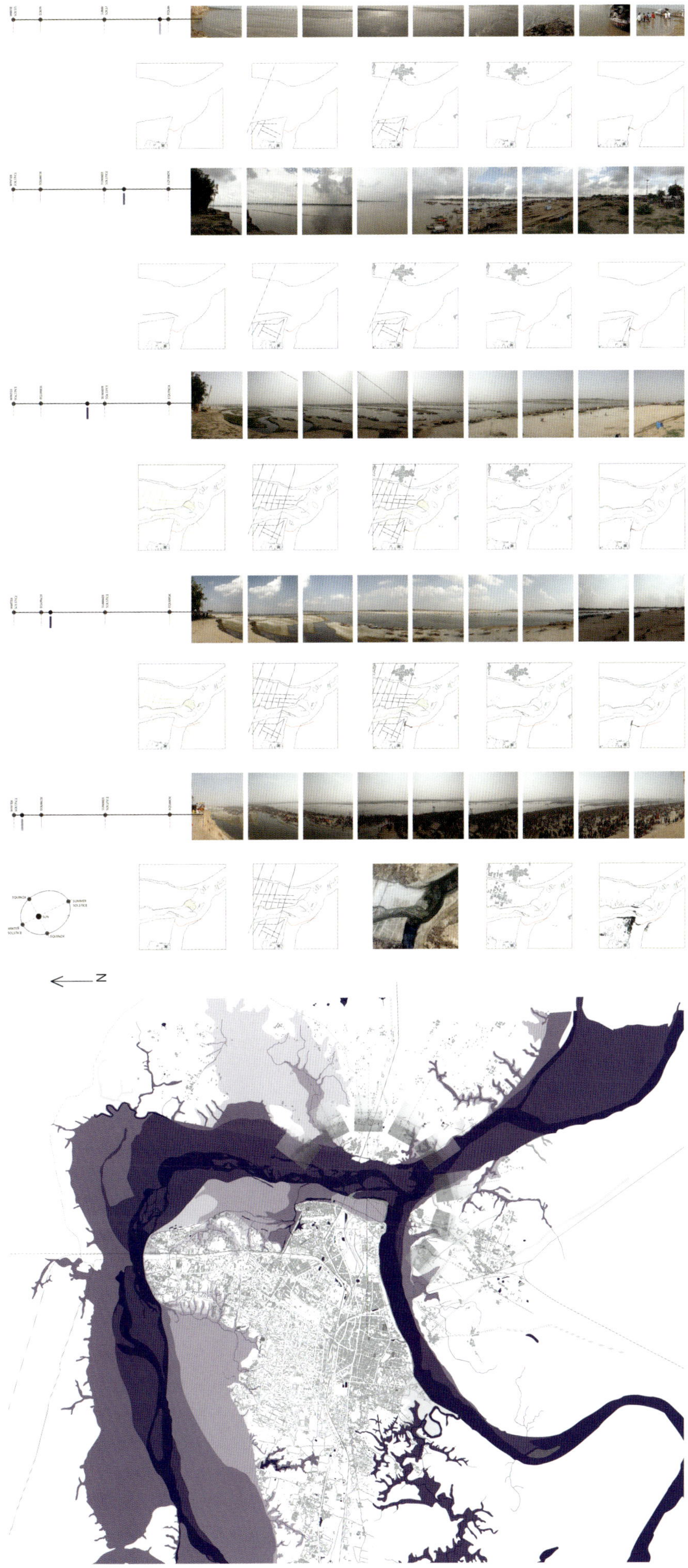

Measures the ways in which the Ganges and Jamuna rivers expand and contract around Allahabad throughout a single solar cycle. Events at the Triveni Sangam are mapped, diagrammed, and photographed from the same vantage point to vividly visualize the cyclical changes taking shape.

Tirthraj Prayag Map. Publisher Unknown, circa 2005. (AA).

Tirthraj Prayag Raj Triveni Sangam Illahabad [Allahabad] U.P. Publisher Unknown, circa 2005. (AA).

This process of slow motion change that takes place at the Sangam in Allahabad is a microcosm of the changes that take place all along the Ganges. Seasonal farmers along the river divide the silt deposited after the monsoons into fields. They build temporary thatch homes and dig shallow wells to take advantage of the high water table level. Desolate dirt paths that terminate in the river, when connected by pontoon bridges, turn into bustling corridors and marketplaces. Stalls open and children swim in the river and play cricket on the shoals. The entire Ganges corridor is transformed into a continuous quilt of agropoleis.

Such changes are impossible to capture in a couple of months. They must be measured and drawn in maps that account for dynamic change. What makes the Kumbh Mela so spectacular, besides its density of population and infrastructure, is just how the Sangam offers a dynamic model of urban space in northern India. The cyclical changes of the Sangam offer a view into a microcosm—an ecosystem—at the scale of a small city. Even though its intensity is definitely unparalleled, the slow motion changes that take place here may proffer insights into how to better integrate change and temporality in the design and life of cities throughout the basin. It offers another way to think about how infrastructure might be used to better organize space and integrate different communities and environmental change.*

Like many sacred Hindu sites within India, it is common to find images known as "calendar art" or "bazaar art." The Triveni Sangam is no exception. The majority of these images depict the celestial importance of the sacred confluence with Ganga riding a crocodile while Jamuna sits atop a turtle and Saraswati a swan (or lotus). Terrestrial practices, like bathing or the placement of landmarks, occupy the same picture plane. It is not uncommon for Krishna or Hanuman to also be in the image. Each of these images foregrounds the intersection of terrestrial and celestial spaces. Pilgrims and tourists will often purchase these images to commemorate their journey to the sacred site.

*Portions of this text on the ephemeral nature of the Triveni Sangam were published in "Ancient Geographies—New Frontiers" in *Sarai Reader: Frontiers*, #7 (2007), "Changes of State: Slow-Motion Trauma in the Gangetic Plains of India" in *Architectural Design*, No. 207 (2010) and "Pop up City" in *The Indian Express* (March 11, 2013).

Plan of Allahabad, showing urban settlement and expansion and contraction of the Ganges and Jamuna rivers from 2005-2006--taken from Triveni Sangam: Celestial-Terrestrial Microcosm, 2006.

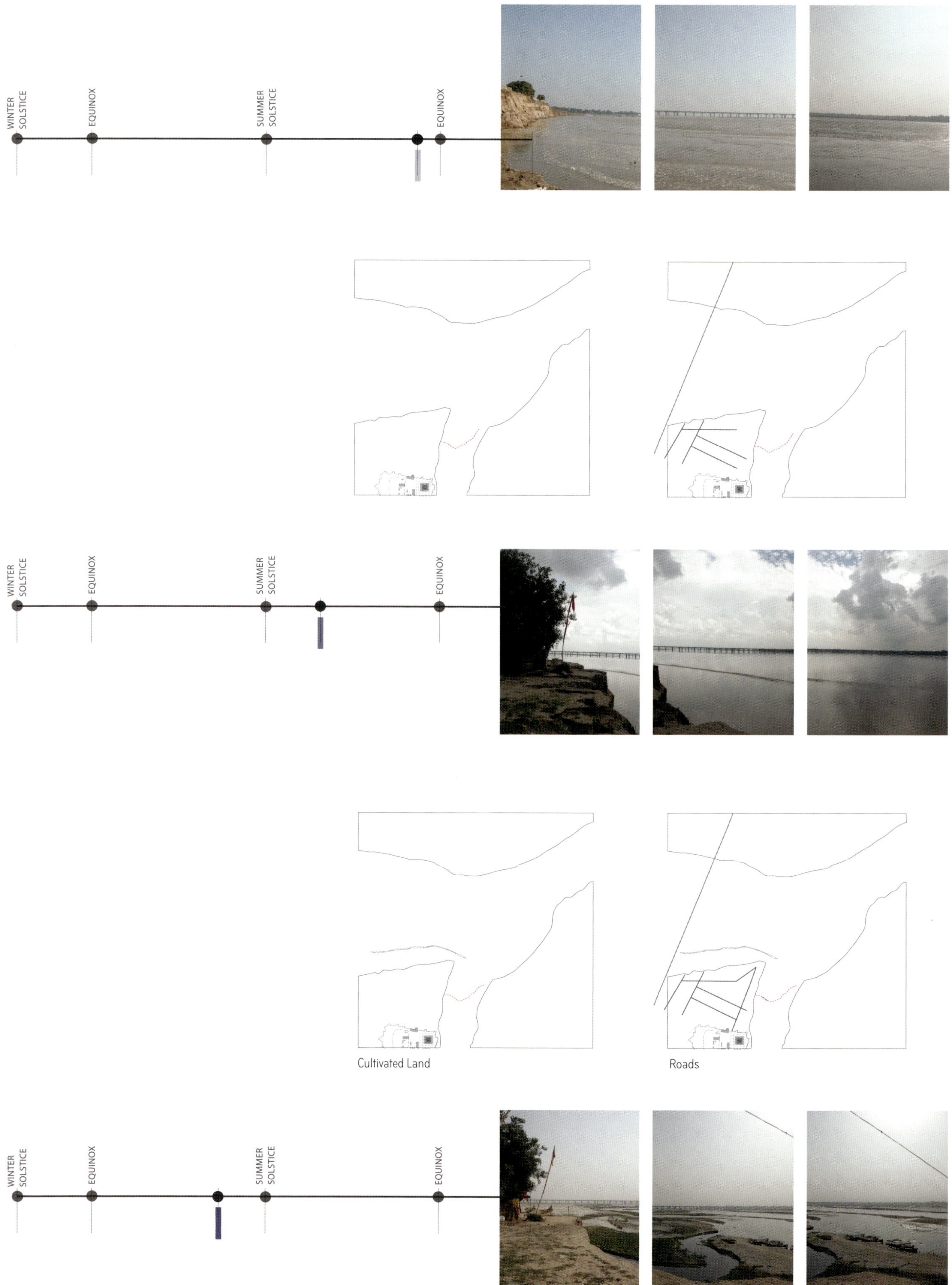

Top portion of drawing, "Triveni Sangam: Celestial-Terrestrial Microcosm," 2006.

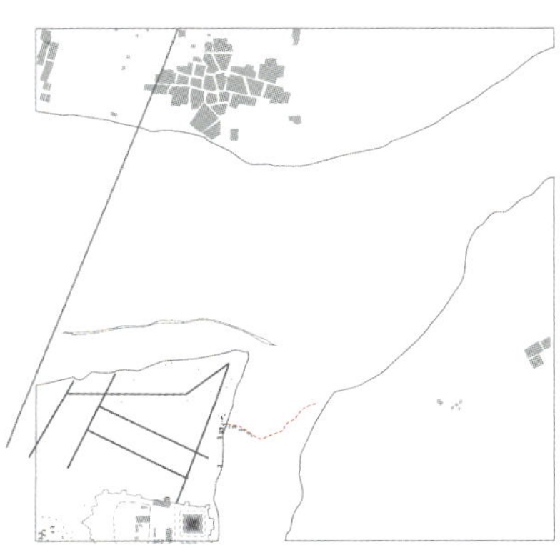

Summation

Settlement

People

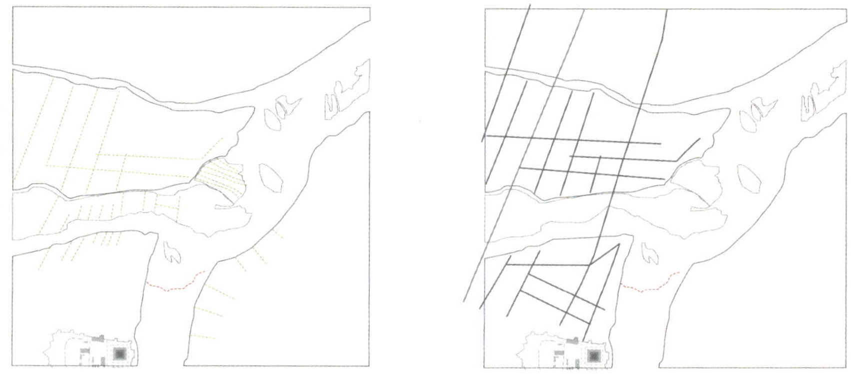

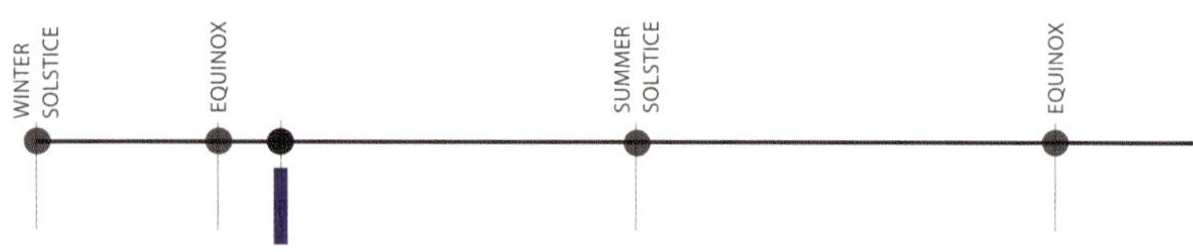

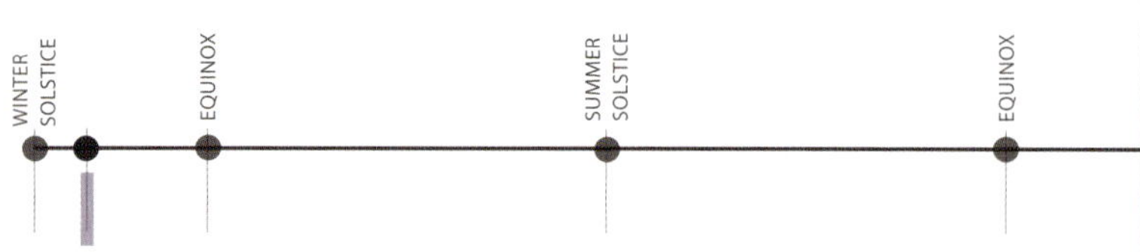

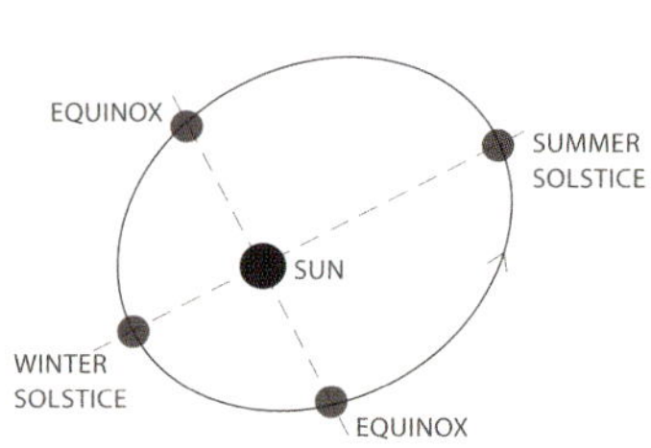

Cultivated Land

Roads

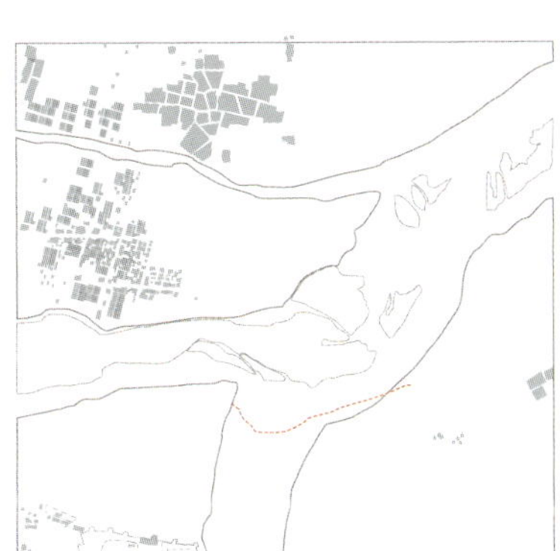

Summation

Settlement

People

1: Panorama of preparations for Magh Mela, December (2005).

3: Panorama where the Ganges and Jamuna Rivers meet–where the blue waters (Jamuna) and brown waters (Ganges) meet is the Triveni Sangam, November (2005).

2: Panorama of bulldozing and levelling ground along with laying temporary plumbing for Magh Mela, December (2005).

4: Panorama just north of the Triveni Sangam, November (2005).

Atop Phahamau Bridges facing Curzon Bridge (East View), Allahabad, July 2006.

Cremation along Ganges beneath Phaphamau Bridge.

Curzon (left) and Phahamau (right) Bridges, (View South) 2006.

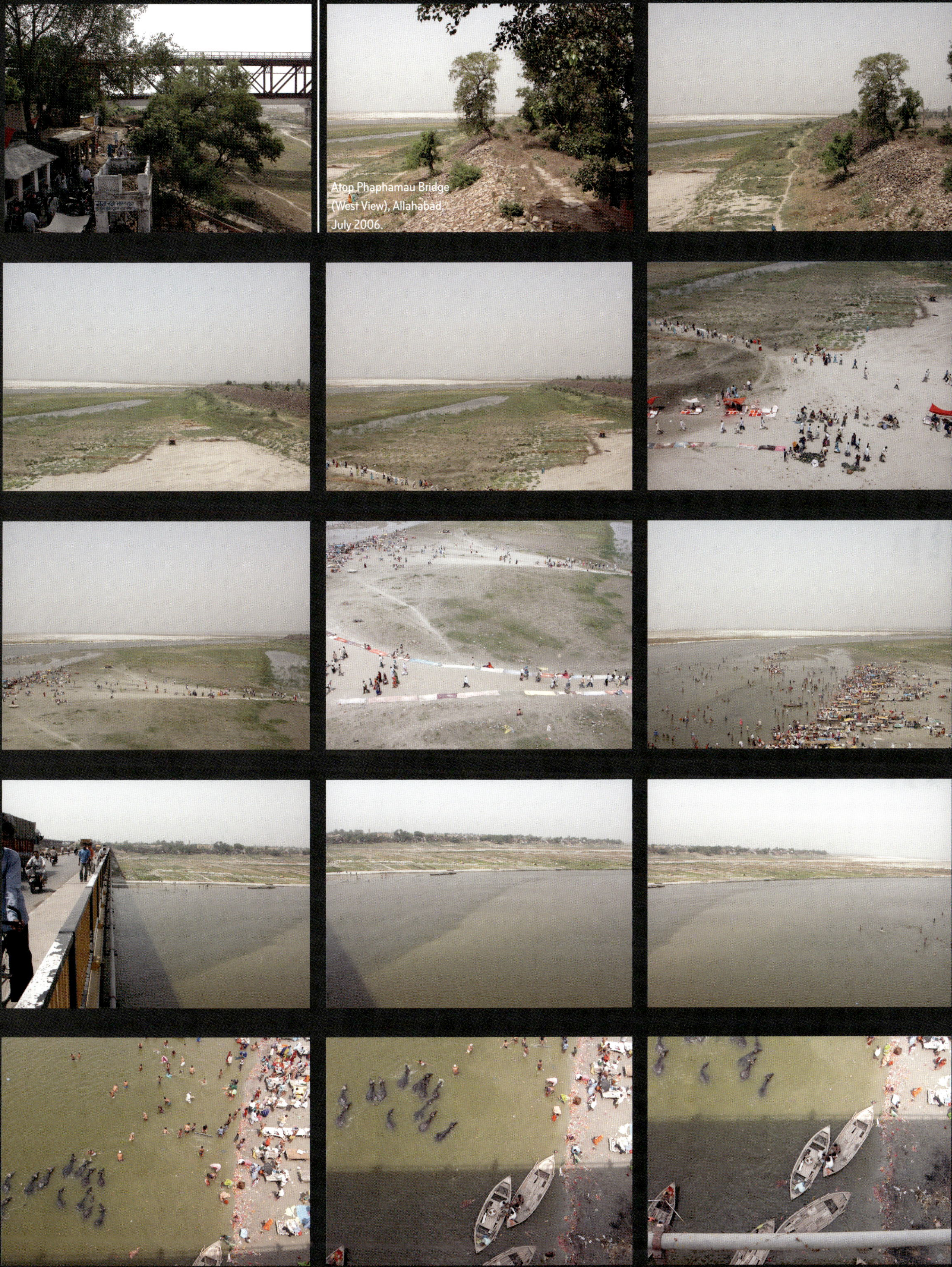

Atop Phaphamau Bridge (West View), Allahabad, July 2006.

Atop Shastri Bridge (View North),
Allahabad, July 2006.

Furrowed Sandbar.

Traces of street and tent grid etched in sandbar from Magh Mela.

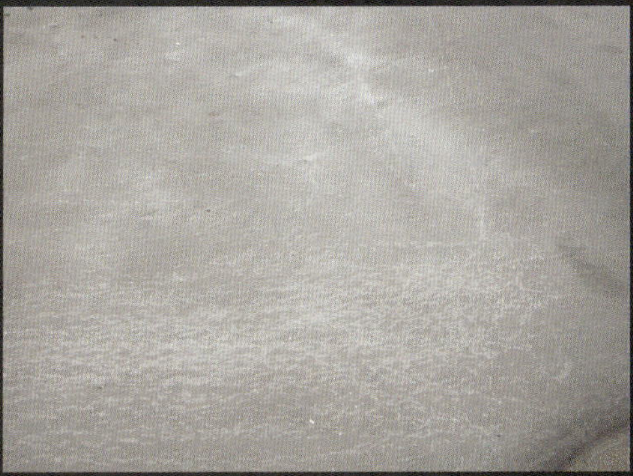

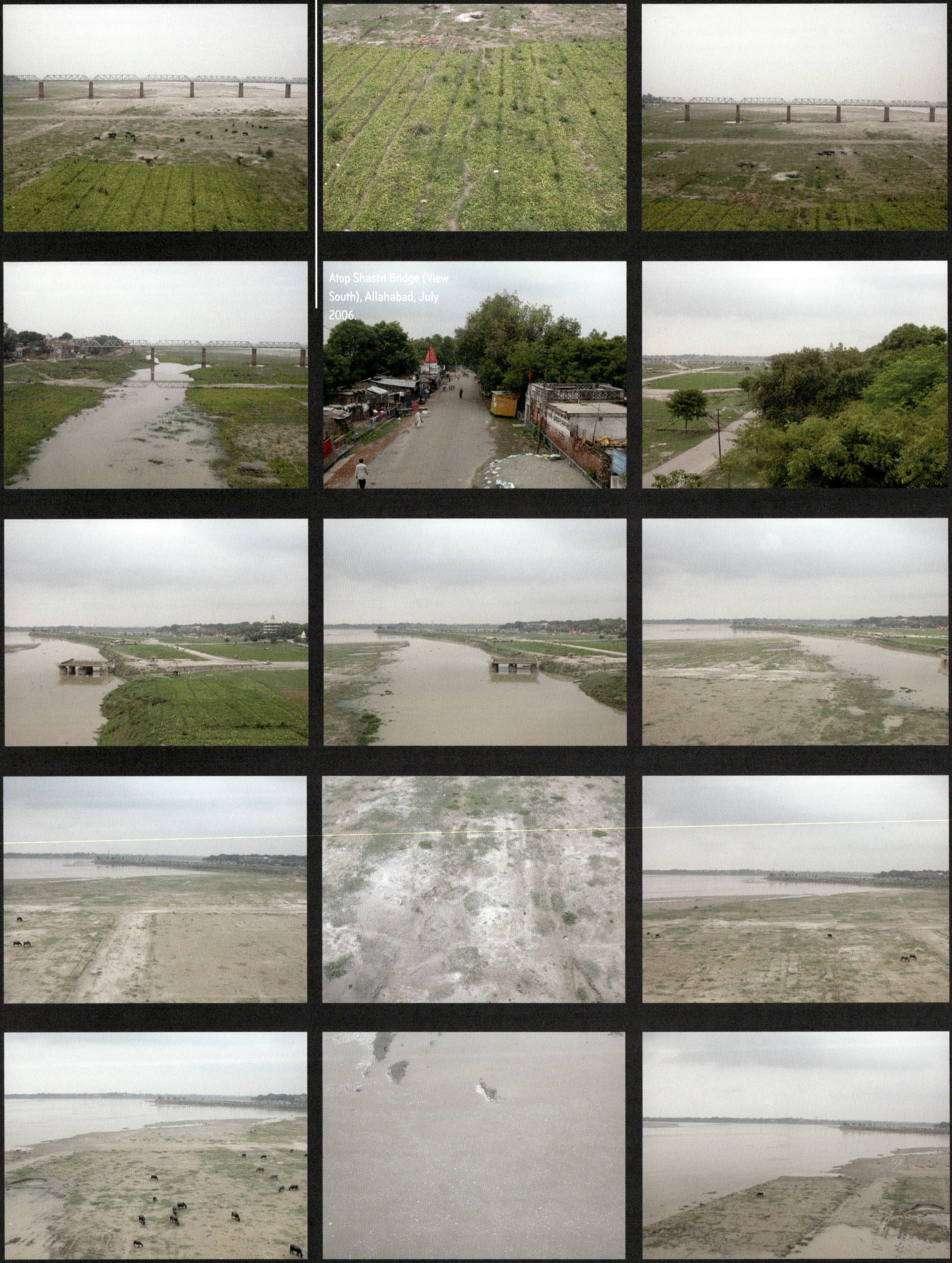

Atop Shastri Bridge (View South), Allahabad, July 2006.

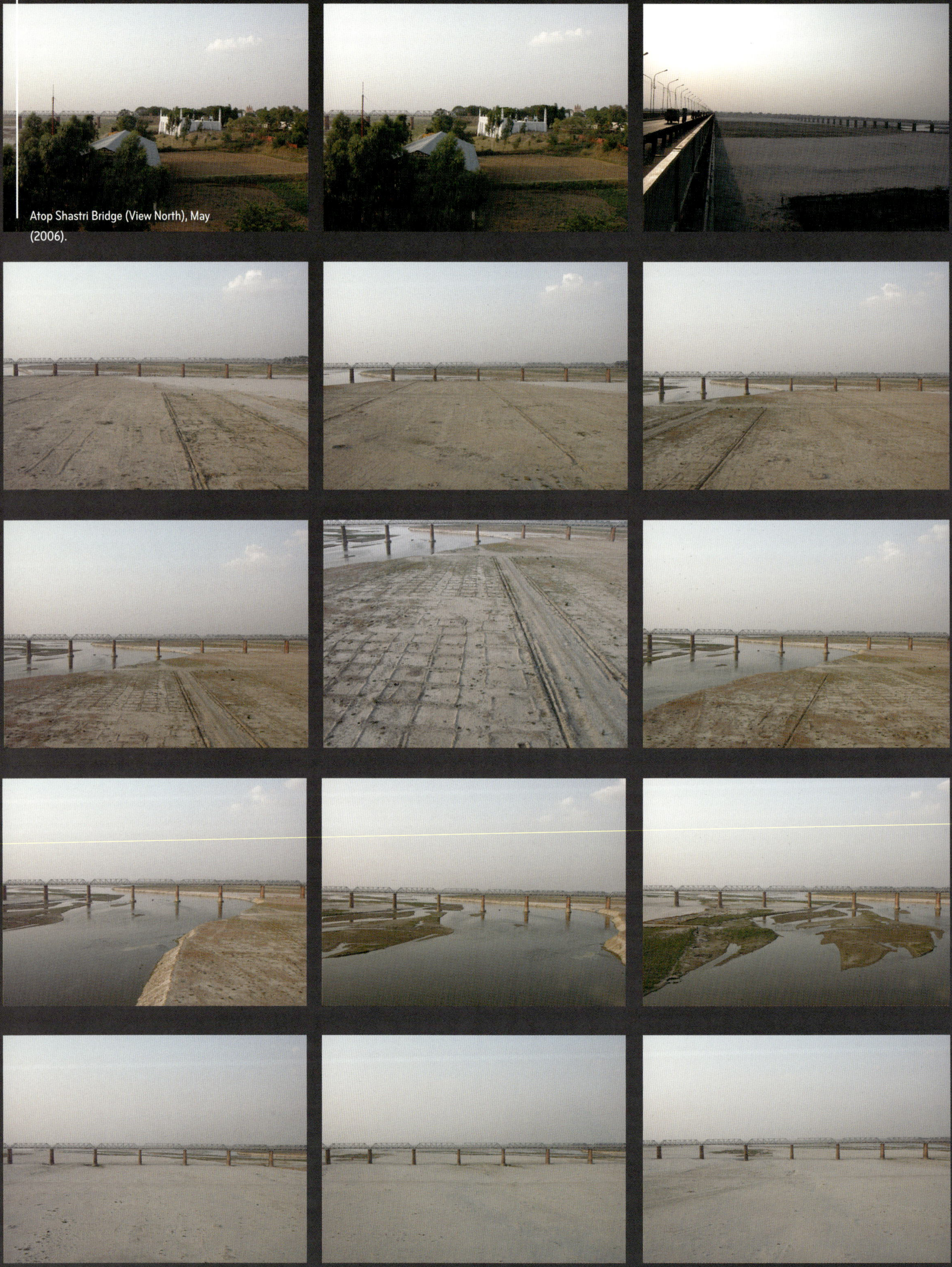

Atop Shastri Bridge (View North), May (2006).

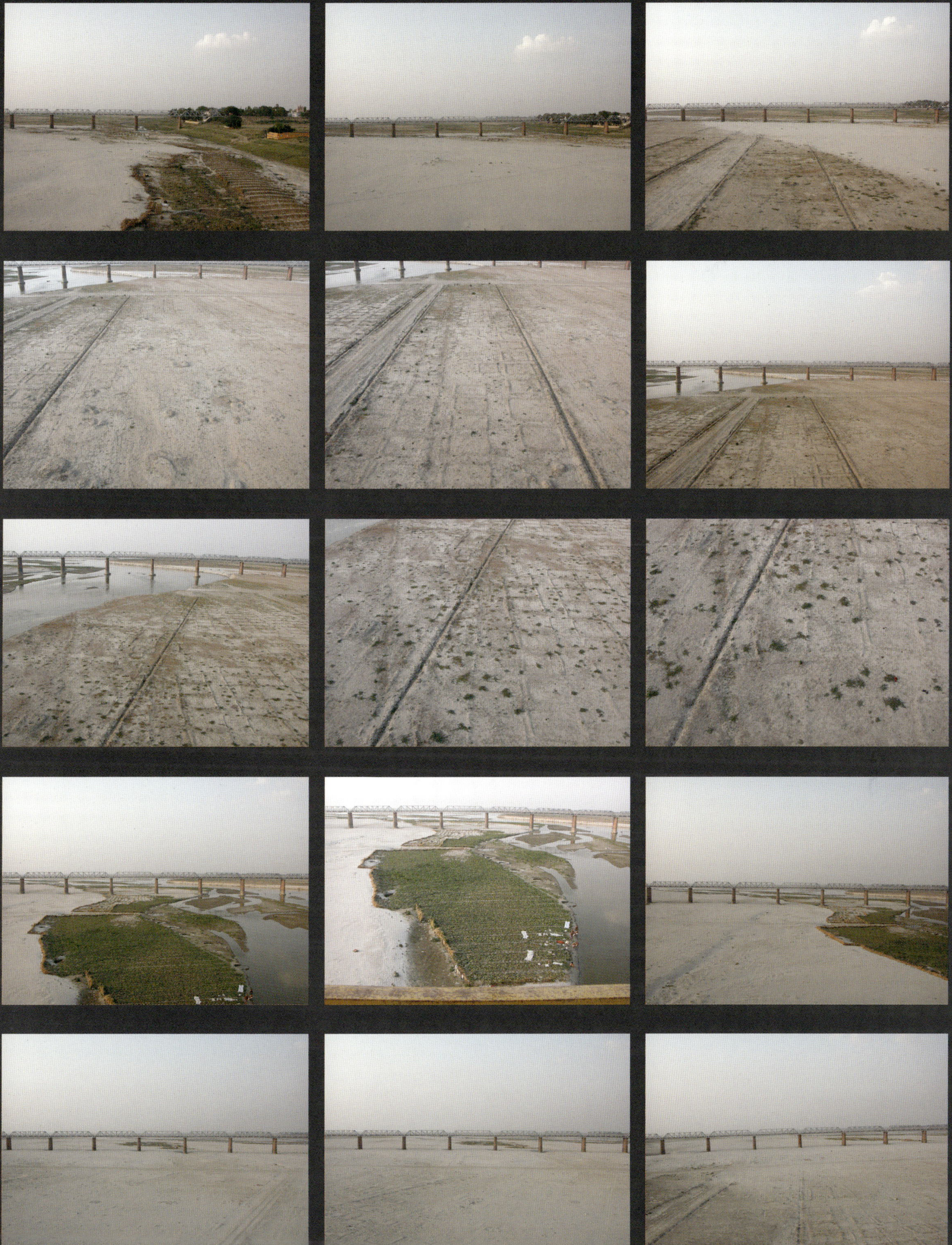

Atop Shastri Bridge (View South), Allahabad, May 2006.

Atop Shastri Bridge (View South),
Kumbh Mela, February (2013).

Atop Shastri Bridge (View North), Kumbh Mela, February (2013).

Panorama north of Sangam, Kumbh Mela, February (2013).

Night Scenes, Kumbh Mela, February (2013).

Spare Pontoons for Bridges, Kumbh Mela, February (2013).

Day Scenes, Kumbh Mela, February (2013).

TRANSECTS: ALLAHABAD TO VARANASI

From its abrupt ninety-degree turn at Allahabad to meet the Jamuna River, the Ganges River meanders en route to the Bay of Bengal. Like any meandering river, it erodes sediments from the outside of a bend and deposits them on the inside. Comparatively speaking, between Allahabad and Varanasi, the river's course has remained relatively stable, producing few oxbow lakes (when a meander is cutoff from the primary stream). Despite its stable center, the bed of the river dynamically expands and contracts with the flows of the southwest monsoon. Throughout the year, the character of the Ganges changes. At times it appears like a line, with clearly defined edges. At other times, the river is like a giant stain or smudge within the landscape. These changes in the character of the river shape how people inhabit its edges. Where rich alluvial soil is deposited annually after the floods, farmers cultivate these areas. They make use of the high-water table level to water *kharif* (monsoon) crops like rice and maize, while others harvest the sand for concrete mixes.

Densely populated, agriculturally productive, and inscribed with canals and railways and dotted with thousands of tubewells, this area has remained historically stable while cyclically indeterminate. A survey of this segment of the river, conducted over a single solar cycle, visually frames the dynamism found throughout the Ganges River Basin.

These sets of maps use the *transect* to expose how water exists as much above the surface of the land as it does below. The transect is a visual technique, devised by Alexander von Humboldt in the late-eighteenth century, and further developed by Patrick Geddes in the late-nineteenth and early-twentieth centuries, the transect draws a swath through the landscape—a sectional sampling of the characteristic land uses. Von Humboldt used it as a tool to better understand microclimates. Geddes, and later Ian McHarg, used it is as both an analytical tool and an instrument of design. More recently, it has been used by New Urbanists to create a consistent gradient between urban and rural. In the case of the Ganges River Basin, cutting transects exposes the many layers of the Ganges Machine. Given the expansive nature of water and transportation infrastructures crisscrossing the Ganges River Basin, it offers a promising method of surveying the layered super-surface of the Ganges Machine.

Each of the six transects run perpendicular to the Ganges River, revealing the parallel infrastructures of railways, canals and national highways like the Grand Trunk Road (NH2), as well as how and where they intersect with urban settlement and agriculture. It is imperative to draw how common and private property alters the flow of water and the transformation of space. Tanks and lakes, agricultural fields and domiciles, roadways and railways, as well as topography and drains, all direct, and often surrender, to the flow and collection of water in different ways.

The Composite Transect is the summation of five different, yet overlapping layers of the super-surface:

1. Rivers
2. Tanks and lakes
3. Canals and drains
4. Settlement
5. Railways and roads

Layers of Transect 1. 2006.

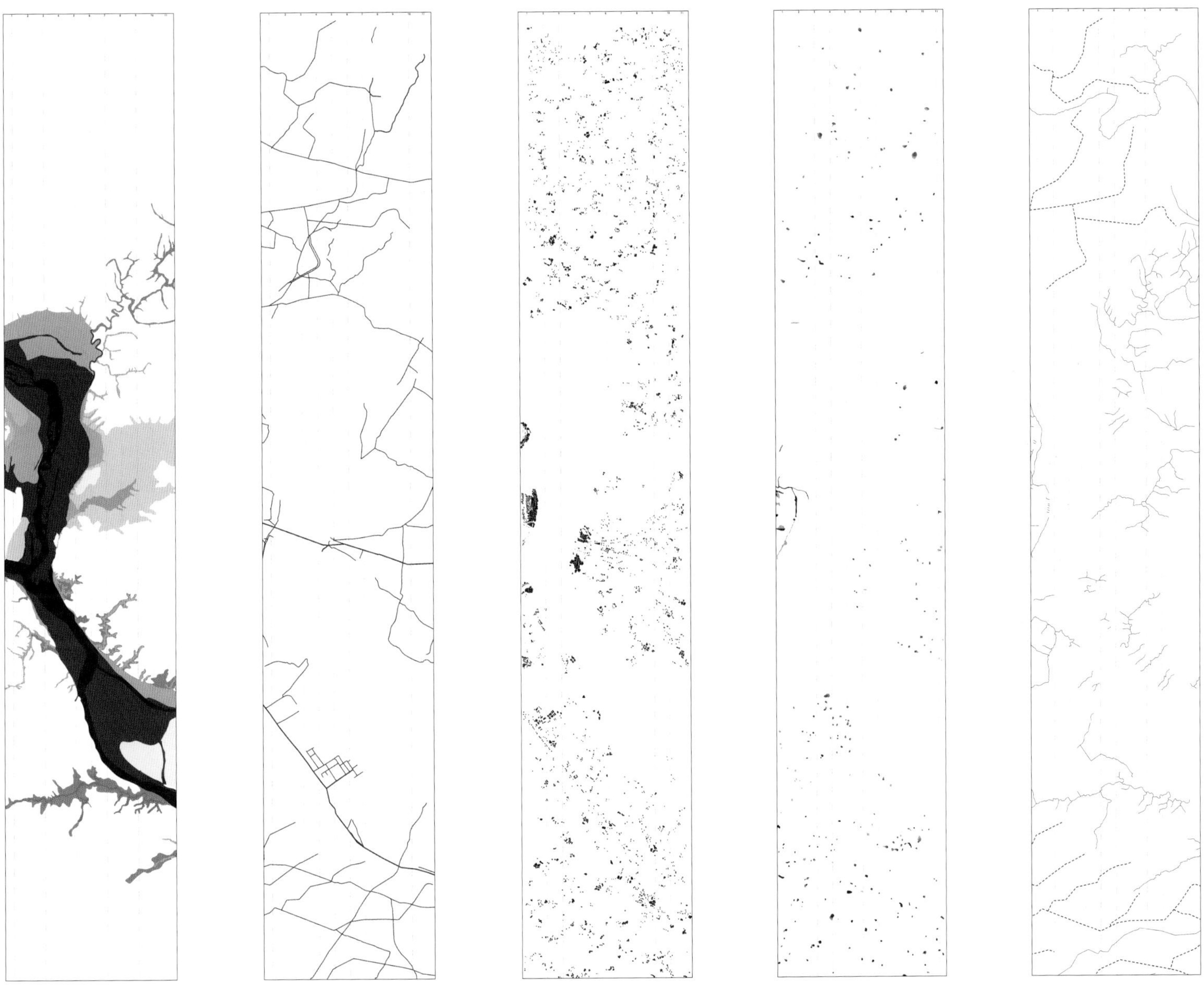

1

2

3

4

5

6

TRANSECTS — ROADS

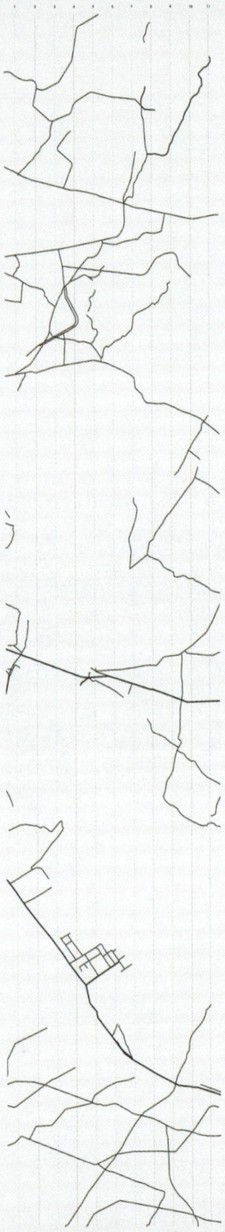

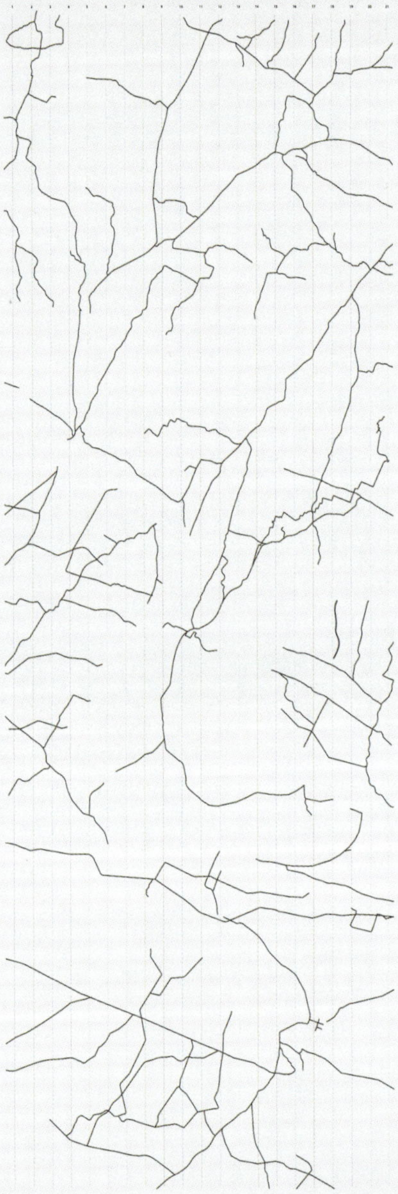

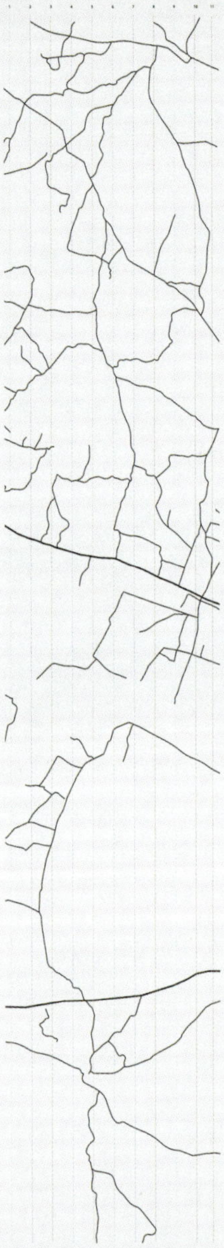

1

2

3

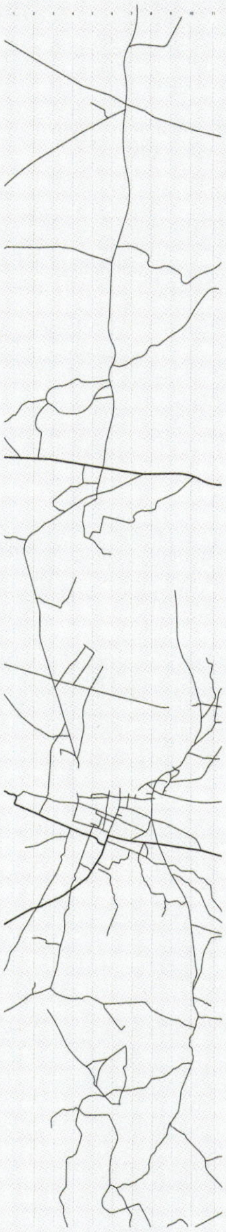

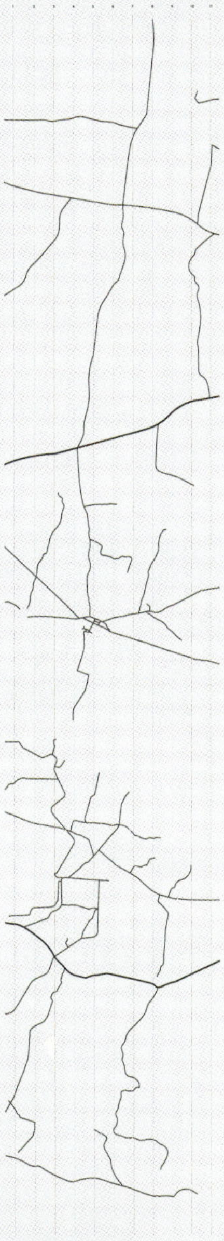

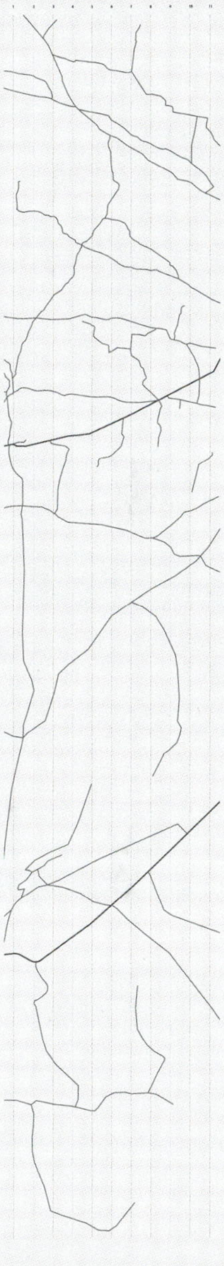

4

5

6

TRANSECTS — SETTLEMENT

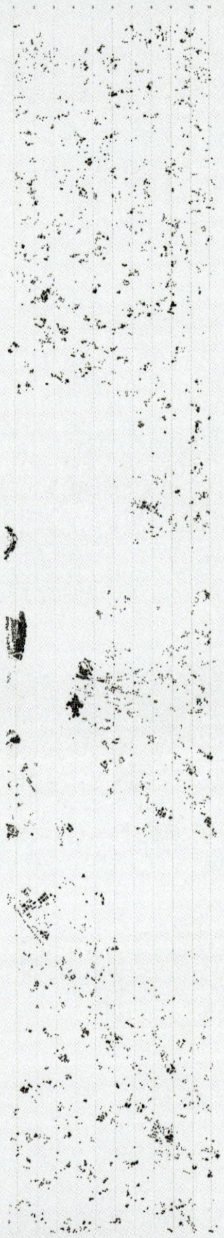

1

2

3

4

5

6

TRANSECTS — TANKS AND LAKES

1

2

3

4

5

6

TRANSECTS — CANALS AND DRAINS

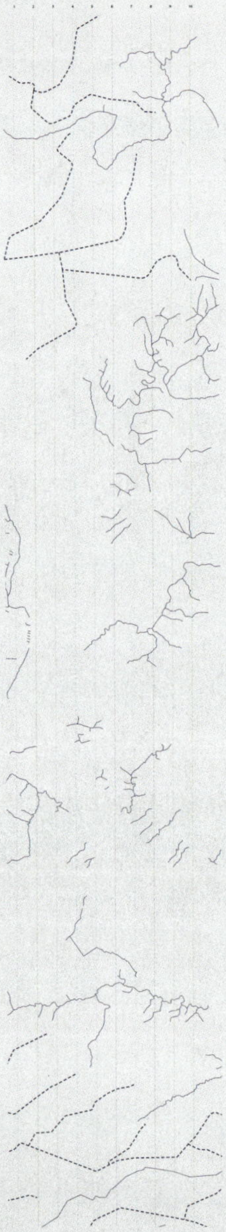

1

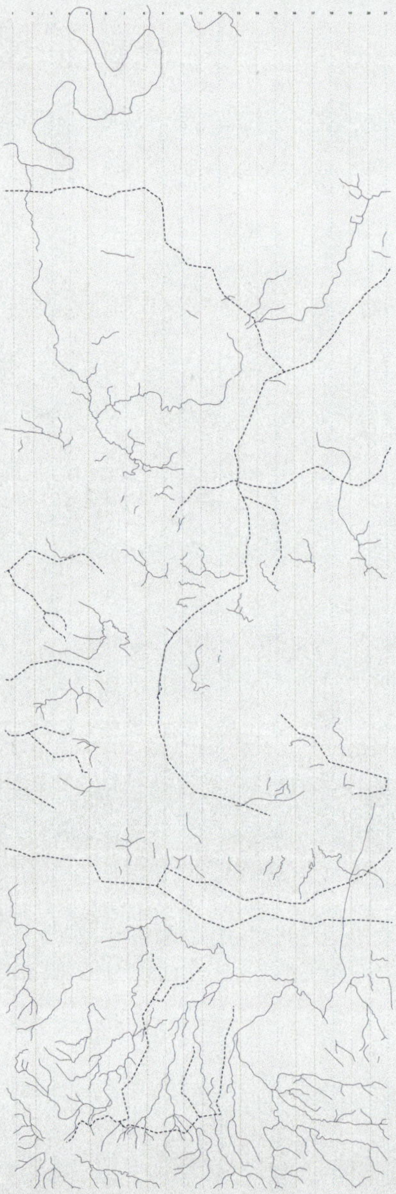

2

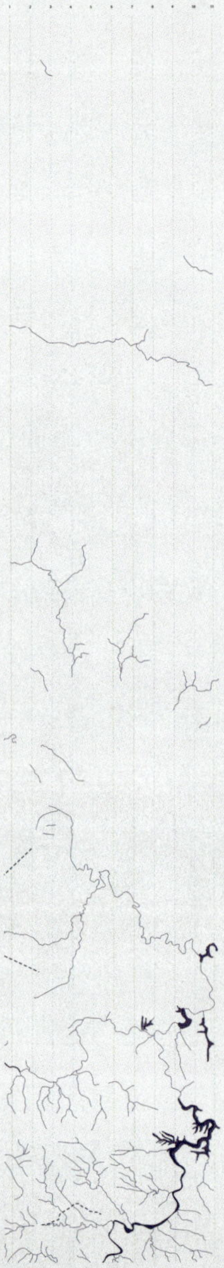

3

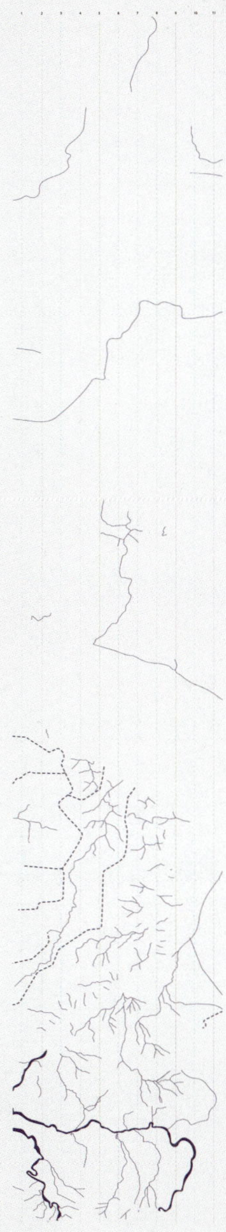

4

5

6

TRANSECTS — SUMMATIONS

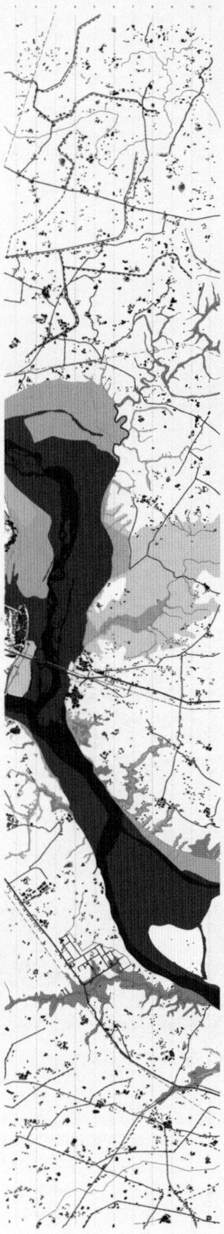

1

2

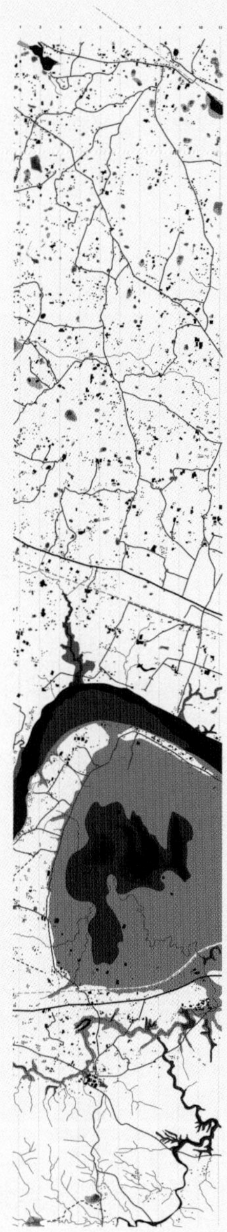

3

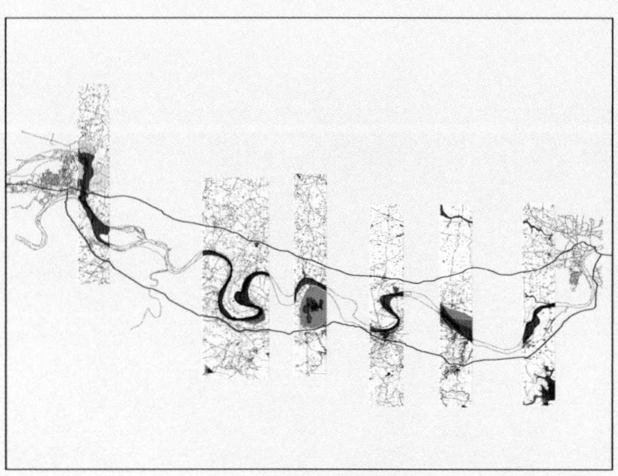

4

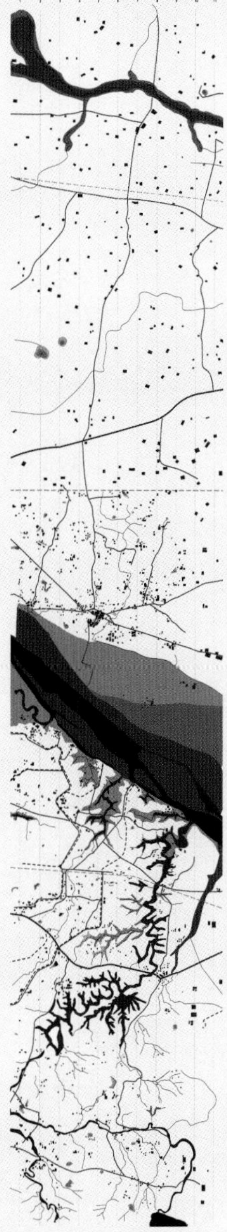

5

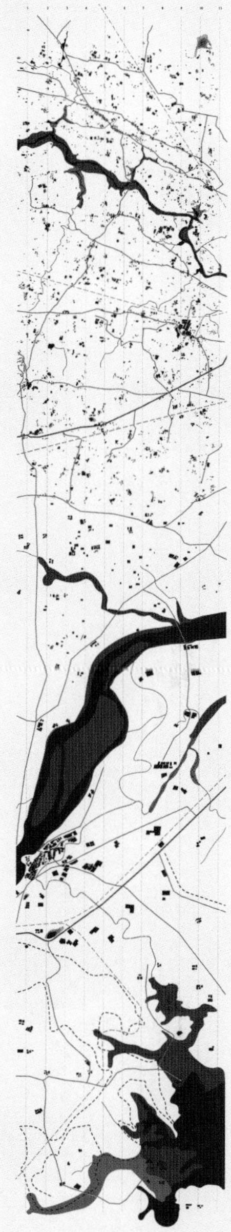

6

All of these layers, when overlaid, reveal a consistency and coherence to the ways in which space is shaped by settlement, agriculture, infrastructure, and the monsoon. A single solar cycle, marked by the expansions and contractions of the riverbed as well as the tanks and lakes, frames the cyclical changes taking place across the region.

TRANSECT VARIATIONS

Building on the composite transects, the following variations of each of the six transects represent two additional ways to visualize these spaces between Allahabad and Varanasi. This is done for each of the six transects. Photographs, taken during March of 2006 aboard a 20-foot (6.1 meter) long boat, document the conditions of each sweep of the river.

1. The composite transect with the distribution of tubewells indexed on the left-side of the drawing.

1.1. An outline of the bed of the river is drawn. The *thalweg* (the deepest point of a river valley) is drawn in red and the vertical sections of the riverbed are drawn in black.

1B. Imagine tracing a line along the thalweg. Next, draw lines perpendicular to the banks of the river that intersect with the thalweg. Rather than seeing the thalweg as a stiff line, think of it as a piece of string with the perpendicular lines as sticks piercing through the string. Pull the string at both ends until it becomes taut. Note how the sticks, once perpendicular to the edge of the river, now radiate about the straightened thalweg—indexing the river's curvature. Cut a vertical section every fifth perpendicular line, drawing not only the profile of the bed but also how the bed profile expands during the southwest monsoon and contracts in the months afterward.

1B.1. A detail of 1B: the expansion and contraction of the bed of the Ganges River in Sections D, E, and F is measured and drawn.

1C. Instead of representing topographic change in vertical section (like 1B), this drawing represents changes in elevation in plan: straightening the thalweg, indexing its curvature, and using the thalweg as a datum line, each white line marks a change in elevation of 1 meter (3.3 ft). The rise and run of the bed of the river and the sweep of the river are drawn in relationship to one another.

1C.1. A detail of 1C: the expansion and contraction of the bed of the Ganges River in Sections D, E, and F is measured and drawn in plan.

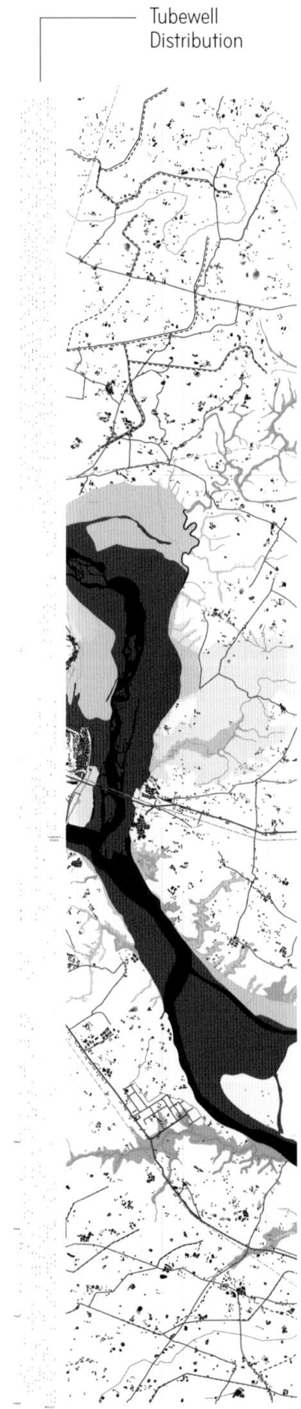

1

* All coordinates were plotted using a handheld Garmin eTrex, 12 Channel GPS in 2005-06.

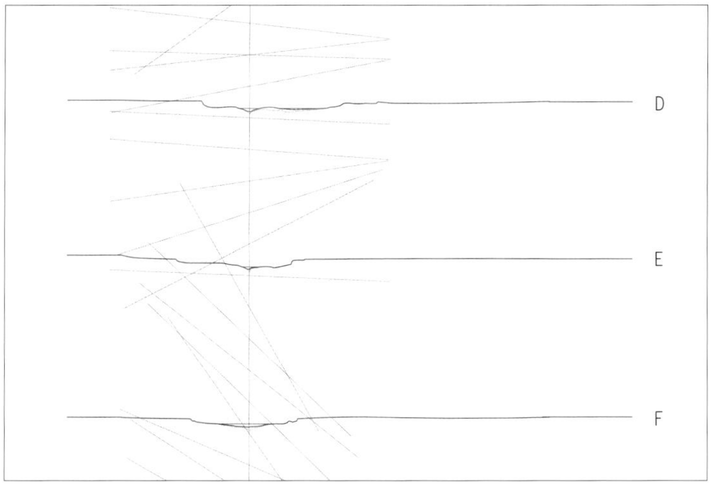

1B.1

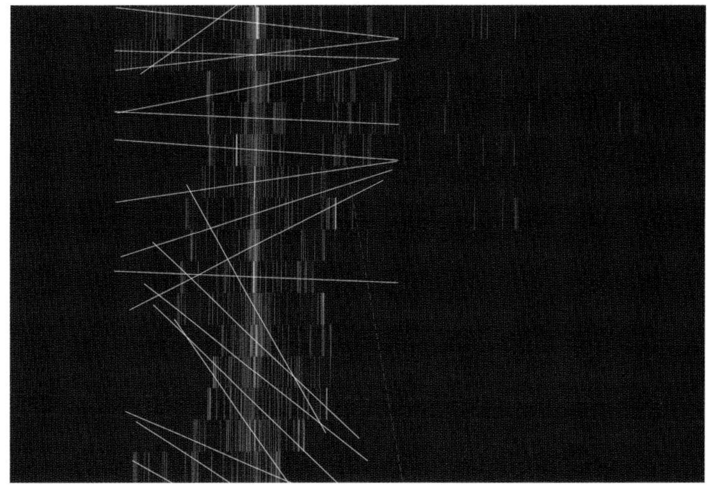

1C.1

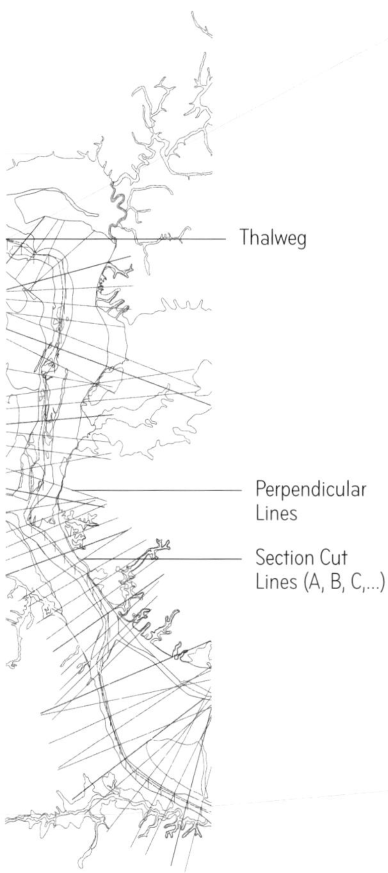

1.1

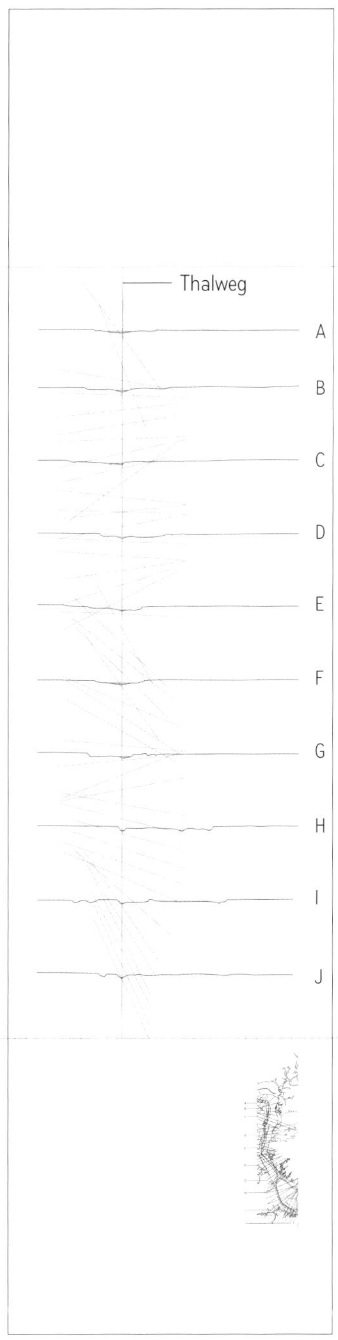

1B

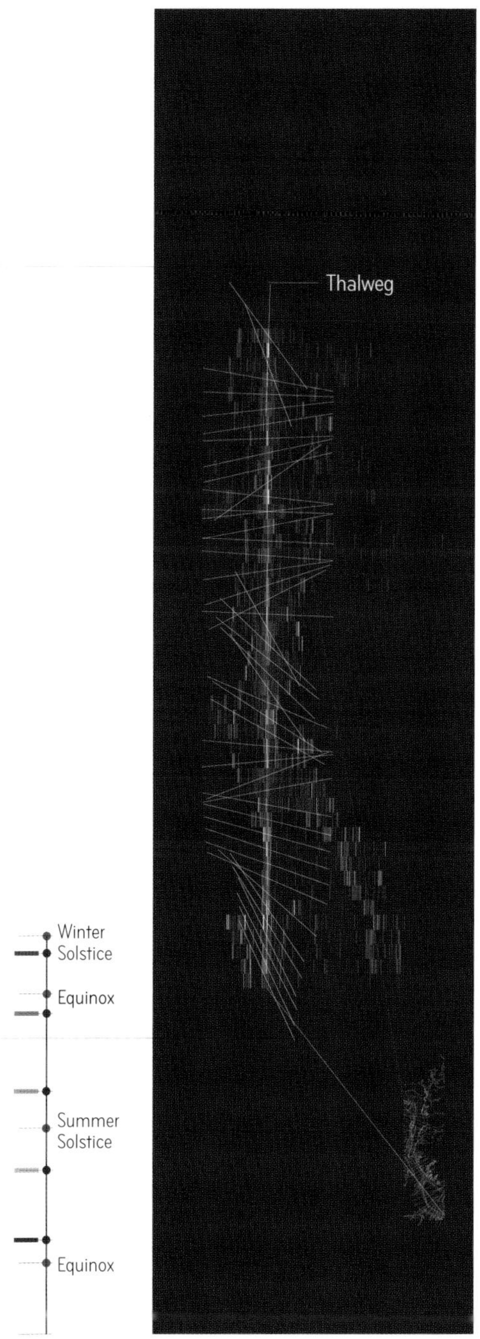

1C

Bamboo Bandalling (used to maintain an adequate depth for river navigation) southeast of Allahabad, March (2006).

TRANSECT 1

TRANSECT 1

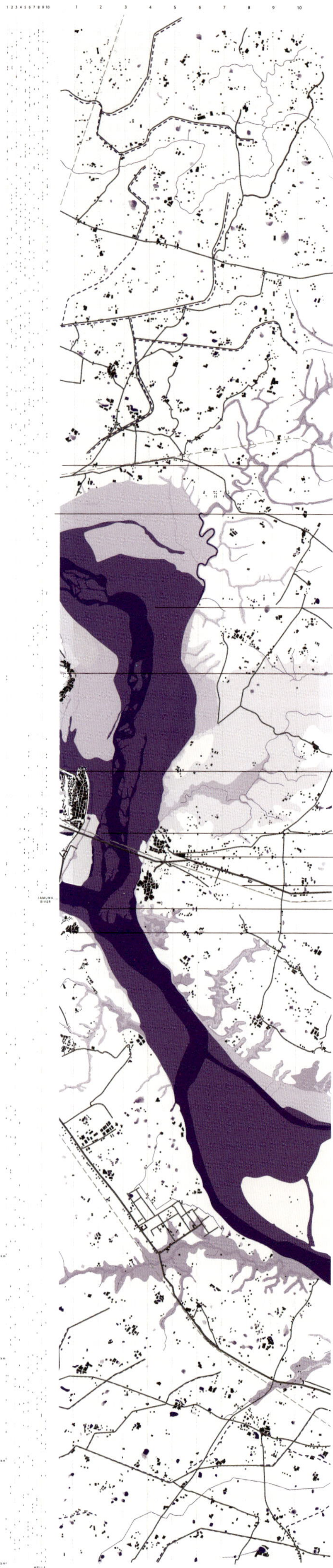

STARTED IN: 1902
COMPLETED IN: 1905
LENGTH: 3,000 FT (914M)
15 PIERS AT DEPTH OF 100 FT (30.4M)
SINGLE LINE OF RAILWAY BETWEEN GIRDERS AND ROADWAY ON TOP

STARTED IN: 16TH CENTURY
COMPLETED IN: 16TH CENTURY
LENGTH: 1.1 MILES (1.8KM)
HEIGHT: APPROX. 50 FT (15M)
BUILT BY MUGHAL EMPEROR AKBAR IN THE 16TH CENTURY, THE BUNDH (EMBANKMENT) PROTECTS ALLAHABAD FROM THE FLOODS OF THE GANGES. A NEW SEWERAGE TREATMENT PLANT IS ON THE NORTH SIDE OF THE EMBANKMENT.

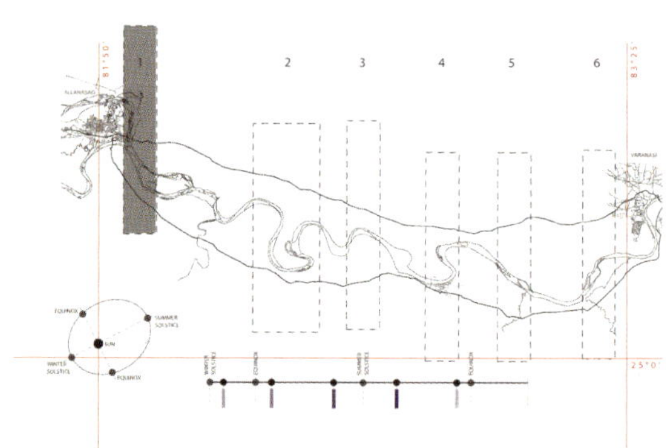

Transect 1

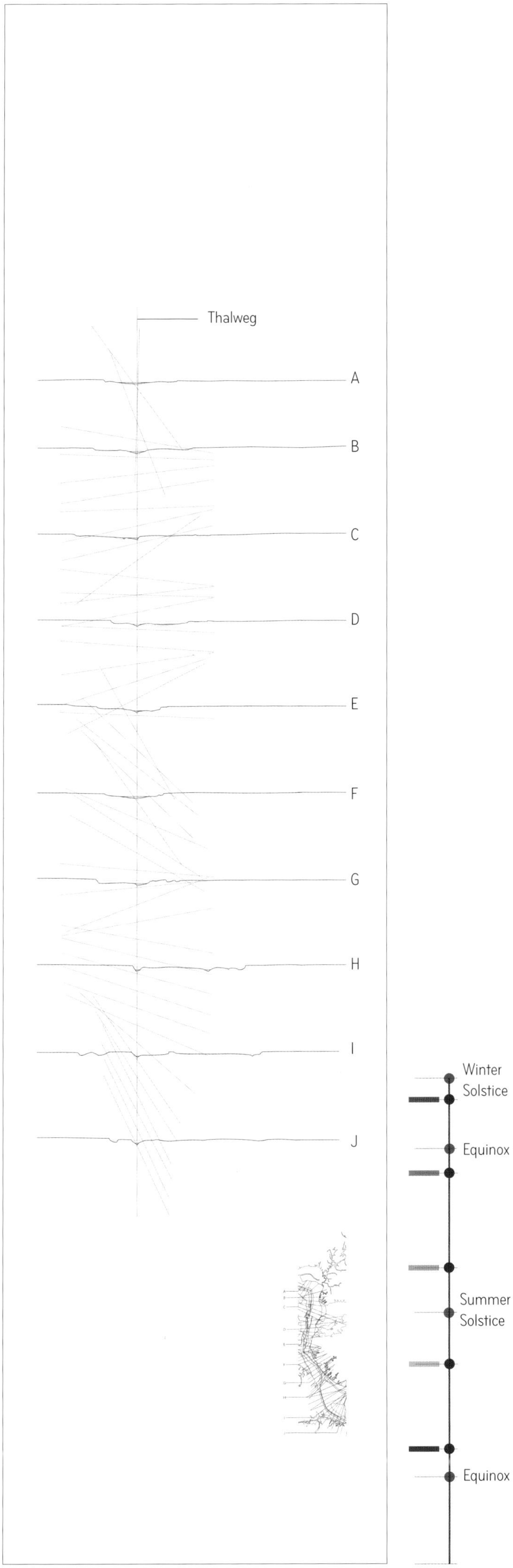
Thalweg
A
B
C
D
E
F
G
H
I
J
Winter Solstice
Equinox
Summer Solstice
Equinox

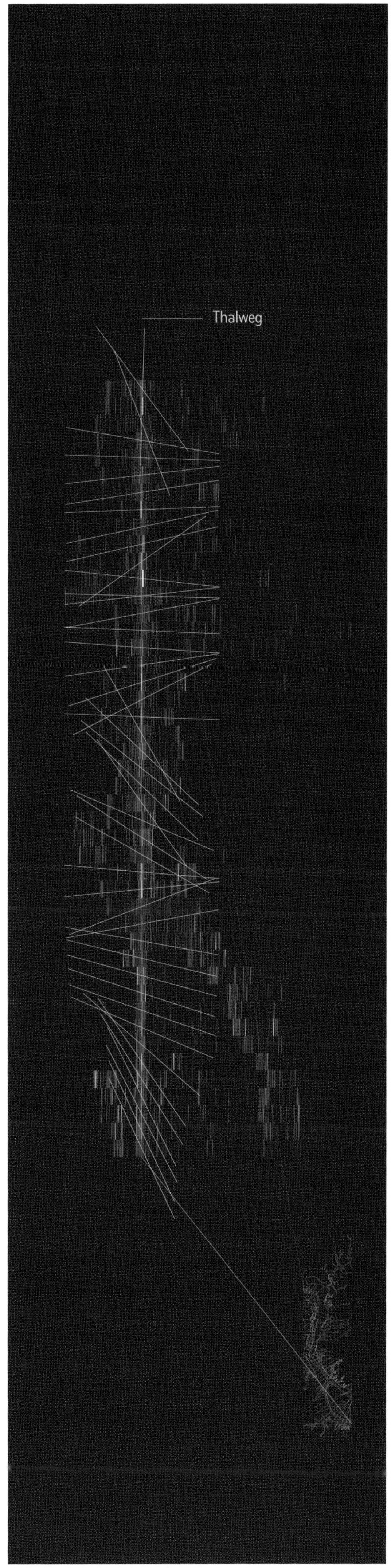
Thalweg

TRANSECT 1

Atop Naini Bridge (View West) above Jamuna, Allahabad, May 2006.

Atop Naini Bridge (View East),
Allahabad, May 2006.

T44: N 25° 25.061', E 81° 53.358', March (2006).

T46: N 25° 24.993', E 81° 53.966'.

T50: N 25° 24.187', E 81° 54.534'.

T51: N 25° 23.805', E 81° 54.742'.

T52: N 25° 23.464', E 81° 54.858'.

"Mining" Sand South of Allahabad.

T45: N 25° 25.062', E 81° 53.519'

T47: N 25° 24.844', E 81° 54.180'

T48: N 25° 24.704', E 81° 54.294'

T54: N 25° 22.505', E 81° 55.561'

T54 (continued).

T57: N 25° 19.705', E 81° 57.228', March (2006).

T61: N 25° 19.280', E 81° 59.549'.
T64: N 25° 19.882', E 82°01.030'.

T65: N 25°20.455', E 82°01.656', March (2006).

T66: N 25°20.957', E 82°01.874'.

T67: N 25°21.239', E 81°02.267'.

T69: N 25°21.098', E 82°03.140'.

T68: N 25°21.113', E 82°02.898'

T70: N 25°20.463', E 82°04.112'

T73: N 25°17.286', E 82°04.628', March (2006).

T77: N 25°16.202', E 82°05.661'.

T78: N 25°16.245', E 82°06.250'.

Tons and Ganges River Confluence.

T75: N 25°16.511', E 82°04.990'.

Banks of Ganges River, March (2006).

TRANSECT 2

STARTED IN: 1978
COMPLETED IN: 2003
ULTIMATE IRRIGATION POTENTIAL: 65.42 (TH HA)
POTENTIAL CREATED: 37.23
GROSS IRRIGATED AREA: 65.415

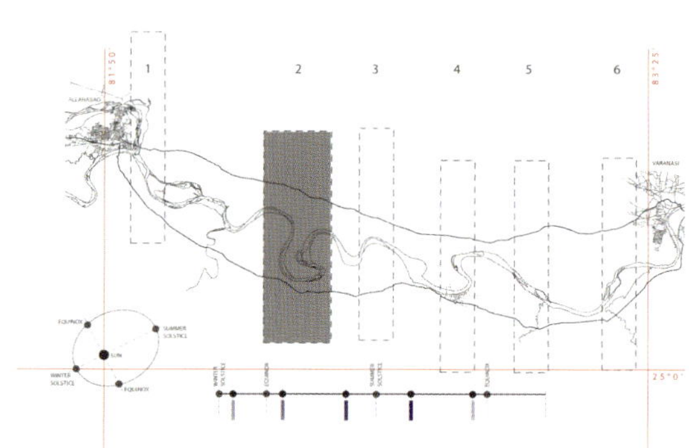

Transect 2

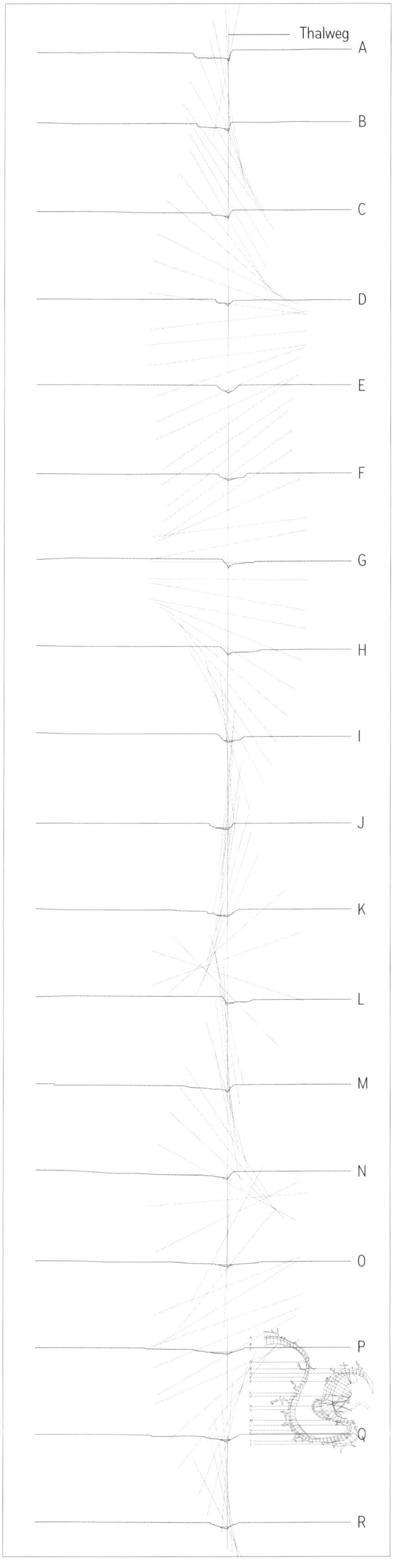
Thalweg
A
B
C
D
E
F
G
H
I
J
K
L
M
N
O
P
Q
R

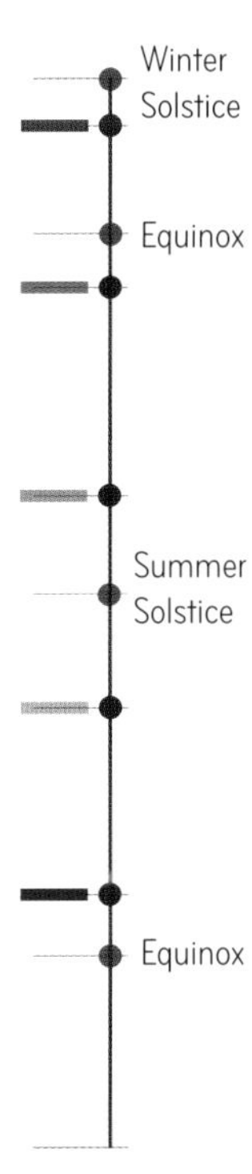
Winter Solstice
Equinox
Summer Solstice
Equinox

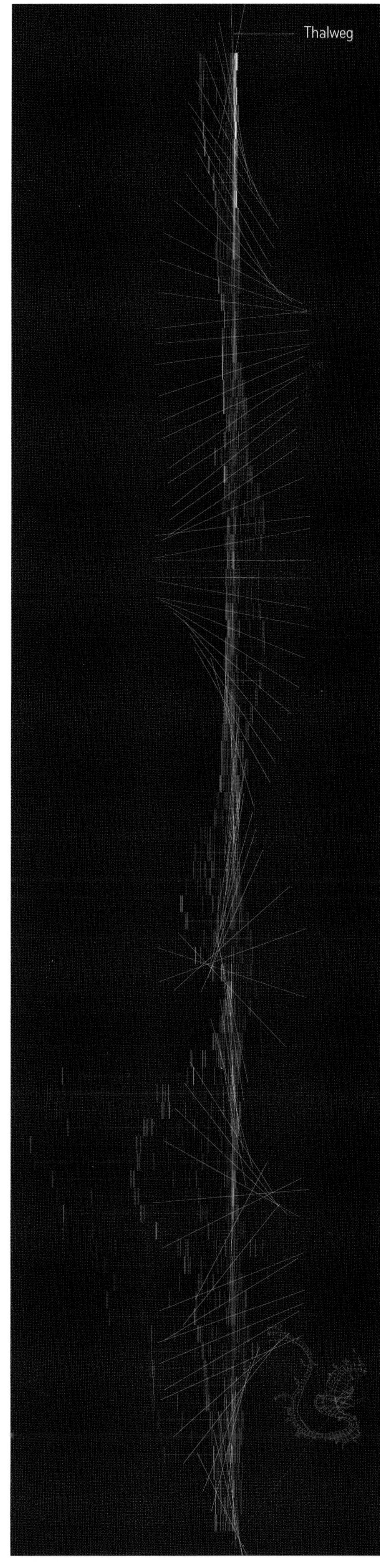
Thalweg

TRANSECT 2

Gyanpur Pump Canal.

T79: N 25°17.433', E 82°13.300', March (2006).

T84: N 25°13.945', E 82°11.921'.

T85: N 25°13.563', E 82°11.777'.

T87: N 25°13.318', E 82°11.740'.

T88: N 25°13.070', E 82°11.753'.

T89: N 25°12.870', E 82°12.171'

T92: N 25°12.116', E 82°11.914', March (2006).

T94: N 25°10.982', E 82°12.679'.

T96: N 25°10.719', E 82°14.731'.

T95: N 25°10.639', E 82°14.309'.

T97: N 25°11.026', E 82°15.943'.

T100: N 25°11.853', E 82°17.049'.

T101: N 25°12.069', E 82°16.937'.

T105: N 25°12.509, E 82°14.674', March (2006).

T107: N 25°14.033', E 82°14.314'.

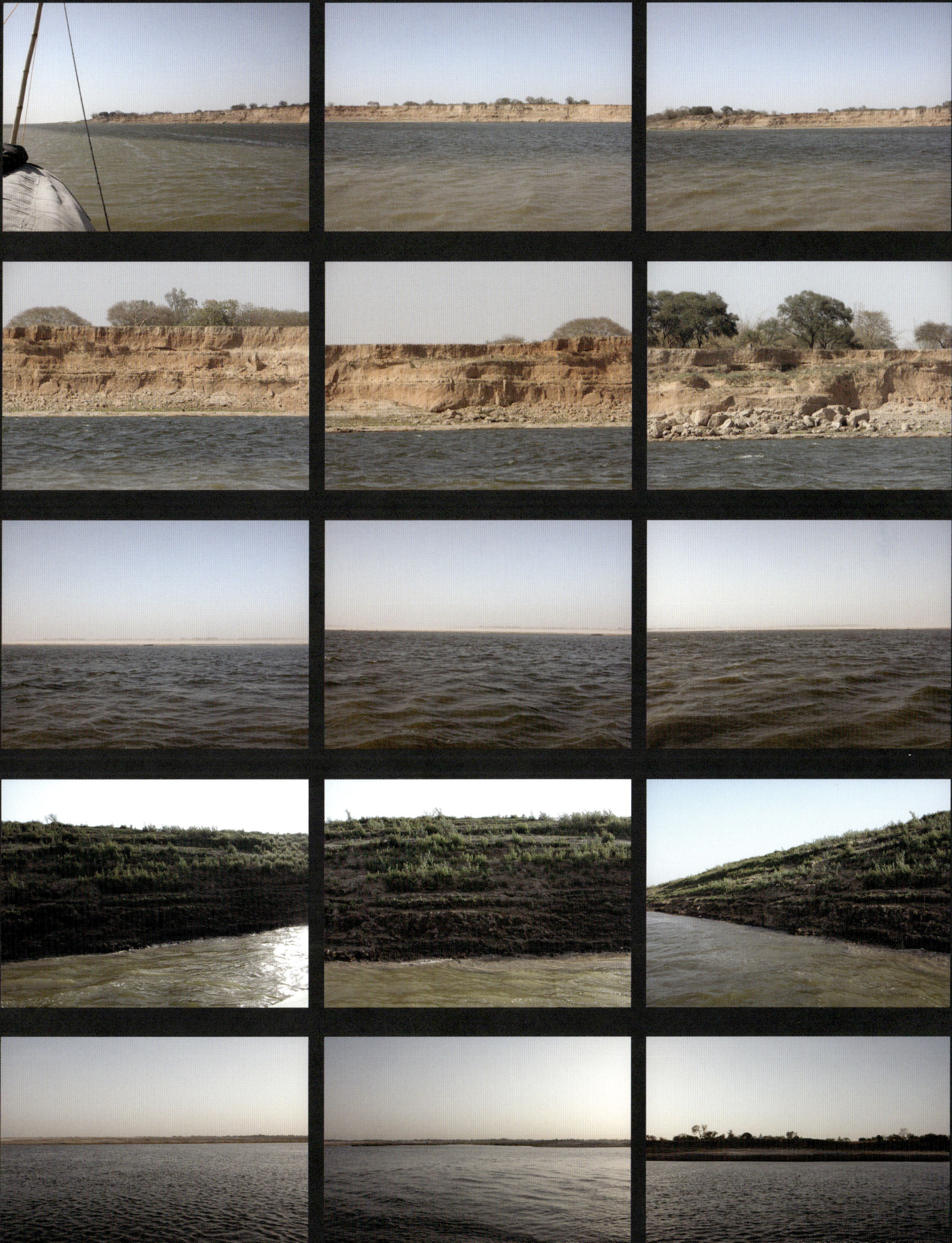

T109: N 25°13.954', E 82°14.597'.

T110: N 25°13.983', E 82°14.715'.

T111: N 25°14.109', E 82°14.953', March (2006).

T114: N 25°16.241', E 82°16.007'.

T115: N 25°16.193', E 82°18.168'.

T113: N 25°15.632', E 82°15.499'.

TRANSECT 2 - DETAIL 1

By zooming in and delaminating the surface of the transect, it is possible to see the depth of the section as well as map the choreography of water and agriculture across common and private property. What crops are grown during the kharif season (May-November) and the rabi season (November-May) are part of this choreography, as are the number and distribution of gravity flow canals and lift irrigation implements (tubewells, mobile engines and pumps). 2010.

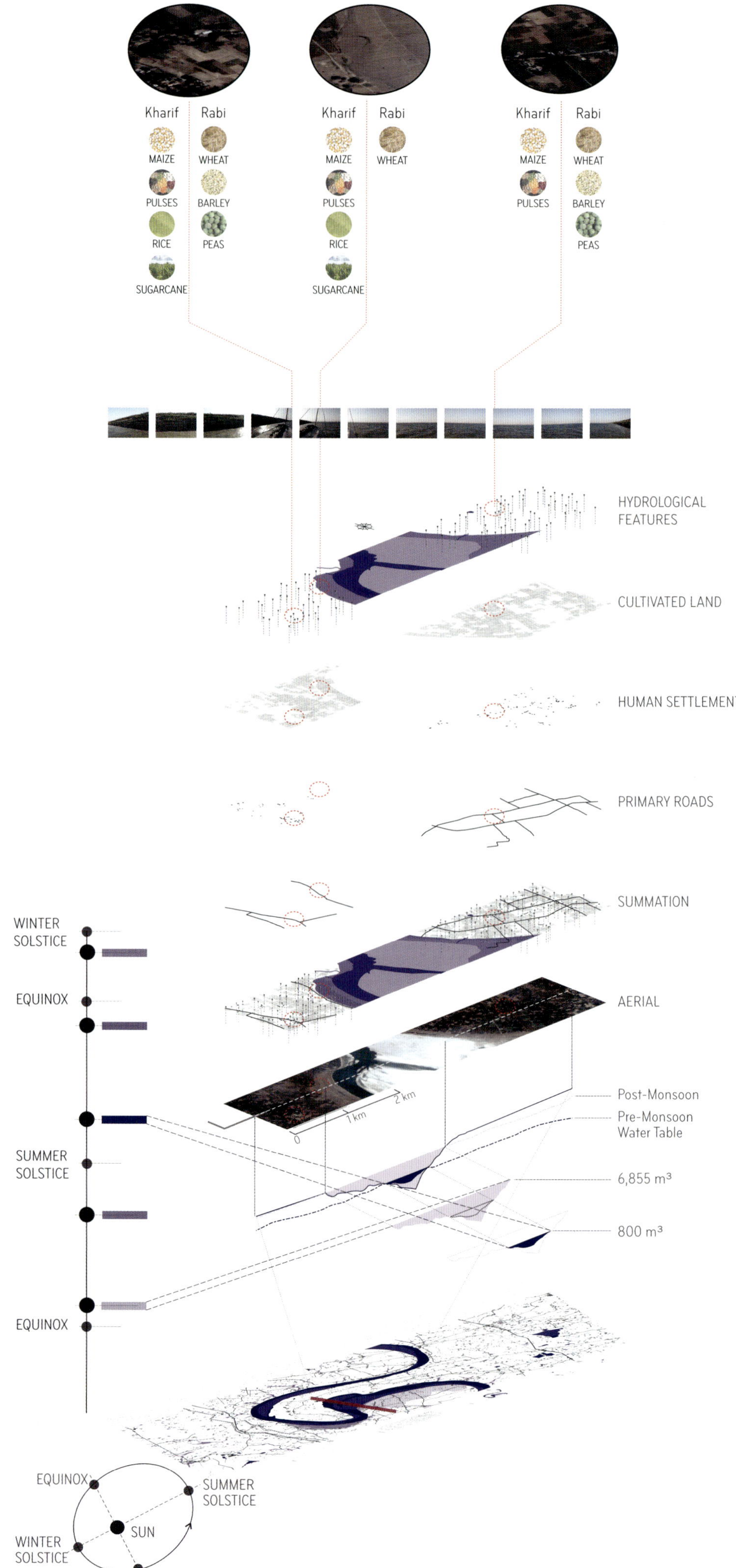

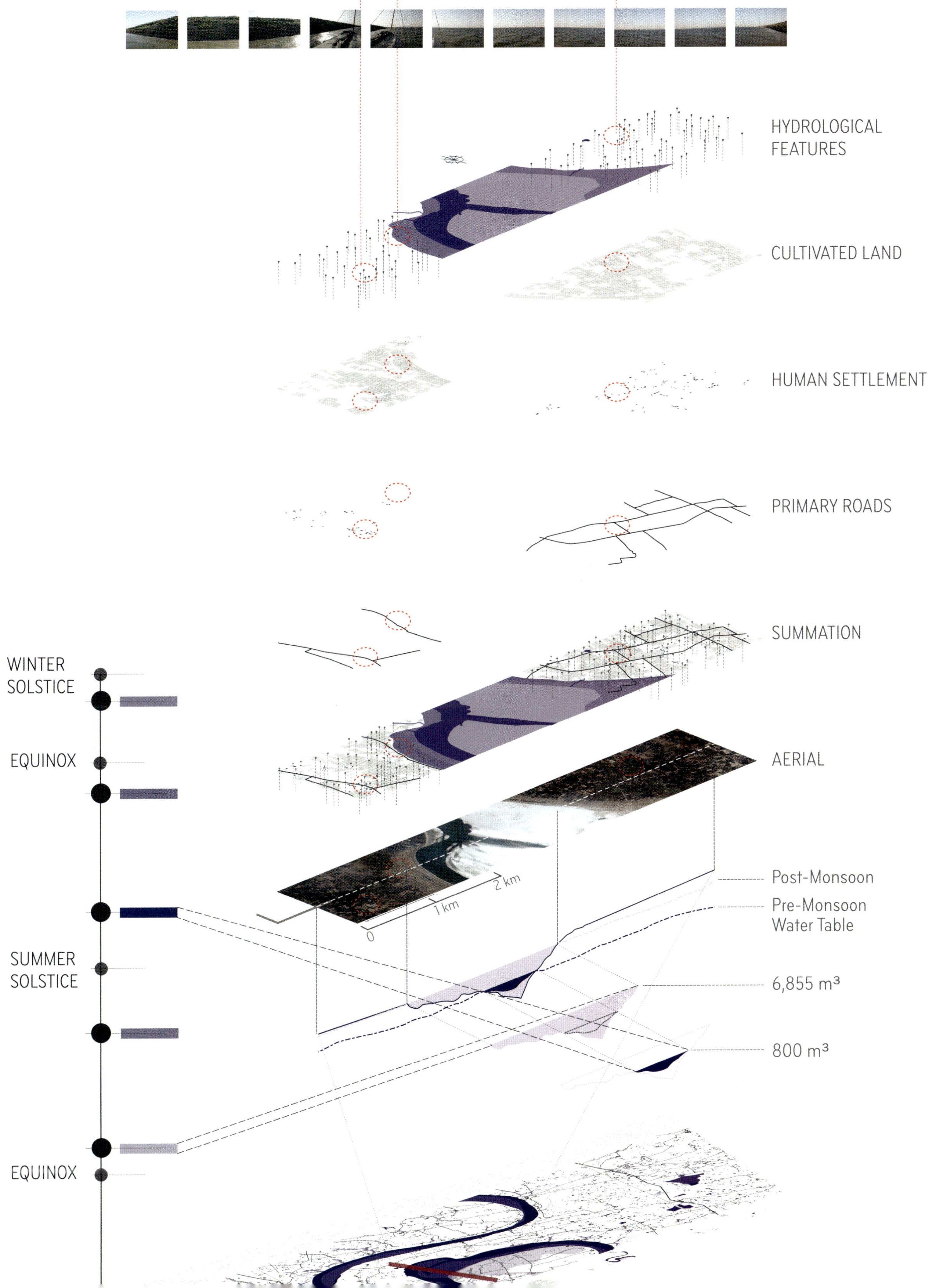

TRANSECT 2 – DETAIL 1

Along Left Bank of Ganges River, south of Gopiganj, March 2006.

TRANSECT 3

Grand Trunk Road (NH2) at Gopiganj

Ganges River

ONE LEG OF THE GOLDEN QUADRILATERAL (GQ) PROJECT
STARTED IN: 2000
COMPLETED IN: 2012
GQ LENGTH: 5,846 KM (3,633 MILES)

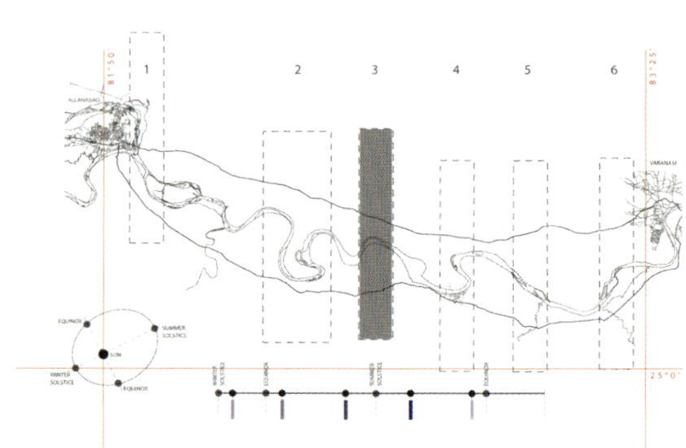

Transect 3

TRANSECT 3

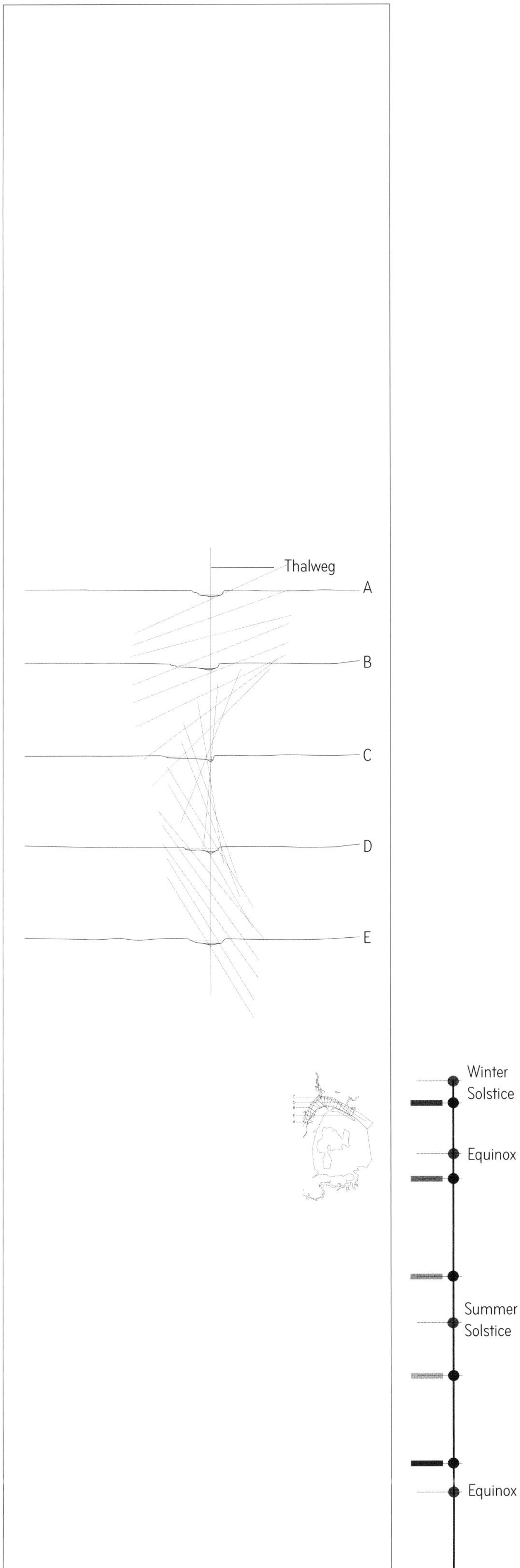
Thalweg
A
B
C
D
E
Winter
Solstice
Equinox
Summer
Solstice
Equinox

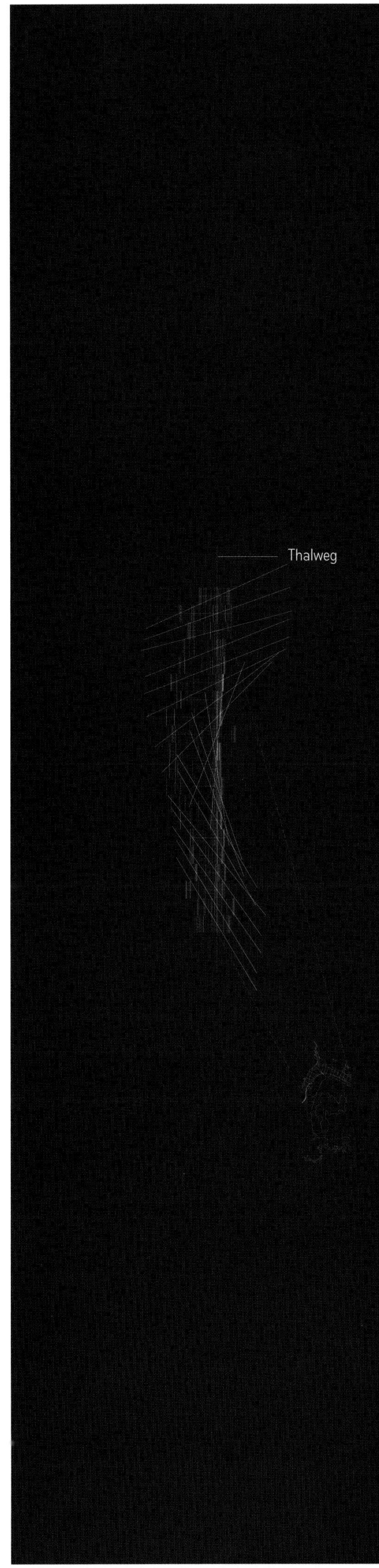
Thalweg

T117: N 25°12.601' E 82°19.722', March (2006).

T120: N 25°12.751', E 82°21.706'.

T122: N 25°13.366', E 82°21.885'.

T118: N 25°12.638', E 82°21.453'.

T123: N 25°13.741', E 82°21.981'.

T124: N 25°15.510', E 82°22.849'.

T126: N 25°15.432', E 82°24.270'.

T130: N 25°12.687', E 82°27.338'.

T132: N 25°11.222', E 82°27.888'.

T133: N 25°10.512', E 82°28.475'.

T125: N 25°15.611', E 82°23.041', March (2006).

T131: N 25°12.156', E 82°27.546'.

T135: N 25°10.442', E 82°29.535'.

Along the Ganges River at Mirzapur, March (2006).

TRANSECT 4

GHATS AT MIRZAPUR

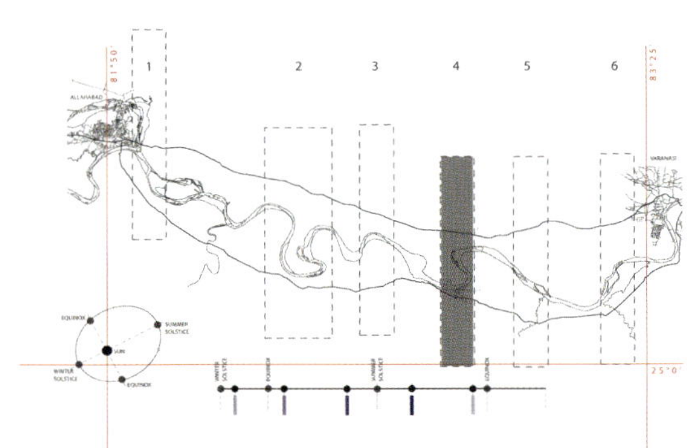

Transect 4

TRANSECT 4

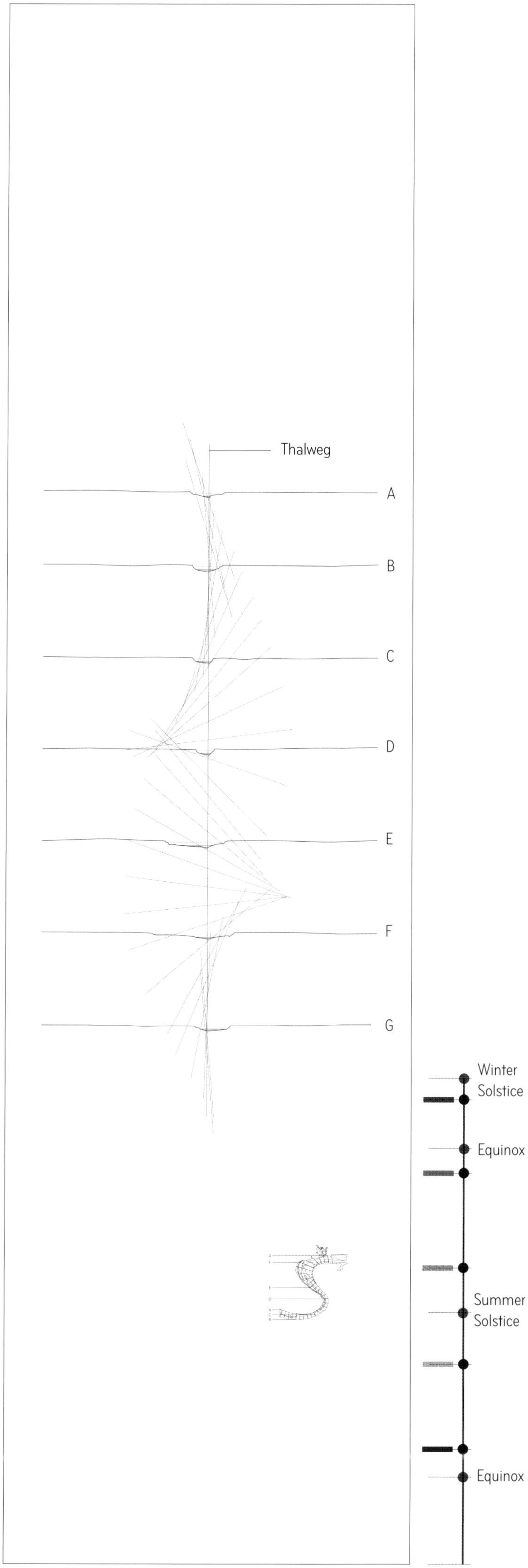
Thalweg
A
B
C
D
E
F
G
Winter Solstice
Equinox
Summer Solstice
Equinox

Thalweg

T136: N 25°09.128', E 82°33.285', March (2006).

T138 N 25°09.125', E 82°33.619'.

T141: N 25°13.408', E 82°33.532'.

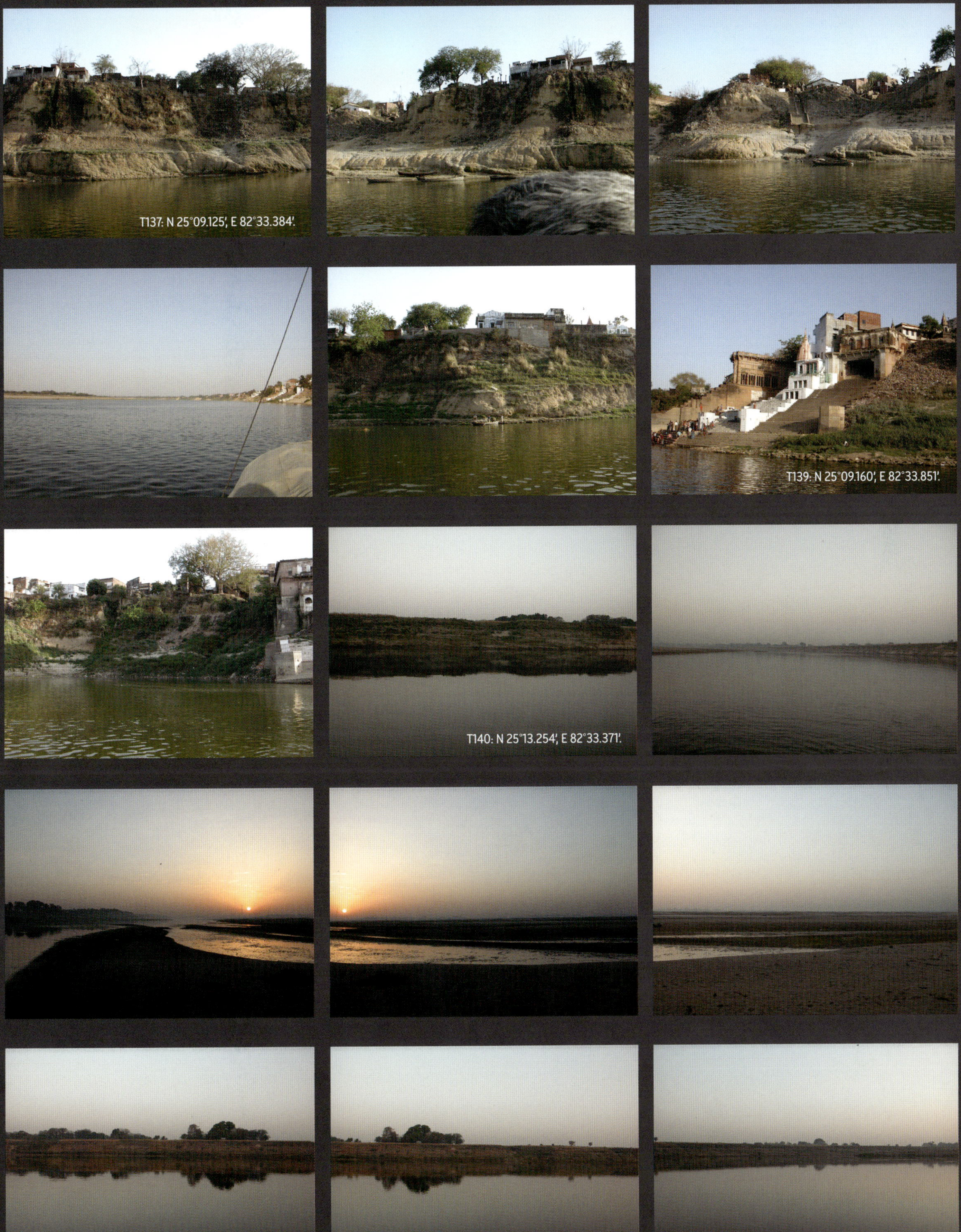

Ghats at Mirzapur along Ganges.

T142: N 25°13.625', E 82°35.259', March (2006).

T148: N 25°13.768', E 82°38.377'.

T150: N 25°13.126', E 82°39.140'.

T145: N 25°13.901', E 82°35.892'.

T149: N 25°13.560', E 82°38.814'.

T151: N 25°12.754', E 82°39.306'.

T152: N 25°12.523', E 82°39.603'.

Along Ganges River southeast of Mirzapur, March (2006).

TRANSECT 5

TRANSECT 5

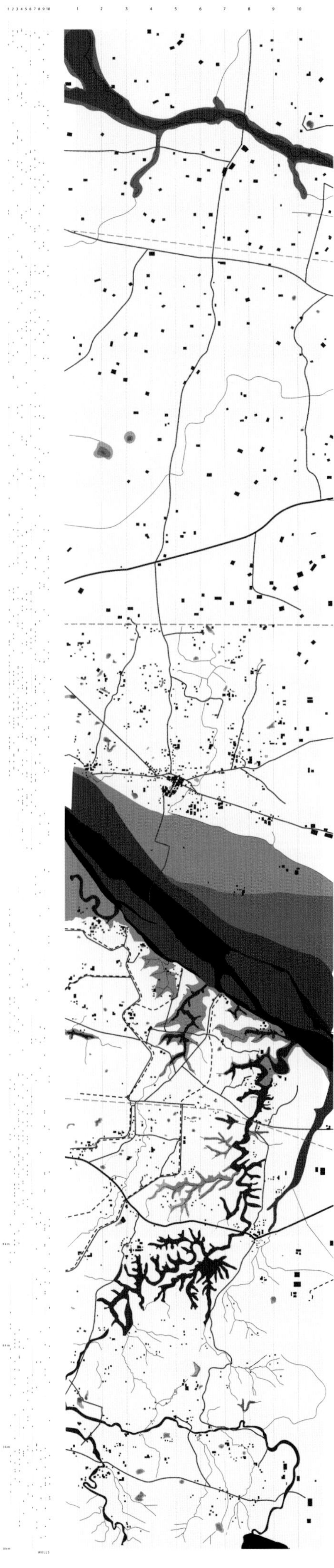

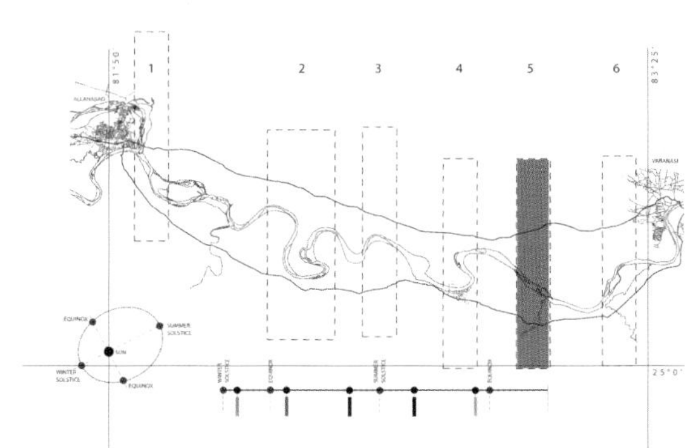

Transect 5

TRANSECT 5

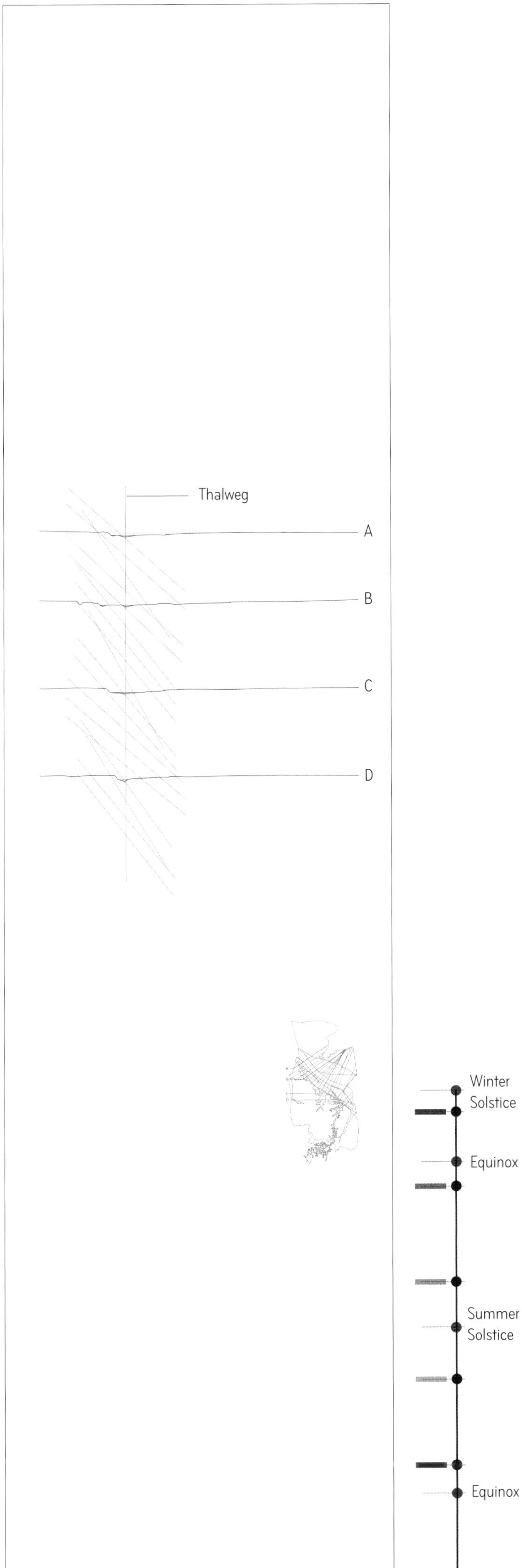
Thalweg
A
B
C
D
Winter
Solstice
Equinox
Summer
Solstice
Equinox

Thalweg

T154: N 25°12.142', E 82°40.343', March (2006).

T156: N 25°11.900', E 82°40.810'.

T158: N 25°10.974', E 82°42.098'.

T162: N 25°09.263', E 82°44.020'.

T164: N 25°08.592', E 82°44.717'.

T155: N 25°12.030', E 82°40.582'.

T157: N 25°11.644', E 82°41.138'.

T161: N 25°09.891', E 82°43.452'.

TRANSECT 6

Floating Pump house West of Varanasi, March (2006).

TRANSECT 6

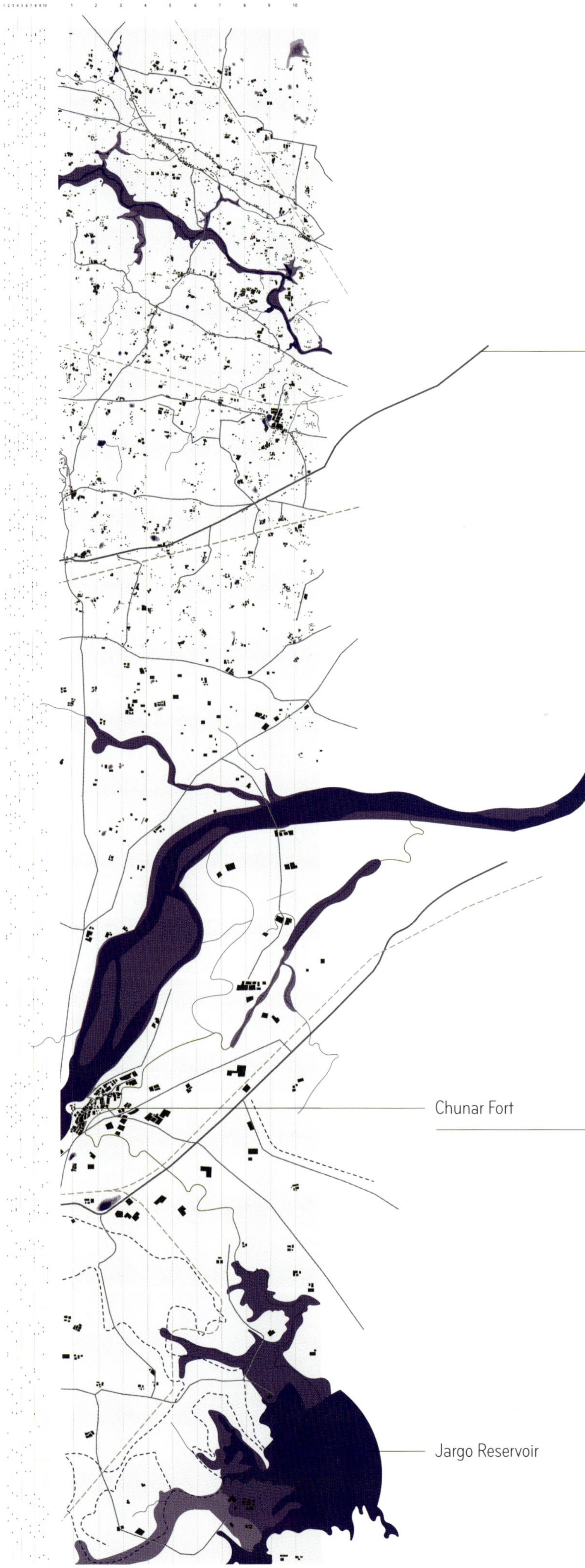

STARTED IN: 1969 (APPROX.)
COMPLETED IN: 1974 (APPROX.)
RENOVATED IN: 2000-03
ULTIMATE IRRIGATION POTENTIAL: 73.145 (TH HA)
POWER REQUIRED: 17,480 KVA

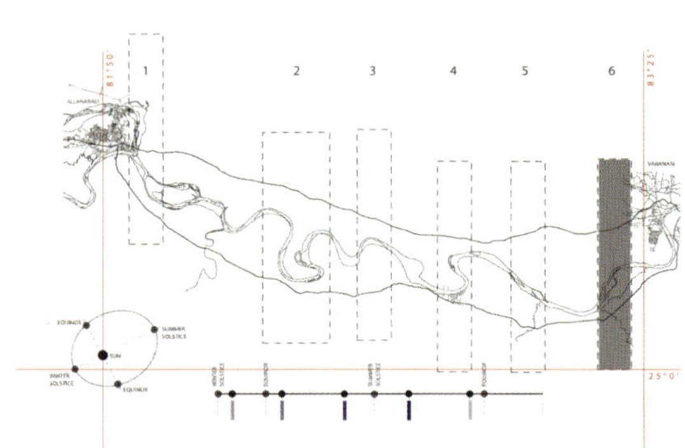

Transect 6

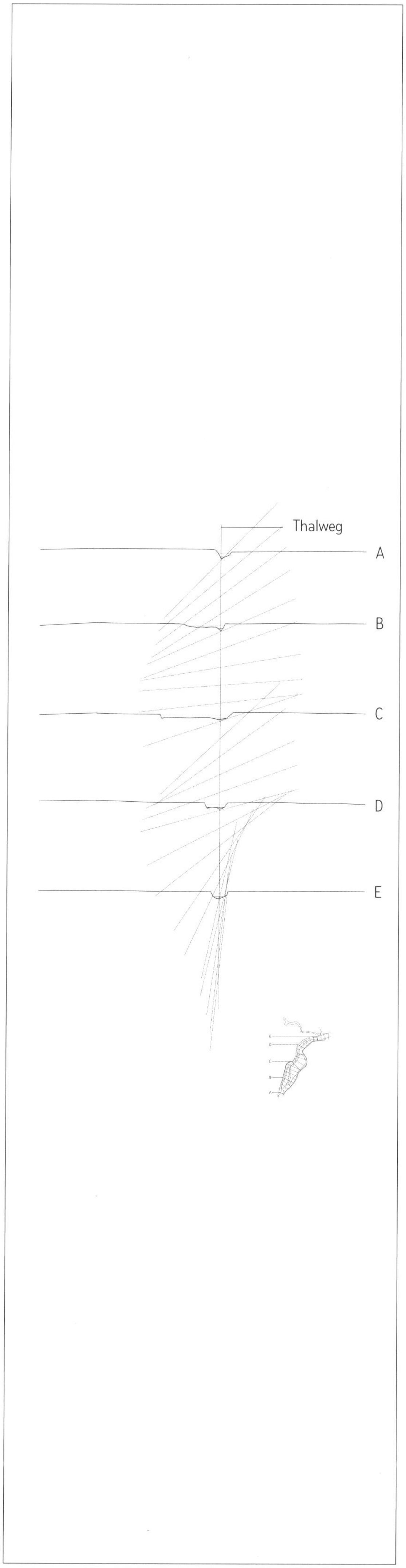
Thalweg
A
B
C
D
E

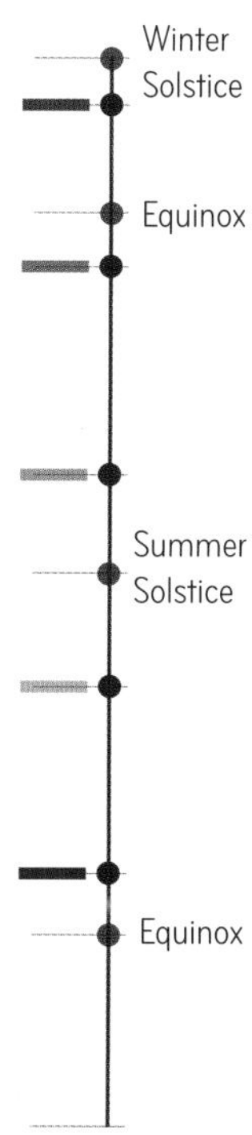
Winter
Solstice
Equinox
Summer
Solstice
Equinox

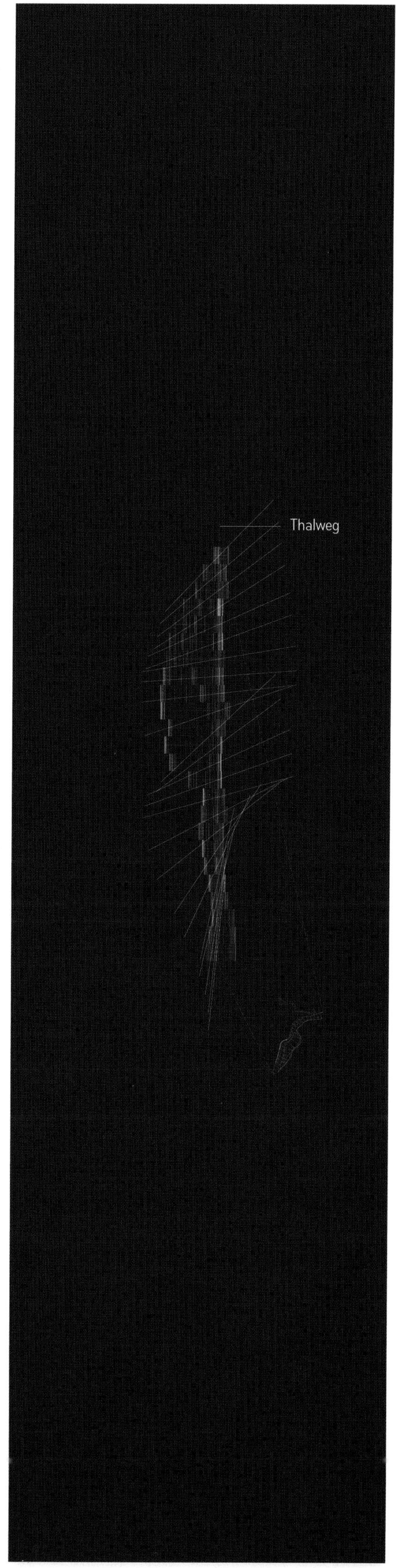
Thalweg

TRANSECT 6

T165: N 25°06.912', E 82°49.375', March (2006).

T166: N 25°06.853', E 82°50.322'.

T167: N 25°06.807', E 82°51.074'.

T168: N 25°07.138', E 82°52.208'.

Chunar Fort

T171: N 25°09.554', E 82°52.774', March (2006).

T174: N 25°12.255', E 82°54.922'.

T176: N 25°12.444', E 82°56.616'.

Diesel Powered Pum

T177: N 25°12.241', E 82°58.324'.

Narainpur Pumping Station.

T179: N 25°12.338', E 83°00.353'.

T172: N 25°10.285', E 82°52.991'.

T173: N 25°10.332', E 82°53.301'.

T175: N 25°12.340', E 82°55.260'.

Floating Diesel Pump House.

T178: N 25°12.159', E 82°59.373'.

NARAINPUR PUMPING CANAL

A hybrid between the pump and the canal, Narainpur Pump Canal diverts 1,800 cusecs (51 cumecs) of water. It is a yellow ochre concrete structure stained brown by the flooding of the Ganges, with steel intake pipes that are only visible above the meander of the river when the water level is low. Said to be the largest pump canal system in Asia, it operates 14 pumps with two backups, irrigating an area of 73,415 ha (283.5 square miles). Made of poured-in-place concrete, the Pump House was begun and completed during the Fourth Five Year Plan (1969-74).

Location: Near village of Narainpur, in Mirzapur District (U.P.)
Districts Served: Chandauli, Mirzapur, and Varanasi (U.P.)
Completed: Fourth Five Year Plan (1969-74)
Renovated: 2000-03
Irrigable Area: 73.145 Thha (hectares)
Power Required: 17,480 KVA
Pumps: 14 Working
(2014).

TRANSECT 6 — NARAINPUR

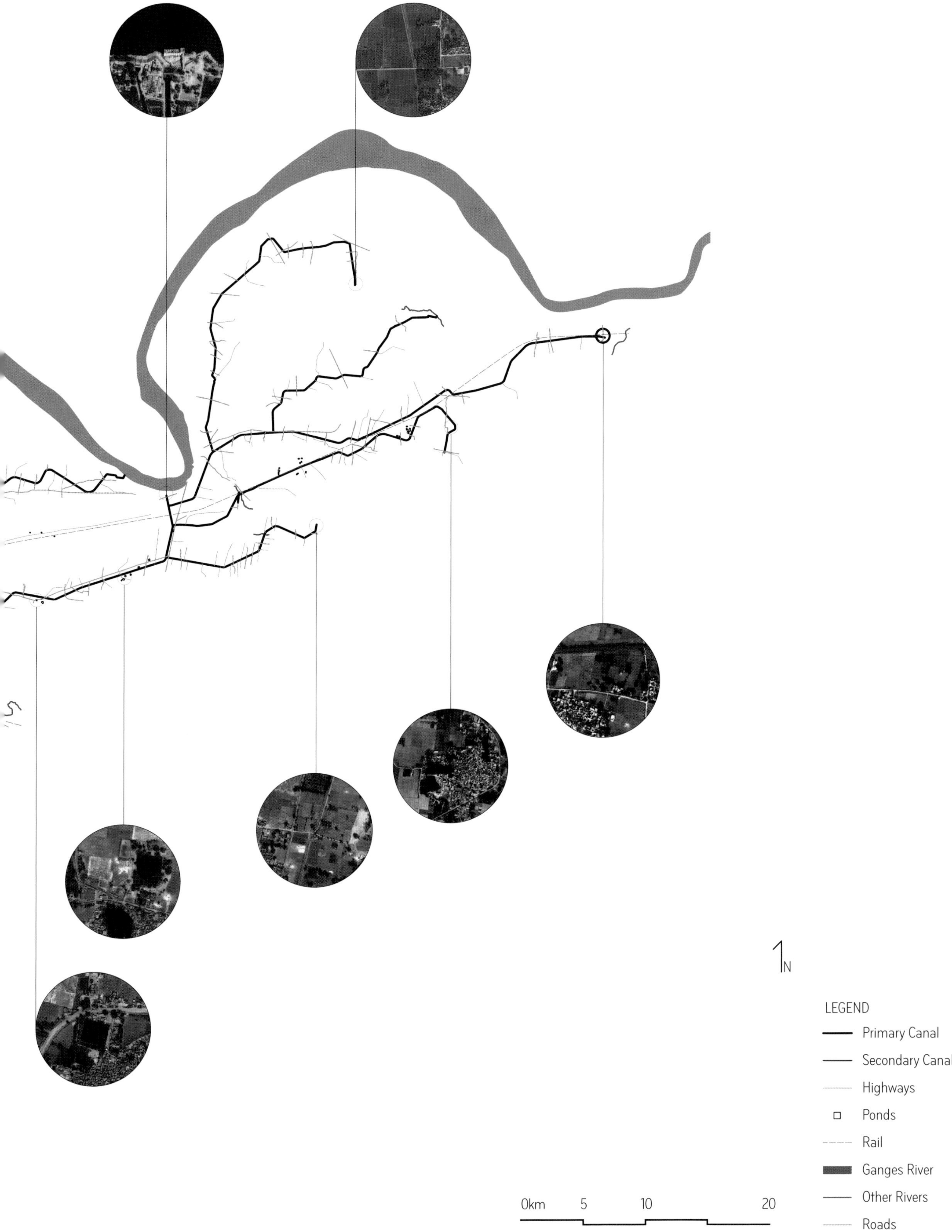

1: Panorama at Narainpur Pumping Canal at the 14 pumps in August, 2008.

2: Panorama downstream from the Narainpur Pump House in August, 2008.

3: Panorama further downstream of Narainpur Pump House with escapes and minor distribution canals in August, 2008.

Lieutenant-Colonel Forrest. "Sacred Tank and Pagodas near Benares." **A Picturesque Tour Along the Rivers Ganges and Jumna in India**. London: R. Ackermann, 1842. (Princeton University).

VARANASI

Benares [Varanasi] is likewise one of the most ancient cities of India...Some of these ghauts [ghats] are very fine edifices...Indeed the whole face of the city towards the river is one continued line of these ghauts, which, exclusive of the ornament they are to Benares, are highly commodious and useful in the facility for bathing which they present to its vast population...—Lieutenant Colonel Forrest, A Picturesque Tour Along the Rivers Ganges and Jumna in India (1824)

The cliffs of Varanasi (often called Benares or Banaras) are concealed beneath some 3 miles (4.8km) of over 100 distinct sets of steps (*ghats*). Crowned by palaces, guesthouses, and temples, these steps lead to the waters of the Ganges River. Like separate floors of a building, they abut one another, but maintain considerable autonomy in terms of how they are designed and used. A ghat descending from a palace and used for bathing might share an edge with a ghat for water buffalos on one side, while on the other side a religious ritual or prayer (*puja*) is taking place. A trip up and down the river by boat is like viewing a succession of vertical section cuts. Such heterogeneity of design and use, along with the flooding of the Ganges and the temples half under water and caked in silt, are what have always made Varanasi picturesque to travelers.

If strips of ghats crowned by buildings define the urban-riparian edge of Varanasi, then the courtyard and *kunds* (tanks or pools) define the historic core of the city. Courtyard buildings, mostly shops on the ground floor and residences up above, characterize most of the urban fabric of the historic core. These residential voids, combined with the once-copious number of tanks and pools, created a very different grain to the strips of ghats along the river. Today, most of the courtyard structures remain while the kunds are quickly disappearing. Grand Trunk Road (NH2) delimits the outer reaches of the historic core. Beyond the highway, the city grows more diffuse. As seen from the plan of the city in the late 1970s, hundreds—if not thousands—of tubewells draw heavily from groundwater resources (see Varanasi—Morphological Development drawing). These tubewells, combined with the growing paucity of land, create a dense yet largely disparate urban archipelago.

Lieutenant-Colonel Forrest. "City of Benares from the Ganges." **A Picturesque Tour Along the Rivers Ganges and Jumna in India**. London: R. Ackermann, 1842. (Princeton University).

"Flood in Ganges. Benares." Saeed Bros, Benares. Circa 1920. (AA).

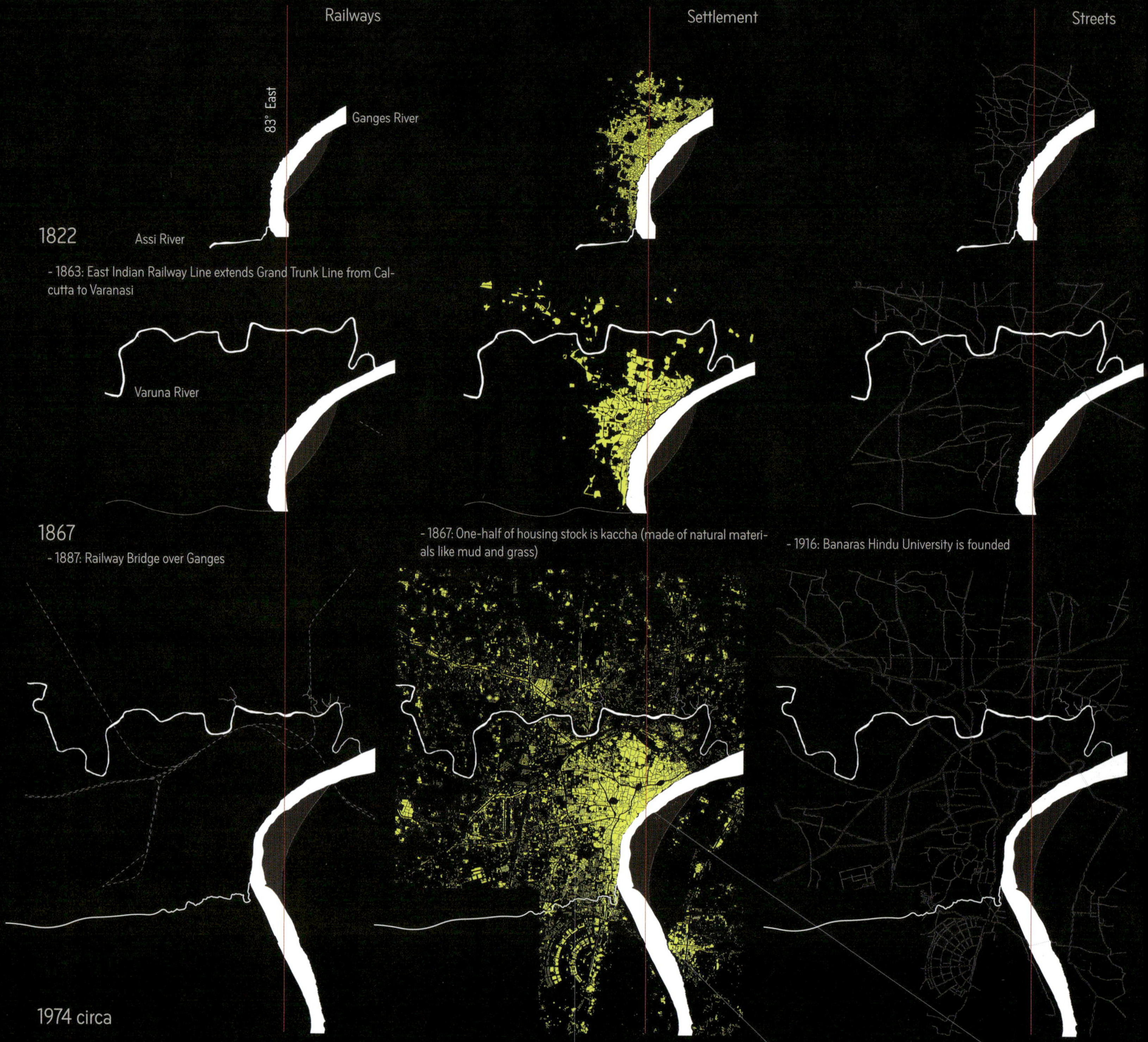

Varanasi hugs the left bank of the Ganges River at a moment where the river flows due north. Known for its temples and ghats (steps) and dense living quarters, the city has been a major pilgrimage center for over 2,000 years. James Prinsep, an English scholar, produced the first modern census of the city in 1822. Prinsep's map and writings described a city composed of water: bordered by the Ganges to the east and myriad kunds (tanks or pools) adjacent to temples and orchards filled the city. Drains and paths leading from these kunds to the Ganges defined principal thoroughfares that exist today. Bounded to the north by the Varuna River, to the south by the Assi River (though now more a drain than river), the Ganges to the east and Grand Trunk Road (NH2) to the west, the city has rapidly grown westward since the 1950s. With most of the historic kunds filled in, the city is today speckled by thousands of tubewells.

Note: 1822 and 1867 drawings based on maps in R. L. Singh, **Banaras: A Study in Urban Geography** (Banaras : Nand Kishore, 1955); 1979 map based on 1979 Survey of India Map.

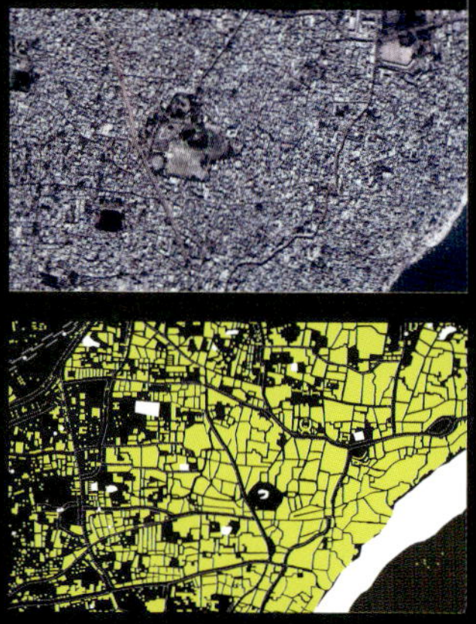

Historic Core (with Goudalia Crossing)

Assi Ghat (caked in silt post-monsoon)

Near Goudalia Crossing

VARANASI – MORPHOLOGICAL DEVELOPMENT

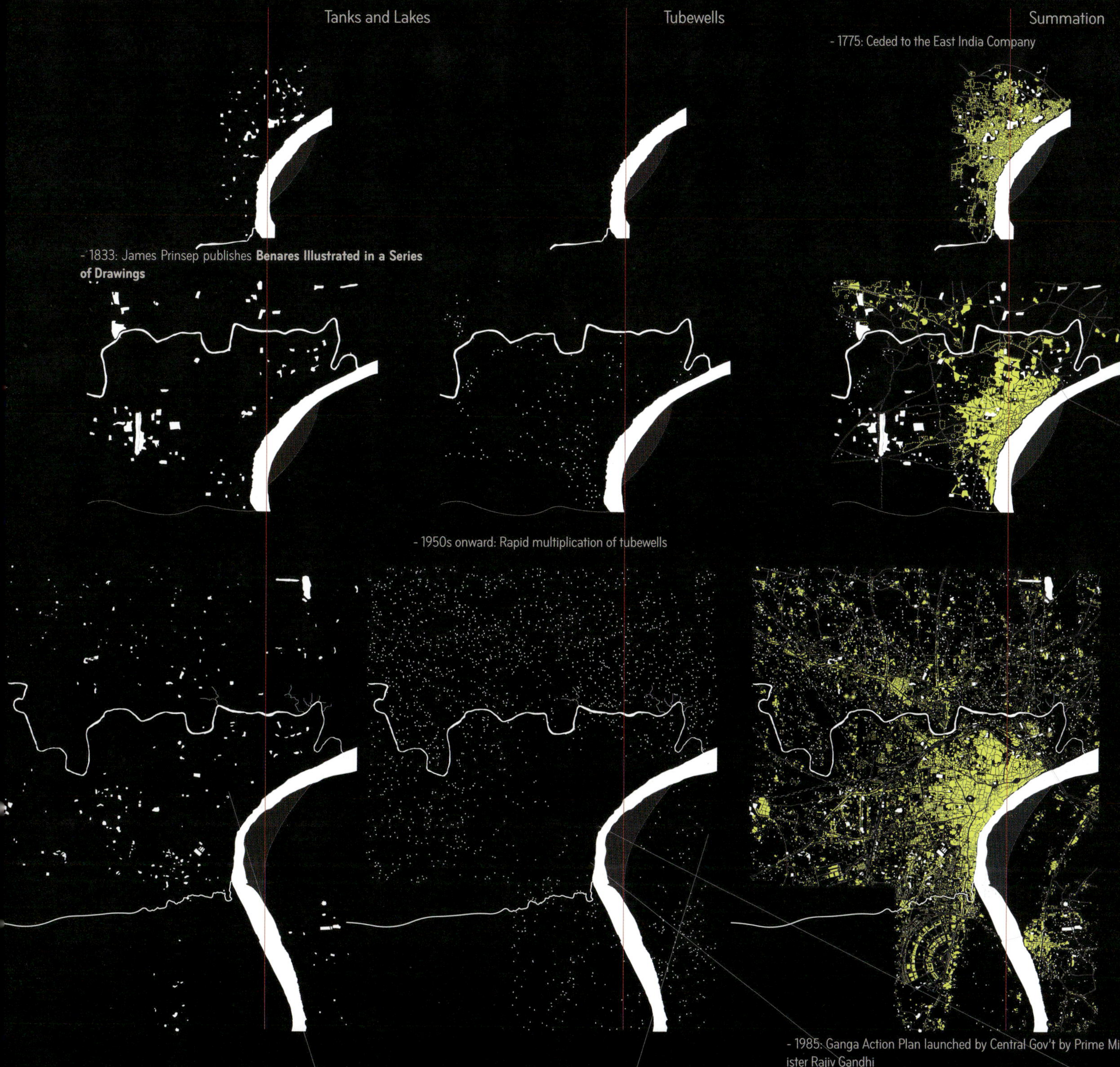

1: Panorama at Meer Ghat October 2, 2005.

4: Panorama at Meer Ghat October 9, 2005.

5: Panorama at Meer Ghat October 10, 2005.

VARANASI – MEER GHAT

2: Panorama at Meer Ghat
October 3, 2005.

3: Panorama at Meer Ghat
October 5, 2005.

VARANASI — MEER GHAT

7: Panorama at Meer Ghat October 12, 2005.

9: Panorama at Meer Ghat November 16, 2005.

10: Panorama at Meer Ghat June 14, 2006.

6: Panorama at Meer Ghat
October 11, 2005.

VARANASI — MEER GHAT

8: Panorama at Meer Ghat
October 13, 2005.

Puja Pandal built to look like Taj Mahal, October (2005).

Puja Pandal, October (2005).

Path to temporary installations and Puja Pandals, October (2005).

Durga and idols en route to Ganges River, October (2005).

Durga and idols en route to Ganges River, October (2005).

DURGA PUJA VARANASI

In the autumn, one of two Nine Night Cycles in Hinduism is observed in Varanasi. Commonly known as Durga Puja, the entire city is transformed in recognition of goddess Durga's triumph over the demon Manisha. Large puja pandals (temporary structures to revere the goddess) are built and incorporated into a set of permanent structures like Durga Kund to venerate Durga. Devotees travel to each of these structures, navigating a sacred geography outlined by neon lights. These lights adorn buildings and streets, announcing paths to the elaborately adorned monumental edifices. Inside these spaces scenes from Durga's victories are recreated through dioramas and paintings.

Ishwar Gangai, June (2006).

Pichas Mochan Tal, June (2006).

Pitarkunda Tal, June (2006).

Soma Kund, June (2006).

Ram Kund June (2006).

KUNDS OF VARANASI

Varanasi, once known as a city of orchards and kunds (tanks or pools), is one U.P.'s most populous cities. Few of the tanks recorded in nineteenth century accounts exist today. Important infrastructures to temper the deluge of the southwest monsoon and replenish aquifers, kunds were often rimmed with important temples and seminaries. Those attached to temples, like Durga Kund, are well-preserved and used throughout the year. Others, such as Soma Kund, fill with water when it rains, but otherwise collect refuse and serve as grazing grounds for cows and goats throughout most of the year. Sill others, such as Ram Kund and Company Gardens tank are covered in algae much of the year. Today, those not attached to temples or religious sites are fast disappearing due to urban growth and land scarcity. Filled, plotted, and sold, these former kunds are often plagued by waterlogging and flooding.

Durga Kund, June (2006).

Lakshmi Kund, June (2006).

Bakaria Kund, June (2006).

Machodary Park, June (2006).

Company Gardens, June (2006).

KUNDS OF VARANASI

Varanasi, June (2006).

Varanasi, October (2005).

Right Bank of Ganges, opposite Varanasi, November (2005).

Varanasi, June (2006).

Varanasi, October (2005).

Varanasi, October (2005).

GHATS OF VARANASI

Varanasi, September (2005).

Varanasi, September (2005).

Varanasi, June (2006).

Varanasi, October (2005).

Varanasi, October (2005).

GHATS OF VARANASI

Ghats, September (2005).

Survey of Ghats north of Assi Ghat, July (2011)

Clearing of silt from ghats, September (2005).

Meer Ghat, October 04 (2005).

Meer Ghat, October 05 (2005).

Meer Ghat, October 09 (2005)

Meer Ghat, Auspicious Bathing Hour, October 17 (2005).

GHATS OF VARANASI

Meer Ghat, October 10 (2005).

Meer Ghat, October 11 (2005).

Meer Ghat, October 14 (2005).

INTRODUCTION

1. See Govind Sadashiv Ghurye, *Anatomy of a Rururban Community* (Bombay: Popular Prakashan, 1963).
2. Roger Revelle and V. Lakshminarayana, "The Ganges Water Machine," *Science*, New Series, Vol. 188, No. 4188, Food Issue (May 9, 1975), 611.
3. Roger Revelle and Theodore Herman, "Some Possibilities for International Development of the Ganges-Brahmaputra River Basins," in *Water for the human environment: proceedings of the first World Congress on Water Resources, held September 24-28, 1973*, in Chicago, Illinois, U.S.A. (Champaign: International Water Resources Association, 1973), 154—163.
4. Joseph Conrad, "Geography and Some Explorers," *The National Geographic Magazine*, Vol. XLV, No. 3 (March, 1924), 241–356.
5. James Johnson, *An Account of a Voyage to India, China &c. in His Majesty's Ship Caroline* (London: Richard Phillips, 1806), 32.

PART 1

1. Indira Gandhi, "Foreword,"*Gaṅgā*, C. Sivaramurti, (New Delhi: Orient Longman, 1976), i.
2. John Watson McCrindle, trans. *Ancient India as described by Megasthenês and Arrian* (Calcutta: Thacker, 1877), 68.
3. Strabo, *The Geography of Strabo, Volume 15*, with an English translation by Horace Leonard Jones (London: Loeb Library, 1917 –33), 19.
4. See Susan Gole, *India Within the Ganges* (New Delhi: Jayaprints, 1983); Aaron Arrowsmith, *A Compendium of Ancient and Modern Geography: For the Use of Eton School* (London: For the Author, 1839).
5. Raghubir Singh, *The Ganges* (London: Thames and Hudson, 1992), 11.
6. Diana Eck, "Gangā: the Goddess in Hindu Sacred Geography," *The Divine Consort: Rādhā and the goddesses of India*, eds. John Stratton Hawley and Donna Marie Wulff (Berkeley: Berkeley Religious Studies Series, 1982), 166—67.
7. Felix V. Raper, "Narrative of A Survey For the Purpose of Discovering The Sources of the Ganges," *Asiatic Researches or, Transaction of the Society Instituted in Bengal, For Inquiring into the History and Antiquities, the Arts, Sciences, and Literature, of Asia*, Vol. 11 (London: T. Maiden and W. Wilson, 1812), 506.
8. Ibid, 505.
9. Saivites (worshippers of Siva) consider this point the abode of Siva (Har) while Vaisnavites (followers of Vishnu) call it Haridwar, the home of Vishnu (or Hari).
10. Mildred Archer, "India and the Illustrated Book," in *A Journey to Hindoostan: Graphic Art of British India, 1780-1860*, catalogue and exhibition by Thomas P. Bruhn (Storrs: William Benton Museum of Art, University of Connecticut, 1987), 7.
11. James Baillie Fraser, *Journal of a Tour Through Part of the Snowy Range of the Himālā Mountains, and to the Sources of the Rivers Jumna and Ganges* (London: Rodwell and Martin, 1820), 469.
12. Ibid, 370—371.
13. Ibid, v.
14. William Hodges, *Travels in India, 1780, 1781, 1782 & 1783* (London: by Author, 1793), iii—iv.
15. William Fraser Tytler, "Notice respecting the Journey to the Sources of the Jumna and the Ganges, by J. B. Fraser, Esq." *The Edinburgh Philosophical Journal*, Vol. I, edited by David Brewster and Robert Jameson (Edinburgh: Archibald Constable and Company, 1819), 128.
16. James Baillie Fraser, *Journal of a Tour Through Part of the Snowy Range of the Himālā Mountains* (1820), 1.
17. Ibid, 114.
18. Ibid, 107-108.
19. Ibid, 109-110.
20. Ibid, 110.
21. Ibid, 138.
22. Mildred Archer, "James Baillie Fraser (1783—1856)," in *A Journey to Hindoostan: Graphic Art of British India, 1780-1860* (1987), 63.
23. Mildred Archer and Ronald Lightbrown, *India Observed: India as Viewed by British Artists, 1760-1860* (London: Victoria and Albert Museum in association with Trefoil Books, 1982), 100.
24. Thomas Skinner, *Excursions in India; Including a Walk Over The Himalaya Mountains, to the Sources of the Jumna and the Ganges*, Vol. 2 (London: Henry Colburn and Richard Bentley, 1832), 141—142.
25. R. J. Wasson, "A sediment budget for the Ganga-Brahmaputra catchment," *Current Science*, Vol. 84, No. 8, (25 April 2003), 1041—1047.
26. See Rajiv Sinha, "Why do Gangetic rivers aggrade or degrade?" *Current Science*, Vol. 89, No. 5,

(10 September 2005), 836–840.

27. Jules Verne, *The Demon of Cawnpore*, trans. A. D. Kingston (New York: Charles Scribner's and Sons, 1881), 88-89.
28. William John McGee, *The Flood Plains of Rivers* (New York: Publishers' Printing Company, 1891), 229.
29. Proby T. Cautley, *Ganges Canal. A disquisition on the heads of the Ganges and Jumna canals, North-western Provinces, in reply to strictures by Major-General Sir Arthur Cotton* (London: Private Circulation, 1864), 28.
30. Government of India, *Settlement Report of the District of Moozuffernugger: Including a Report on the Permanent Settlement of the Western Pergunnahs of the District* (Allahabad: North-Western Provinces Government Press, 1873), 45.
31. Verne, 177.
32. Ibid, 175.
33. Ibid, 73.
34. Ibid, 177.
35. Ibid, 177.
36. Ibid, 177–178.
37. James Rennell, "An Account of the Ganges and Burrampooter Rivers" *Philosophical Transactions of the Royal Society of London*, Vol. LXXI, Part I (1781), 103; also to be found in James Rennell, *Memoir of a Map of Hindoostan: or the Mogul Empire: with an introduction, illustrative of the Geography and Present Division of that Country: And a Map of the Countries Situated between the Head of the Indus, and the Caspian Sea* (London: M. Brown, 1788), 268.
38. Rennell (1781), 103.
39. Matthew Edney, *Mapping an Empire: The Geographical Construction of British India, 1765–1784* (Chicago: University of Chicago Press, 1997), 8.
40. Ibid, 8.
41. J. Long, *Selections from the Unpublished Records of Government for the years 1748 to I767 Inclusive Relating Mainly to The Social Condition of Bengal, with a Map of Calcutta in 1784* (Calcutta: Office of Superintendent of Government Printing, I869), 487.
42. Clements Robert Markham, *Major James Rennell and the Rise of Modern English Geography* (London: Cassell and Company, 1895), 50.
43. For more on the history of the traverse or route survey, see Matthew Edney, *Mapping an Empire* (1997), 91–96.
44. J. Long, *Selections from the Unpublished Records of Government for the years 1748 to I767* (I869), l.
45. Markham, 47.
46. Ibid, 9.
47. This map was originally published in 1777 by Andrew Dury for the East India Company and was eventually republished in London by Robert Sayer in 1786. The 1786 version is what is reprinted in this book.
48. Henry Thomas De La Bache, *A Geological Manual*, 3rd Edition (London: Charles Knight, 1833), 74-75.
49. Charles Lyell, *Principles of Geology; or, The Modern Changes of the Earth and Its Inhabitants Considered as Illustrative of Geology* (New York: D. Appleton & Co., 1865), 277–78, 293.

NOTES

PART 2

1. Proby Cautley, *Report of the Ganges Canal Works: from their commencement until the opening of the canal in 1854*, Vol. I (London: Smith, Elder and Co., 1860), 3.
2. Charles Eliot Norton, "The Opening of the Ganges Canal," *The North American Review*, (Oct. 1855), 539.
3. Cautley, (1860), 26.
4. Ibid, 18.
5. "The Ganges Canal," *Supplement to the Delhi Gazette* (April 8, 1854), reprinted in C. W. Grant, *Indian Irrigation: Being a Short Description of the System of Artificial Irrigation and Navigation in India* (London: Smith, Eldon & Co., 1854), x.
6. Cautley (1860), 23-25.
7. Ibid, 29.
8. "Opening of the Ganges Canal," *The Illustrated London News* (Supplement, Aug. 5, 1854), 118.
9. Norton (1855), 535.
10. Ibid, 539.

11. C. W. Grant, *Indian Irrigation: Being a Short Description of the System of Artificial Irrigation and Navigation in India* (London: Smith, Eldon & Co., 1854), 35.
12. Norton (1855), 535.
13. "The Ganges Canal," *Supplement to the Delhi Gazette* (April 8, 1854), reprinted in C. W. Grant, *Indian Irrigation: Being a Short Description of the System of Artificial Irrigation and Navigation in India* (London: Smith, Eldon & Co., 1854), ix.
14. Charles Eliot Norton "Canals of Irrigation in India," *The North American Review*, (October 1853), 463.
15. Ibid, 464-465.
16. Norton (1855), 541.
17. Ibid, 543.
18. Karl Marx, "The British Rule in India," *New York Daily Tribune*, (June 25, 1853), reprinted in *Karl Marx and Friedrich Engels On Colonialism* (Moscow: Foreign Languages Pub. House, 1960), 31-36.
19. George Smith, *Twelve Indian Statesmen*, Second Edition, (London: John Murray, 1898), 275.
20. William Wilson Hunter, *Rulers of India: The Marquess of Dalhousie* (Oxford: Clarendon Press, 1890), 11.
21. Ibid, 275-76.
22. H. G. K., "The Opening of the Ganges Canal," *Blackwood's Edinburgh Magazine*, Vol. 76 (July—Dec., 1854), 475.
23. "The Great Ganges Canal," *The Civil Engineer and Architect's Journal*, Vol. XVII (1854), 330.
24. Charles Eliot Norton, "The Opening of the Ganges Canal," *The North American Review*, (Oct. 1855), 540.
25. Ian Stone, *Canal Irrigation in British India: Perspectives on Technological Change in a Peasant Economy* (Cambridge: Cambridge University Press, 1984), 18.
26. Sharad K. Jain, Pushpendra K. Agarwal, Vijay P. Singh, *Hydrology and Water Resources of India* (Dodrecht: Springer, 2007), 367.
27. James Crofton, *Report on the Ganges Canal*, Vol. I (Calcutta: Public Works Department, 1865), 81.
28. *The Encyclopaedia Britannica: a dictionary of arts, sciences and literature*, Vol. XVII (New York: Henry G. Allen Company, 1890), 572
29. Doab comes from Persian, literally translated, means "land between two rivers."
30. "The Ganges Canal," *Supplement to the Delhi Gazette* (April 8, 1854), reprinted in C. W. Grant, *Indian Irrigation: Being a Short Description of the System of Artificial Irrigation and Navigation in India* (London: Smith, Eldon & Co., 1854), x.
31. Richard Baird Smith, *A Short Account of the Ganges Canal, With a Description of Some of the Works* (Roorkee: Thomason College Press, 1870), 5.
32. "Opening of the Ganges Canal," *The Illustrated London News* (Supplement, Aug. 5,1854), 118; Richard Baird Smith, *A Short Account of the Ganges Canal, With a Description of Some of the Principal Works* (Roorkee: Thomason College Press, 1870), 2.
33. John William Kaye, *The Administration of the East India Company; A History of Indian Progress*, Second Edition (London: Richard Bentley, 1853), 276.
34. Richard Baird Smith, "The Cauvery, Kistnah, and Godavery," *The Athenaeum: Journal of Literature, Science, and the Fine Arts*, No. 1514 (Nov. 1, 1868), 1332.
35. Cautley (1860), 5.
36. See Ian Stone (1984), 13; George Walter MacGeorge, *Ways and Works in India: Being an Account of the Public Works in that Country from the Earliest Times up to the Present Day* (Westminster: Archibald Constable and Company, 1894), 126.
37. William Elliot Morton, *Notes on the levels of the Eastern Jumna canal* (Agra: The Secundra Orphan Press, 1852), 5.
38. Henry Meredith Vibart, *Addiscombe, Its Heroes and Men of Note* (Westminster: Archibald Constable and Co., 1894), 8.
39. Ibid, 333.
40. Prior to 1809, forty engineers and artillerymen received training at Woolwich Royal Military Academy at Woolwich; however, the Company needed far more trained to cadets to command their growing forces in India after defeating the Marathas in the early-nineteenth century. Given the fixed quota at Woolwich, the Company purchased the former estate of the Earl of Liverpool to house their new seminary. See Henry Meredith Vibart (1894), 15.
41. R. H. Phillimore, *Historical Records of The Survey of India*, Vol. II (Dehra Dun: Survey of India, 1950), 442.
42. Mildred Archer and Ronald Lightbrown, *India Observed: India as Viewed by British Artists, 1760-1860* (London: Victoria and Albert Museum in association with Trefoil Books, 1982), 92.
43. Mildred Archer, "An artist engineer—Colonel Robert Smith in India (1805-1830)" *The Connoisseur*, Vol. 179 (February, 1972), 81.

44. Ibid, 79.
45. See Russell Dionne and Roy MacLeod, "Science and Policy in British India, 1858–1914, Perspectives on a Persistent Belief," *Asie du Sud: Traditions et Changements* (Paris: Centre National de la Recherche Scientifique, 1978), 59.
46. Alfred Deakin, *Irrigated India: An Australian View of India and Ceylon, Their Irrigation and Agriculture* (London: W. Thacker, 1893), 312.
47. See Joyce Brown, "Sir Proby Cautley, Pioneer of Indian Irrigation," *History of Technology*, Vol. III, edited by A. Rupert Hall and Norman Smith (London: Mansell, 1978), 46; Proby Cautley, "A description of the use of wells for foundations," *Journal of the Asiatic Society of Bengal*, Vol. 8, Issue 88 (1839), 327-340.
48. William Elliot Morton (1852), 5.
49. Ibid, 6–7.
50. Cautley (1839), 48.
51. Romesh Dutt, *The Economic History of India in the Victorian Age: From the Accession of Queen Victoria in 1837 to the Commencement of the Twentieth Century* (London: Kegan Paul, Trench, Trübner & Co., 1906), 168-69.
52. Cautley (1860), 4.
53. Ibid, 29.
54. John William Kaye, *The Administration of the East India Company; A History of Its Progress*, Second Edition (London: Richard Bentley, 1853), 288.
55. Ibid, 289.
56. Cautley (1860), 33.
57. Richard Baird Smith, *Italian Irrigation: A Report on the Agricultural Canals of Piedmont and Lombardy* (London: W. H. Allen & Co., 1852), 370.
58. Kaye (1853), 289-290.
59. Cautley, (1860), 33.
60. Ibid, 33-37.
61. Kaye (1853), 289.
62. Cautley, (1860), 39.
63. Ibid, 63.
64. Ibid, 40.
65. Ibid, 99.
66. Ibid, 99.
67. Ibid, 100.
68. Ibid, 100.
69. Ibid, 103.
70. Ignorance of developments in Italian irrigation, namely hydraulic architecture, is most peculiar, especially given that Major-General John Garstin, Acting Chief-Engineer of the Bengal Establishment, translated Paul Frisi's *A Treatise on Rivers and Torrents; with The Method of Regulating Their Course and Channels* in 1818. What is more, Garstin dedicates the translation to Governor-General of India Warren Hastings. In his preface, Frisi notes, "Hydraulic Architecture arose, advanced, and almost attained perfection, in Italy, where they have written on every point connected with the theory of torrents and rivers, the conducting and distribution of clear and turbid waters, the slopes, the directions, and the variations of channels; and, in a word, on the whole range of Hydrometry and Hydraulics" (v).
71. Two years later, Captain Richard Baird Smith, commissioned by the Company to travel to Lombardy and Piedmont, reported on all fiscal and engineering matters of canals in these two provinces. He would publish his highly referenced *Italian Irrigation: A Report on the Agricultural Canals of Piedmont and Lombardy* in 1852.
72. Cautley (1860), 104.
73. Ibid, 423.
74. J. S. Beresford, "Memorandum on the Irrigation Duty of Water, And the Principles on Which Its Increase Depends," *Professional Papers on Indian Engineering*, Vol. V, edited by A. M. Lang (Roorkee: Thomason College Press, 1876). 410–411.
75. Cautley (1860), 124.
76. Flynn (1892), 127.
77. Cautley (1860), 125.
78. William Wilson Hunter, *The Imperial Gazetteer of India*, Vol. V (London: Trübner & Co., 1877), 333.
79. Cautley (1860), 130.
80. Macgeorge (1894), 148.
81. R. P. Tickell, "Headworks of the Ganges Canal," *Minutes of the Proceedings of the Institution of Civil*

Engineers, With Other Selected and Abstracted Papers, Vol. LIV (London: Published by the Institution, 1878), 294-95.

82. Robert Burton Buckley, *The Irrigation Works of India*, Second Edition (London: E & F. N. Spon, Ltd., 1905), 134; Alfred Deakin, *Irrigated India: An Australian View of India and Ceylon, Their Irrigation and Agriculture* (London: W. Thacker, 1893), 181.
83. Bholanauth Chunder, *The Travels of a Hindoo to Various Parts of Bengal and Upper India*, Vol. I (London: N. Trübner & Co., 1869), 345.
84. Paramand, *Mahāmanā Madan Mohan Malaviya: An Historical Biography* (Banaras: Banaras Hindu University, 1985), 245.
85. Ibid, 246.
86. Ibid, 256-57.
87. See Kelly D. Alley, *On the Banks of the Ganga: When Wastewater Meets a Sacred River* (Ann Arbor: University of Michigan Press, 2002), 114-116.
88. Pushpendra K. Agarwal, Vijay P. Singh, *Hydrology and Water Resources of India* (Dodrecht: Springer, 2007), 369.
89. Planning Commission of Government of India, "Final Report: Pilot Study on Water Use Efficiency for Upper Ganga Canal (Uttar Pradesh)" (February, 2003), Chapter 1, 2.
90. Deakin (1893), 183.
91. Ibid, 184.
92. Flynn (1892), 170.
93. Richard Baird Smith, "The Ganges Canal," in Charles William Grant, *Indian Irrigation Being a Short Description of the System of Artificial Irrigation and Canal Navigation in India: with Proposal for Carrying the Same Into Effect by Private Enterprise* (London: Smith, Elder, & Co., 1854), xv.
94. Stone (1984), 41.
95. Deakin (1893), 184.
96. *The Roorkee Treatise on Civil Engineering in India*, Volume 2, compiled by Julius George Medley, third edition enlarged and improved and edited by A. M. Lang (Roorkee: Thomason College Press, 1877), 522.
97. Jagdish Mohan and A. K. Sharma, "Operational Problems of Upper Ganga Canal," *CBIP Symposium on Operation and Maintenance of Canal Systems* (New Delhi: 1980), 122.
98. S. J. Burnell, "On the Recently-Erected Bridges of Western Europe," *Examples of the Architecture of the Victorian Age and Monthly Review of the World's Architectural Progress*, Vol. 1 (London: Darton and Hodge, 1862), 138.
99. G. R. Burnell, *The Rudiments of Hydraulic Engineering*, Part II (London: John Weale, 1859), 211.
100. See Javed Majeed, "Comparativism and references to Rome in British imperial attitudes to India," *Roman Presences: Receptions of Rome in European Culture, 1789-1945*, edited by Catherine Edwards (Cambridge: Cambridge University Press, 1999), 88-109.
101. Deakin (1893), 26.
102. "Lord Hardinge's Administration," *The Calcutta Review*, Vol. 8 (1847), 527; originally printed in *London Quarterly Review*, No. 155 (June 1846).
103. Julius George Medley, *A Year's Campaigning in India, From March, 1857, To March, 1858* (London: W. Thacker and Co., 1858), 121.
104. George R. Bramhall, "What Civilization Owes the Architect and Civil Engineer," *Scientific American: Supplement*, Vol. 20, No. 11 (October 17, 1885), 8157. As late as 1966, it was said that the Solani Aqueduct was "Unequalled in its dimensions by any work executed anywhere in the world at that time, this crossing is a magnificent monument to the outstanding engineering skill of its builders."
105. International Commission on Irrigation and Drainage, *Proceedings [of the] special session on new and promising developments in the field of irrigation, drainage and flood control, New Delhi, 1966* (New Delhi, 1966), 584.
106. Deakin (1893), 184.
107. World Bank, *Report and Recommendation of the President of the International Development Association to the Executive Directors on a Proposed IDA Credit in an Amount of SDR 117.5 million to India for the Upper Ganga Irrigation Modernization Project* (May 3, 1984), 14.
108. Ibid, 15.
109. Ibid, iii.
110. World Bank, *Implementation Completion Report, India: Upper Ganga Irrigation Modernization Report* (Credit 1483-IN) (June 7, 1995), iii.
111. Hunter (1877), 86.
112. Joachim Hayward Stocqueler, *The Hand-Book of British India: A Guide to the Stranger, Traveler, Resi-*

dent, And All Who May Have Business With or Pertaining to India, Third Edition (London: W. H. Allen and Co., 1854), 421.

113. Lieutenant Price is credited with overseeing the Solani Aqueduct and for Price's Bazaar in Roorkee.
114. Grant (1854), xvii.
115. *Thomason Civil Engineering College, Calendar: 1871-71* (Roorkee: Thomason Civil Engineering College Press, 1871), 13-15.
116. *Account of Roorkee College, Established for the Instruction of Civil Engineers, with a Scheme for its Enlargement* (Agra: Secundra Orphan Press, 1851), 20.
117. Ibid, 20.
118. Joachim Hayward Stocqueler, *The Hand-Book of British India: A Guide to the Stranger, the Traveller, the Resident, and All Who May Have Business With Or Appertaining To India* (London: W. M. H. Allen, And Co., 1854), 418—424.
119. Dionne and Macleod (1978), 60.
120. William Willcocks, *Sixty Years in the East* (Edinburgh: William Blackwood and Sons, 1935), 33.
121. *General Report on Public Instruction in the North Western Provinces of the Bengal Presidency, For 1851-52* (Agra: Secundra Orphan Press, 1853), 174-76.
122. Ibid, 71.
123. Ibid, 132.
124. Arun Kumar, "Colonial Requirements and Engineering Education: The Public Works Department, 1847-1947," *Technology and the Raj: Western Technology and Technical Transfers to India, 1700-1947*, edited by Roy MacLeod and Deepak Kumar (New Delhi: Sage, 1995), 216-229.
125. Stocqueler (1854), 422.
126. Ibid, 422.
127. "Roorkee," *The Illustrated London News* (Oct. 10, 1857), 357.
128. Ibid, 357.
129. Hunter (1877), 86.
130. Cautley (1860), 166.
131. Ibid, 192.
132. Ibid, 244.
133. Ibid, 246.
134. Ibid, 28.
135. Ibid, 67.
136. Ibid, 67.
137. Ibid, 423.
138. *The Roorkee Treatise on Civil Engineering in India* (1877), 526.
139. William Elliot Morton, *Professional Papers printed at the Thomason Civil Engineering College, Roorkee No. II: On Rajbuhas* (Roorkee: Thomason College Press, 1856), i.
140. Cautley (1860), 427.
141. Ibid, 449.
142. Morton (1856), 5.
143. Ibid, 5.
144. Ibid, 7. Morton, Superintendent Engineer of the EJC, penned the introduction to the first edition on rajbuhas on December 29, 1852—just a month after Cautley gave his orders for the construction of rajbuhas along the Ganges Canal.
145. Cautley (1860), 439.
146. Cautley (1860), 379.
147. *Thomason Civil Engineering College, Calendar: 1871-71* (1871), 194.
148. *Report on the Administration of the N.-W. Provinces for the Year 1867-68* (Allahabad: Government Press, 1868), 58.
149. For more on this, see Michael Lewis, "The Personal Equation: Political Economy and Social Technology on India's Canals, 1850-1930," *Modern Asian Studies*, Vol. 41, No. 5 (Sep., 2007), 967-994.
150. Cautley (1860), 3.
151. Julia A. Stone, *Illustrated India: Its Princes and People: Upper, Central, and Farther India, Up the Ganges and Down the Indus* (Hartford: American Publishing Company, 1877), 239.
152. Cautley (1860), 299.
153. Ibid, 299.
154. Ibid, 299.
155. Ibid, 431.
156. Colonel Sykes, "Notes on Public Works" *Journal of the Statistical Society of London*, Vol. 21, No. 2

(Jun, 1858), 125.

157. "The Military Board-- Ordnance Commissariat-- New Secretariat of Public Works" *Indian Mail and Register of Intelligence for British and Foreign India, China, and All Parts of the East*, Vol. XIII (January—December, 1855), 158.
158. George Julius Medley, *India and Indian Engineering: Three Lectures Delivered at the Royal Engineering Institute, Chatham in July, 1872* (London: E. & F. N. Spon, 1873), 20.
159. Richard Baird Smith, *Report on the Famine of 1860-61* (Calcutta: Home Department, 1861), 24.
160. Brown (1978), 80.
161. Colin Campbell Scott-Moncrieff, *Report of the Indian Irrigation Commission, 1901-03* (Calcutta: Office of the Superintendent of Government Printing, India, 1903), 87.
162. Macgeorge (1894), 161.
163. *Imperial Gazetteer of India: Provincial Series; United Provinces of Agra and Oudh*, Vol. I (Calcutta: Superintendent of Government Printing, 1908), 208.
164. Medley (1873), 101.
165. William Wilson Hunter, *The Imperial Gazetteer of India: Provincial series*, Volume XXIV (London: Trübner & Co., 1877), 208.
166. Flynn (1892), 4.
167. Ibid, 4.
168. Ibid, 185.
169. Herbert M. Wilson, *Pumping Water for Irrigation* (Washington, D.C.: Government Printing Office, 1896), 18.
170. Herbert M. Wilson, *Irrigation in India* (Washington, D.C.: Government Printing Office, 1892), 409; See James Wescoat, "Water Rights in South Asia and the United States: Comparative Perspectives, 1873-2000," *Land, Property and the Environment*, ed. John F. Richards (Oakland: ICS, 2001), 313.
171. Herbert M. Wilson, "American and Indian Irrigation Works," *The Irrigation Age* (1894), 107.
172. Ibid, 107.
173. Ibid, 109.
174. Deakin (1893), 21.
175. Ibid, 21.
176. Dadabhai Naoroji, *Poverty of India* (London: Vincent Brooks, Day and Sons, 1878), 40.
177. Ibid, 50.
178. Ibid, 36-38.
179. Arthur Cotton, *The Madras Famine: With Appendix Containing a Letter from Miss Florence Nightingale, and other papers* (London: Simpkin, Marshall & Co., Stationers' Hall Court, 1877).
180. William Digby, *The Famine Campaign in Southern India (Madras and Bombay Presidencies and Province of Mysore) 1876-1878*, Vol. I (London: Longmans, Green, and Co., 1878), 6-7.
181. John Strachey and Richard Strachey, *The Finances and Public Works of India from 1869 to 1881* (London: K. Paul, Trench & Co., 1882), 171.
182. Ibid, 171.
183. See Dionne and MacLeod (1978), 65.
184. Dadabhai Naoroji, *Poverty and Un-British Rule in India* (London: Swan Sonnenschein & Co., 1901), 212.
185. Romesh Dutt, *The Economic History of India in the Victorian Age: From the Accession of Queen Victoria in 1837 to the Commencement of the Twentieth Century*, Third Edition (London: Keegan Paul, Trench, Trübner & Co., 1908), 174.
186. John M. Hurd, "Railways," *The Cambridge Economic History of India*, Vol. 2, c. 1757 to 1970, ed. Dharma Kumar and Meghnad Desai (Cambridge: Cambridge University Press, 1983), 742.
187. Elizabeth Whitcombe, "Irrigation," *The Cambridge Economic History of India*, Vol. 2, c. 1757 to 1970, ed. Dharma Kumar and Meghnad Desai (Cambridge: Cambridge University Press, 1983), 704.
188. Dutt (1908), 374-75.
189. Ibid, 174.
190. See David Arnold, *Science, Technology and Medicine in Colonial India* (New York: Cambridge University Press, 2000), 111.
191. Dutt (1908), 376.
192. Mohandas K. Gandhi, *Hind Swaraj or Indian Home Rule* (Ahmedabad: Navajivan Publishing House, 1939), 41-42.
193. Colin Campbell Scott-Moncrieff, *Report of the Indian Irrigation Commission, 1901-03* (Calcutta: Office of the Superintendent of Government Printing, India, 1903), 33.
194. "Sir Colin C. Scott-Moncrieff," *The Spectator* (December 1, 1917), 7.
195. Scott-Moncrieff (1903), 87.

196. Ibid, 75.
197. See Elizabeth Whitcombe, "Irrigation," *The Cambridge Economic History of India*, Vol. 2, c. 1757 to 1970, ed. Dharma Kumar and Meghnad Desai (Cambridge: Cambridge University Press, 1983), 677-737; See Stone (1984).
198. Scott-Moncrieff (1903), 60, 73.
199. Colin Campbell Scott-Moncrieff, "Opening Address," *Nature* (Sept. 7, 1905), 466.
200. Ibid, 466.
201. Monier Williams, *Modern India and the Indians: being a series of impressions, notes, and essays* (London: Trübner & Co, Ludgate Hill, 1879), 333—334.

PART 3

1. William Stampe, "East and West in Co-operative Action: A Himalayan Dream," *The Asiatic Review*, Vol. XLIII, No. 154 (April, 1947), 97.
2. Ibid, 97.
3. Ibid, 97.
4. Stephen Leggett, "The Amritsar Hydro-Electric Irrigation Installation," *Minutes of Proceedings*, Vol. 212, Issue 1921 (Jan. 1, 1921), 66.
5. Bernard Darley, "The Development of Irrigation in India," *Journal of the Royal Society of Arts*, Vol. 90, No. 4602 (Dec. 12th, 1941), 50.
6. William Stampe, *The Ganges Canal Hydro-Electric Scheme with Its Associated State Tube-Well Irrigation Project* (Allahabad: Superintendent, Printing and Stationary, United Provinces, 1937), 1.
7. Stampe (1947), 98.
8. Ibid, 100.
9. Ibid, 108.
10. Ibid, 99.
11. Ibid, 100.
12. Ibid, 108.
13. Sam Higginbottom, "Tube-Well Irrigation," *The Allahabad Farmer*, Vol. XI, No. 5, (September 1937), 246.
14. See "Sir William Leonard Stampe, C.I.E.," *Civil Engineering and Public Works Review*, Vol. 47, No. 547, (January 1952), 37.
15. Sanjay Pahuja, Catherine Tovey, Stephen Foster and Hector Garduno, *Deep Wells and Prudence: Towards Pragmatic Action for Addressing Groundwater Overexploitation in India* (Washington, D.C.: The World Bank, 2010), ix.
16. B. D. Dhawan, "Utilisation of Ground Water Resources: Public versus Private Tubewells," *Economic and Political Weekly*, Vol. 9, No. 39 (Sep. 28, 1974), A76.
17. Duleep Danton and Andy J. Marr, "Conjunctive Use of Surface and Groundwater in A Water-Abundant River Basin," 17th *International River Symposium, Conference Proceedings* (Brisbane: International River Symposium, 2005), 4.
18. Sir William Stampe summoned poetry to exalt the tubewell, but never wrote verse in honor of the technology.
19. B. D. Dhawan, *Development of Tubewell Irrigation in India* (New Delhi: Agricole Publishing Academy, 1982), 46-47.
20. Government of India, "Provisional Population Totals—Uttar Pradesh—Data Sheet, Census of India 2011" (Lucknow: Ministry of Home Affairs, 2011).
21. George C. Taylor, Jr., *Historical Review of the International Water-Resources Program of the U.S. Geological Survey*, 1940-70, Geological Survey Professional Paper 911 (Washington, D.C.: United States Government Printing Office, 1976), 68.
22. Ibid, 68.
23. Ibid, 69.
24. Robert J. McMahon, "Food as a Diplomatic Weapon: The India Wheat Loan of 1951," *Pacific Historical Review*, Vol. 56, No. 3 (Aug., 1987), 349-377.
25. Karl Wittfogel, *Oriental Despotism: A Comparative Study of Total Power* (New Haven: Yale University Press, 1957).
26. Ford Foundation, Agricultural Production Team; Ministry of Food and Agriculture and Ministry of Community Development and Cooperation, *Report on India's Food Crisis & Steps to Meet It* (New Delhi: Govt. of India, Ministry of Food and Agriculture and Ministry of Community Development and Cooperation, 1959), 17.

27. Planning Commission, *PEO Study No. 284, High-Yielding Varieties Programme* in India, Part I (New Delhi: Government of India, 1971), 1.
28. Sukhamoy Chakravarty, *Development Planning: the Indian Experience* (New York: Oxford University Press, 1987), 22.
29. See Tushaar Shah, *Wells and Welfare in the Ganga Basin: Public Policy and Private Initiative in Eastern Uttar Pradesh, India*, Research Report 54 (Colombo: International Water Management Institute, 2001).
30. See World Bank, *Staff Appraisal Report: India, Second Uttar Pradesh Public Tubewells Project*, Report No. 4167-IN (Washington D.C.: World Bank, 1983).
31. B. R. Tomlinson, *The Economy of Modern India, 1860-1970* (New York: Cambridge University Press, 1993), 206.
32. Andre Gunder Frank, "Reflections on Green, Red and White Revolutions in India," *Economic and Political Weekly*, Vol. 8, No. 3 (Jan. 20, 1973), 119.
33. Ibid, 119.
34. Joseph W. Elder, "Land Consolidation in an Indian Village: A Case Study of the Consolidation of Holdings Act in Uttar Pradesh," *Economic Development and Cultural Change*, Vol. 11, No. 1 (Oct., 1962), 16-40.
35. B. D. Dhawan, *Development of Tubewell Irrigation in India* (New Delhi: Agricole Publishing Academy, 1982), 46-47.
36. B. D. Dhawan, "Tubewell Irrigation in the Gangetic Plains," *Economic and Political Weekly*, Vol. 12, No. 39 (Sep. 24, 1977), A91.
37. Shah (2001), 17.
38. Tushaar Shah, *Taming the Anarchy: Groundwater Governance in South Asia* (Washington, D.C.: Resources for the Future, 2009), 240; See Marcus Moench, "When the Well Runs Dry but Livelihood Continues: Adaptive Responses to Groundwater Depletion and Strategies for Mitigating the Associated Impacts," *Agricultural Groundwater Revolution: Opportunities and Threats to Development*, edited by Mark Giordano and K. G. Vilholth (Cambridge: CABI Publishing, 2007), 173-192.
39. Shah (2001), 24.
40. Niranjan Pant, "Control of and Access to Groundwater in UP," *Economic and Political Weekly*, Vol. 40, No. 26 (Jun. 25 - Jul. 1, 2005), 2673.
41. Shah (2009), 45.
42. Shah (2001), 5.

INDEX

ACKNOWLEDGEMENTS

The completion of this book would not have been possible without the encouragement of my family, both above and along the Equator. They have always supported the peculiar mode of work (lifestyle) that I developed to undertake such a protracted project.

The research, both in the field and in archives, would never have been possible without the support of a J. William Fulbright Fellowship to India (2005-06), a grant from the Ford Foundation (2006-07), as well as a Tata Study Grant (2008) and a Frederick Sheldon Traveling Fellowship (2009-10), both from Harvard University. It was at Allahabad University where I had the privilege to know Prof. Harkiesh N. Misra, then Chair of the Dept. of Geography. Prof. Misra generously hosted me at the institution. He encouraged my work on the Ganges and provided insights and feedback that helped shape my thinking about this work. He and his family made me feel most welcome and at home in Allahabad. It was also at Allahabad where I came to know Ajay Sharma and Sabir Anwar. Their generosity and friendship were exceptional. And while I met Dr. Robert Phillips and Cathy Phillips née Brooks in Delhi, I did not get to know them until they arrived at the Allahabad University Guest House one week after I did. Having never been to India before, I relied on their knowledge and years of experience. I remain indebted to their kindness and unwavering friendship.

Over the past nine years, I have made countless trips by boat down the Ganges. I was fortunate to have Michael Scahill and Gabriel Gomez agree to join me on my first, and only attempt, to travel up river from Varanasi to Allahabad. While the trip up river was a bust, we cemented lasting friendships that extend well beyond the steps of Meer Ghat in Varanasi. Christopher Holland and Navin Tripathi joined me on one of my most colorful and memorable trips between Allahabad and Varanasi in March of 2006. They managed to keep me from falling over the side of the boat when a Ganges River dolphin jumped up near our boat and scouted places to stop and document. Though Dr. Frederick Coolidge was unable to join us as we made our way slowly down the Ganges, he was there in spirit. Sarah Mehta's moral support and humor throughout my stay and beyond has been without equal. Girish Kaul, Varrtika Mudaliar, and S. K. Bharathi were invaluable in keeping things running smoothly throughout my time in India as a Fulbright Fellow. Lauren Portada, a rare mixture of visual acumen and deep imagination, always reminded me what was important in a set of working drawings. Dustin Smith and Bobby Poulson-Houser both accompanied me to the *Kumbh Mela* in Allahabad in 2013. Never losing sight of me, they kept an eye on the time and my back while I was dangling from bridges and standing in the middle of traffic.

Lynnette Widder, Margaret Lewis, Steven Whitten, and Stefanie Victor, all read versions of my initial proposal in 2004. As they will recall, I proposed to finish this work within eighteen months as opposed to nine years. Had they not helped me limit the scope of the project, this work may have taken an additional decade or two. Ganesan Balachander's encouragement and support from the Ford Foundation was crucial to making this work a reality. Ramesh Jain was an early advocate of this work, ever patient and always believing in its utility, I hope the book lives up to his expectations.

ACKNOWLEDGEMENTS

Since embarking on this project, I have benefitted from comments and suggestions from John R. Stilgoe, Jorge Silvetti, Rodolfo Machado, Erika Naginski, Woodman Taylor, Justin Fowler, Christina Welsch, and Mario Gandelsonas. Each has given invaluable feedback and encouragement. I could not have completed this work without the support of a remarkable and diverse group of individuals—especially Ivi Diamantopoulou, Cesar Ferro, Alexander Keller, Maynard Hayden León, Jesse Honsa, Dan Hui, Diana Mangaser, Kevin Park, Jesen Tanadi, and Robert Williams.

Portions of this work were published in *Topos*, *Architectural Design*, *Sarai Reader*, *Onsite*, *The Indian Express*, and reviewed in *UnCube*. Jessica Bridger, Adrian Lahoud, Anthony Burke, Charles Rice, Yamini Lohia, and Smriti Vohra turned half formed thoughts and ideas into complete sentences. Likewise, portions of this work were exhibited in solo and group shows in Asia, Europe, as well as North and South America. I must thank Ana María Durán-Calisto, Varrtika Mudaliar, Noel Moreno, Yu Morishita, and Evita Yumul for all of their help.

I could not have wished for more thorough support than that provided by Aleksandr Bierig. A dear friend and scrupulous critic and editor, I am indebted to all of his help and insights. He made critical interventions throughout the development of this book. I must also thank Duncan Corrigall for his help in catching mistakes and making astute changes and suggestions. With great gratitude I thank my publisher, Gordon Goff, for his uncompromising support of this project and his immeasurable patience.

Ever since meeting Rahul Mehrotra in 2010, he has always encouraged and supported me in this project. I thank him for the thoughtful foreword.

Above all, *Ganges Water Machine* owes a great deal of gratitude to Felipe Correa for having listened to me talk about this project for over a decade, seen me change course and expand the scope of the book more times than he would care to remember, and yet is still willing to talk to me. For that I am eternally grateful.